A TEXTBOOK OF MEASUREMENT AND METROLOGY

By

R.K. RAJPUT

M.E. (Heat Power Engg.) Hons., Gold Medallist, Grad. (Mech. Engg. & Elec. Engg.)
M.I.E. (India); M.S.E.S.I.; M.I.S.T.E.; C.E. (India)
Recipient of: "Best Teacher Award (Academic)"
"Distinguished Author Award"
"Jawahar Lal Nehru Memorial Gold Medal"
for an outstanding research paper
(Institution of Engineers—India)

Principal ***(Formerly)***
Punjab College of Information Technology, and
Thapar Polytechnic College,
PATIALA.

S.K. KATARIA & SONS®

Publisher of Engineering Books
4885/109, Prakash Mahal, Dr. Subhash Bhargav Lane,
Opposite Delhi Medical Association, Daryaganj, New Delhi–110002 (INDIA)
Phone: +91-11-43551243; *Mobile:* +91-9871775858, 7678425209
e-mail: katariabook@gmail.com, katariabooks@yahoo.com
Website: www.skkatariaandsons.com
(e-books, subscriptions and bulk buying also available on our website)

Published by:

S.K. Kataria & Sons®

4885/109, Prakash Mahal, Dr. Subhash Bhargav Lane,
Opposite Delhi Medical Association, Daryaganj, New Delhi-110002 (INDIA)
Phone: +91-11-43551243; *Mobile:* +91-9871775858, 7678425209
e-mail: katariabook@gmail.com, katariabooks@yahoo.com
Website: www.skkatariaandsons.com

First Edition: 2012; Reprint: 2026

Laser Type Setting by:
SARA ASSIGNMENTS, Delhi.

Printed at:
AL AZIZ BOOK BINDING HOUSE

Preface to The First Edition

This book has been written specifically to meet exhaustively the requirements of the subject **"Measurements and Metrology"** of B.E. examination of "UPTU, Lucknow – UTU, Dehradun and PTU, Jalandhar", strictly according to the latest syllabus.

The book consists of 13 chapters namely:

Part I: ***Mechanical Measurements:*** *1. General concepts; 2. Sensors and transducers; 3. Signal conditioning, transmission and processing; 4. Time related measurements; 5. Measurement of pressure; 6. Strain measurement; 7. Measurement of force, torque and vibrations; 8. Temperature measurements.*

Part II: ***Metrology and Inspection:*** 9. *Introduction to metrology and inspection*; 10. *Linear and angular measurements and comparators*; 11. *Measurement of geometric forms*; 12. *Measurement of screw threads and gears*; 13. *Surface texture/finish*.

The book contains comprehensive treatment of the subject matter under wide range of topics mentioned in the syllabus.

Any suggestions for the improvement of this book will be thankfully acknowledged and incorporated in the next edition.

ER. R.K. RAJPUT
(Author)

Contents

PART II: METROLOGY AND INSPECTION

UNIT I

MECHANICAL MEASUREMENTS

Chapter

1

General Concepts

1.1 INTRODUCTION TO MEASUREMENT AND INSTRUMENTATION

1.1.1 Measurement

Measurement is the act, or the result of a quantitative comparison between a predetermined standard and an unknown magnitude. The procedure and apparatus employed for obtaining the comparison, however, must be provable; the procedure for this is called '*calibration*'.

- Probably it has rightly been said: "*whatever exists, exists in some amount*". The determination of the amount is what measurement is all about. If those things that exist are related to the practice of mechanical engineering, then the determination of their amounts constitutes the subject of *mechanical measurements*.
- *Measurements provide us with a means of describing various phenomena in quantitative terms.* The measurements are not necessarily carried out by *mechanical* means.

The following quantities are typically within the scope of mechanical measurements:

- Pressure
- Temperature
- Displacement
- Fluid flow and related parameters
- Acoustics
- Mass, length, time etc.

In several situations, however, the above quantities are not measured by purely mechanical means, but invariably measured by electrical means by *transducing them into an analogous electrical quantity*.

In modern technology, the measurement of mechanical quantities involves the use of *electrical and electronic techniques.*

Measurand. *The physical quantity or the characteristic condition which is the object of measurement in an instrumentation system is variously termed as "measurand", "measurement variable", "instrumentation* *variable" or "process variable".*

The measurand may be:

- Fundamental quantity, *e.g.,* length, mass and time;
- Derived quantity, *e.g.,* speed, velocity, acceleration, pressure.

The word *measurand* is used to designate the particular physical parameter being observed and quantified; that is, the *input quantity* to the measuring process. The act of measurement produces a *result.* Figure 1.1 shows a *fundamental measuring process.*

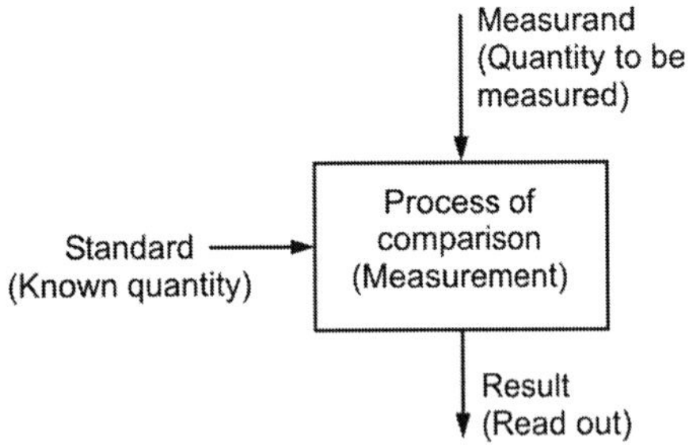

Fig. 1.1 Fundamental measuring process.

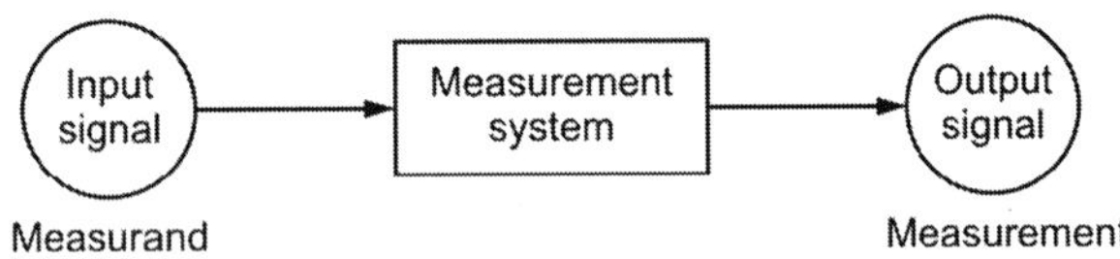

Fig. 1.2

– Whereas the input to a measuring system is known as *measurand* and, the output is called *measurement* (See Fig. 1.2)

The *basic requirements* for getting meaningful result of measurement are:

(*i*) The standard employed for comparison purposes *must be accurately defined* and should be *commonly acceptable.*

(*ii*) The standard must be of the *same character* as the measurand and usually, but not always, is prescribed and defined by a legal or recognized agency or organisation, *e.g.,* Indian Bureau of Standards (IBS) or the International Organisation of Standards (ISO).

(*iii*) The apparatus used and method adopted for the comparison purposes must be *provable.*

- Several measurands of interest *vary with time in a continuous manner over a range of magnitude.*

For example, the speed of the automobile, as it starts from rest has some magnitude at every instant during its motion. A sensor that responds to velocity will produce an output signal having a line variation analogous to the time change in automobile's speed. Such a signal is referred to as **analog signal** *because it is analogous to a continuous physical process.*

The analog signal, which has a value at every instant of time, becomes a **digital signal**.

– *A digital signal is a set of discrete numbers each corresponding to the value of the analog signal at a single specific instant of time.*

– *The* digital signal contains *no information* about the value of the analog signals at times other than sample times.

The mechanical quantities such as pressure, stress, strain, temperatures, fluid-flow rates behave time-wise in an analog manner. However, by converting an analog signal to an equivalent digital signal for the purpose of signal conditioning and/or read out oftenly following distinct *advantages* are obtained:

(*i*) *Noise problems are reduced or sometimes eliminated completely.*

(*ii*) *Data transmission becomes simpler.*

(*iii*) *Computers are designed to process digital information, and direct numerical display or recording is more easily accomplished by manipulating digital quantities.*

1.1.2 Instrumentation

The technology of using instruments to measure and control the physical and chemical properties of materials is called **Instrumentation**.

- When the instruments are used for the *measurement and control of industrial manufacturing conversion or treatment processes*, the term **"process instrumentation"** is used.
- *When the measuring and controlling instruments are combined so that measurements provide impulses for remote automatic action, the result is called a* **control system**.

1.2 SIGNIFICANCE OF MEASUREMENTS

The significance of measurements is discussed below:

- Measurements play a very significant role in every branch of scientific research and engineering processes which include the following:
  - — Control systems;
  - — Process instrumentation;
  - — Data reduction.
- The whole area of automation or automatic controls is based on measurements. The very concept of control is based on the comparison of the actual condition and the desired performance. *The exactness of error depends on the precision and accuracy of measurements made.*
- The measurements confirm the validity of a hypothesis and also add to it its understanding. This eventually leads to new discoveries that require new and sophisticated measuring techniques.
- Through measurements a product can be designed or a process be operated with maximum efficiency, minimum cost, and with desired degree of reliability and maintainability.

1.3 STANDARDS OF MEASUREMENT

As the measurement is a process of comparison, therefore, regardless of our measurement method, we must employ a basis of comparison—*standardized units*. The standard must be precisely defined, and because different systems of units exist, the method of conversion from system to system must be mutually agreed upon.

– It is of significant importance that a relationship between the standards and the read out scale of each measuring system must be established through a process known as *calibration*.

The standards of measurement are often *classified* as follows:

1. Primary standards
2. Secondary standards
3. Working standards.

The **measurement standard** *is the physical embodiment of the unit of measurement as well as that of its submultiple value.*

1. Primary standards

The highest standard of either a base unit or a derived unit is called a ***primary standard.***

- These standards essentially are copies of international prototypes and are kept throughout the world in national standard laboratories and institutions of similar standing.
- These standardized units are:
  - *Quite stable*
  - *Independent*
  - *Invariant*
  - *Not relative but finite.*

While setting up the primary standards, the following points are given due *considerations*:

(*i*) Long time stability of the material.

(*ii*) Low temperature coefficient.

(*iii*) No deterioration in its characteristics with changes in humidity, temperature and other environmental conditions.

(*iv*) Rigidity of construction.

(*v*) Accuracy of machining

(*vi*) Most sophisticated techniques to be used for measuring physical dimensions which determine the accuracy of the standard.

- The "primary standards" *constitute the ultimate basis of reference* and are used for the purpose of "*verification and calibration of secondary standards*".
  - These standards have the "*highest possible accuracy*" but are *very expensive* to own and maintain.
  - The primary standards are *not available for use outside the national laboratories*.

2. Secondary standards

The secondary standards are the *reference calibrated standards, designed and calibrated from the primary standards.*

- These are sent periodically to the national standard laboratories for their calibration (the interval of calibration depends upon the accuracy and the type of standard being maintained).
- These standards are kept by the measurement laboratories and the industrial organisations check and calibrate the general tools for their accuracy and precision.

3. Working standards:

- These standards have an accuracy of *one order lower than that of the secondary standard.*
- These are the normal standards which are *used by the workers and technicians who actually carry out the measurements.*

1.4 S.I. UNITS AND CONVERSION FACTORS

A. Introduction to S.I. Units

S.I., the international system of units is divided into *three* classes:

1. Base units
2. Derived units
3. Supplementary units.

From the scientific point of view division of S.I. unit into these classes is to a certain extra arbitrary, because it is not essential to the physics of the subject. Nevertheless the General Conference considering the advantages of a single, practical, world-wide system for international relations, research and for scientific work, decided to base the international system on a choice of six well-define units given in Table 1.1 below:

TABLE 1.1 S.I. Base units

Quantity	*Name*	*Symbol*
length	metre	m
mass	kilogram	kg
time	second	s
electric current	ampere	A
thermodynamic temperature	kelvin	K
luminous intensity	candela	cd
amount of substance	mole	mol

The second class of S.I. units contains derived units, *i.e.,* units which can be formed by combining base units according to the algebraic relations linking the corresponding quantities. Several of the algebraic expressions in terms of base units that can be replaced by special names and symbols can themselves be used to form other derived units.

Derived units may, therefore, be classified under three headings. Some of them are given in Tables 1.2, 1.3 and 1.4.

TABLE 1.2 Examples of S.I. drived units expressed in terms of base units

Quantity	*S.I. Units*	
	Name	*Symbol*
area	square metre	m^2
volume	cubic metre	m^3
speed, velocity	metre per second	m/s
acceleration	metre per second squared	m/s^2
density, mass density	kilogram per cubic metre	kg/m^3
concentration (of amount of substance)	mole per cubic metre	mol/m^3
activity (radioactive)	1 per second	s^{-1}
specific volume	cubic metre per kilogram	m^3/kg
luminance	candela per square metre	cd/m^2

TABLE 1.3 S.I. Drived units with special names

Quantity	S.I. Units			
	Name	*Symbol*	*Expression in terms of other units*	*Expression in terms of S.I. base units*
frequency	hertz	Hz	—	s^{-1}
force	newton	N	—	$m.kg.s^{-2}$
pressure	pascal	Pa	N/m^2	$m^{-1}.kg.s^{-2}$
energy, work, quantity of hear power	joule	J	N.m	$m^2.kg.s^{-2}$
radiant flux, quantity of electricity	watt	W	J/s	$m^2/kg.s^{-1}$
electric charge	coloumb	C	A.s	s.A
electric tension, electric potential	volt	V	W/A	$m^2.kg.s^{-1}.A^{-1}$
capacitance	farad	F	C/V	$m^{-2}.kg^{-1}.s^4$
electric resistance	ohm	Ω	V/A	$m^{-2}.kg.s^{-3}.A^{-2}$
conductance	siemens	S	A/V	$m^{-2}.kg^{-1}.s^3.A^{-2}$
magnetic flux	weber	Wb	V.S.	$m^2.kg.s^{-2}.A^{-1}$
magnetic flux density	tesla	T	Wb/m^2	$kg.s^{-2}.A^{-1}$
inductance	henry	H	Wb/A	$m^2kg.s^{-2}.A^{-2}$
luminous flux	lumen	lm	—	cd.sr
illuminance	lux	lx	—	m^{-2}.cd.sr.

TABLE 1.4 Examples of S.I. derived units expressed by means of special names

Quantity	S.I. Unitsm		
	Name	*Symbol*	*Expression in terms of S.I. base units*
dynamic viscosity	pascal second	Pa.s	$m^{-1}.kg.s^{-1}$
moment of force	metre newton	N.m	$m^{-2}.kg.s^{-1}$
surface tension	newton per metre	N/m	$kg.s^{-2}$
heat flux density, irradiance	watt per square metre	W/m^2	$kg.s^{-2}$
heat capacity; entropy	joule per kelvin	J/K	$m^{-2}.kg.s^{-1}.K^{-1}$
specific heat capacity, specific entropy	joule per kilogram kelvin	J/(kg.K)	$m^{-2}.s^{-2}.K^{-1}$
specific energy	joule per kilogram	J/kg	$m^{-2}.s^{-2}$
thermal conductivity	watt per metre kelvin	W/(m.K)	$m^{-1}.kg.s^{-3}.K^{-1}$

energy density	joule per cubic metre	J/m^3	$m^{-1}.kg.s^{-2}$
electric field strength	volt per metre	V/m	$m.kg.s^2A^{-1}$
electric charge density	coloumb per cubic metre	C/m^3	$m^{-3}.s.A$
electric flux density	coloumb per square metre	C/m^2	$m^{-2}.s.A$
permittivity	farad per metre	F/m	$m^{-3}.kg^{-1}.s^4.A^2$
current density	ampere per square metre	A/m^2	—
magnetic field strength	ampere per meter	A/m	—
permeability	henry per metre	H/m	$m.kg.s^{-2}.A^{-2}$
molar energy	joule per mole	J/mol	$m^{-2}.kg.s^{-2}.mol^{-1}$
molar heat capacity	joule per mole kelvin	J/(mol.K)	$m^{-2}.kg.s^{-2}.K^{-1}.mol^{-1}$

The S.I. units assigned to third class called "*Supplementary units*" may be regarded either as base units or as derived units. Refer to Table 1.5 and 1.6.

TABLE 1.5 S.I. Supplementary units

Quantity	*S.I. Units*	
	Name	*Symbol*
plane angle	radian	rad
solid angle	steradian	sr

TABLE 1.6 Examples of S.I. derived units formed by using supplementary units

Quantity	*S.I. Units*	
	Name	*Symbol*
angular velocity	radian per second	rad/s
angular acceleration	radian per second squared	rad/s^2
radiant intensity	watt per steradian	W/sr
radiance	watt per square metre steradian	$W\text{-}m^{-2}.sr^{-1}$

TABLE 1.7 S.I. Prefixes

Factor	*Prefix*	*Symbol*	*Factor*	*Prefix*	*Symbol*
10^{12}	tera	T	10^{-1}	deci	d
10^9	giga	G	10^{-2}	centi	c
10^6	mega	M	10^{-3}	milli	m
10^3	kilo	k	10^{-6}	micro	μ
10^2	hecto	h	10^{-9}	nano	n
10^1	deca	da	10^{-12}	pico	p
			10^{-15}	fasnto	f
			10^{-18}	atto	a

B. Definitions

The S.I. *seven* base units and *two* supplementary units are defined below:

(*i*) **Metre.** The metre is the length equal to 1,650, 76373 wavelengths in vacuum of the radiation corresponding to the transition between the levels $2p_{10}$ and $5d_5$ of the krypton-86 atom.

(*ii*) **Kilogram.** One kilogram is equal to the mass of the international prototype of the kilogram.

(*iii*) **Second.** The second is the duration of 9,192, 631, 770 periods of the radiation corresponding to the transition between the two hyperfine levels of the ground state of the cesium-133 atom.

(*iv*) **Ampere.** One ampere is that constant current which, if maintained in two straight parallel conductors of infinite length, of negligible circular cross-section and placed one metre apart in vacuum, would produce between these conductors a force equal to 2×10^{-7} newton per metre of length.

(*v*) **Kelvin.** The kelvin is the fraction $\frac{1}{273.16}$ of thermodynamic temperature of the triple point of water.

(*vi*) **Candela.** The candela is the luminous intensity in the perpendicular direction, of a surface of a $\frac{1}{600,000}$ square metre of a black body at a temperature of freezing platinum under a pressure of 101, 325 newton per square metre.

(*vii*) **Mole.** The mole is the amount of substance of a system which contains as many elementary entities as there are atoms in 0.012 kg, carbon 12. When the mole is used, the elementary entities must be specified and may be atoms, molecules, ions, electrons, other particles or specified groups of such particles.

(*viii*) **Radian.** The radian is the plane angle between two radii of a circle that cut off on the circle an arc equal in length to the radius.

(*ix*) **Steradian.** The steradian is the solid angle which having its vertex in the centre of a sphere, cuts of an area of the surface of the sphere equal to that of a square with sides of length equal to the radius of the sphere.

Some other definitions:

Newton. The newton (N) is a derived unit of force and is defined as the unit of force per second which when acting on a mass of 1 kilogram gives it an acceleration of one metre per second. Since acceleration due to gravity equals 9.81 m/s^2; one kilogram force equals 9.81 newtons.

Joule. The joule (J) is a derived unit of energy, work or quantity of heat and is defined as the work done when a force of one newton acts so as to cause a displacement of one metre. Energy is defined as the capacity to do work. A unit of energy in nuclear physics is the electron volt (eV) which is defined as the energy gained by an electron in rising through a potential difference of one volt.

$$1 \text{ eV} = 1.6021 \times 10^{-19} \text{ J.}$$

Watt. The watt (W) is a unit of power (i.e., rate of doing work)

$$\text{Power in watts} = \frac{\text{Work (or energy) in joules}}{\text{Time in seconds}}$$

Thus 1 watt equals 1 joule/sec.

$$1 \text{ kilowatt-hour (kWh)} = 1000 \text{ watt-hours}$$
$$= 3600000 \text{ joules}$$

Coulomb. The coulomb (C) is the derived unit of charge. It is defined as *the quantity of electricity passing a given point in a circuit when a current of 1A is maintained for 1 second.*

i.e., $$Q = I \cdot t$$

where, Q = charge in coulombs,
I = current in amperes, and
t = time in seconds.

1 coulomb represents 6.24×10^{18} electrons.

Ohm. The ohm (Ω) is the unit of electric resistance and is defined as *the resistance in which a constant current of 1 A generates heat at the rate of 1 watt.*

Siemen. The siemen is a unit of electric conductance (*i.e.,* reciprocal of resistance). If a circuit has a resistance of 5 ohms, its conductance is 0.2 siemens. A more commonly used name for siemen is *mho* (℧).

Volt. The volt is a unit of potential difference and electromotive force. It is defined as *the difference of potential across a resistance of 1 ohm carrying a current of 1 ampere.*

Hertz. The hertz (Hz) is a unit of frequency 1 Hz = 1 cycle per second.

Horse-power. It is a practical unit of mechanical output. BHP (British horse power or brake horse power) equals 746 watts. The metric horse power equals 735.5 watts. To avoid confusion between BHP and metric horse power, the mechanical output of machines in S.I. units, is expressed in watts or kilowatts.

C. Salient Features of S.I. Units

The salient features of S.I. units are as follows:

1. It is a coherent system of units, *i.e.,* product or quotient of any two base quantities results in a unit resultant quantity. For example, unit length divided by unit time gives unit velocity.
2. It is a rationalised system of units, applicable to both, magnetism and electricity.
3. It is a non-gravational system of units. It clearly distinguishes between the units of mass and weight (force) which are kilogram and newton respectively.
4. All the units of the system can be derived from the base and supplementary units.
5. The decimal relationship between units of same quantity makes possible to express any small or large quantity as a power of 10.
6. For any quantity there is one and only one S.I. unit. For example, Joule is the unit of energy of all forms such as mechanical, heat, chemical, electrical and nuclear. However, kWh will also continue to be used as unit of electrical energy.

Advantages of S.I. Units

1. Units for many different quantities are related through a series of simple and basic relationship.
2. Being an absolute system, it avoids the use of factor 'g', *i.e.,* acceleration due to gravity in several expressions in physics and engineering which had been a nuisance in all numericals in physics and engineering.

3. Being a rationalised system, it ensures all the advantages of rationalised MKSA system in the fields of electricity, magnetism, electrical engineering and electrons.
4. Joule is the only sole unit of energy of all forms and watt is the sole unit of power hence a lot of labour is saved in calculations.
5. It is a coherent system of units and involves only decimal co-efficients. Hence it is very convenient and quick system for calculations.
6. In electricity, all the practical units like volt, ohm, ampere, henry, farad, coulomb, joule and watt accepted in industry and laboratories all over the world for well over a century have become absolute in their own right in the S.I. system without the need for any more practical units.

Disadvantages:

1. The non-S.I. time units 'minute' and 'hour' will still continue to be used until the clocks and watches are all changed to kilo seconds and mega seconds etc.
2. The base unit kilogram (kg) includes a prefix, which creates an ambiguity in the use of multipliers with gram.
3. SI units for energy, power and pressure (*i.e.*, joule, watt and pascal) are too small to be expressed in science and technology and, therefore, in such cases the use of larger units, such as MJ, kW, kPa will have to be made.
4. There are difficulties with regard to developing new S.I. units for apparent and reactive energy while joule is the accepted unit for active energy in S.I. systems.

D. Conversion Factors

1. Force:

$$\begin{aligned} 1 \text{ newton} &= \text{kg.m/sec}^2 = 0.012 \text{ kgf} \\ 1 \text{ kgf} &= 9.81 \text{ N} \end{aligned}$$

2. Pressure:

$$\begin{aligned} 1 \text{ bar} &= 750.05 \text{ mm Hg} = 0.9869 \text{ atm} = 10^5 \text{ N/m}^2 \\ &= 103 \text{ kg/m-sec}^2 \\ 1 \text{ N/m} &= 1 \text{ pascal} = 10^{-5} \text{ bar} = 10^{-2} \text{ kg/m-sec}^2 \\ 1 \text{ atm} &= 760 \text{ mm Hg} = 1.03 \text{ kgf/cm}^2 = 1.01325 \text{ bar} \\ &= 1.01325 \times 10^5 \text{ N/m}^2 \end{aligned}$$

3. Work, Energy or Heat:

$$\begin{aligned} 1 \text{ Joule} &= 1 \text{ newton metre} = 1 \text{ watt-sec} \\ &= 2.7778 \times 10^{-7} \text{ kW-h} = 0.239 \text{ cal} \\ &= 0.239 \times 10^{-3} \text{ kcal} \\ 1 \text{ cal} &= 4.184 \text{ joule} \\ &= 1.1622 \times 10^{-6} \text{ kWh} \\ 1 \text{ kcal} &= 4.184 \times 10^3 \text{ joule} = 427 \text{ kgf m} \\ &= 1.1622 \times 10^{-3} \text{ kWh} \end{aligned}$$

$$1 \text{ kWh} = 8.6042 \times 10^3 \text{ cal} = 860.42 \text{ kcal}$$
$$= 3.6 \times 10^6 \text{ joules}$$
$$1 \text{ kgf-m} = \left(\frac{1}{427}\right) \text{kcal} = 9.81 \text{ joules}$$

4. Power:

$$1 \text{ watt} = 1 \text{ joule/sec} = 0.86 \text{ kcal/h}$$
$$1 \text{ h.p.} = 75 \text{ m kgf/sec.} = 0.1757 \text{ kcal/sec.}$$
$$= 735.5 \text{ watt}$$
$$1 \text{ kW} = 1000 \text{ watts} = 860 \text{ kcal/h}$$

5. Specific beat:

$$1 \text{ kcal/kg-°K} = 4184 \text{ joules/kg-K}$$

6. Thermal conductivity:

$$1 \text{ watt/m-K} = 0.8598 \text{ kcal/h-m-°C}$$
$$1 \text{ kcal/h-m-°C} = 1.16123 \text{ watt/m-K}$$
$$= 1.16123 \text{ joules/s-m-K.}$$

7. Heat transfer co-efficient:

$$1 \text{ watt/m}^2\text{-K} = 0.86 \text{ kcal/m}^2\text{-h-°C.}$$
$$1 \text{ kcal/m}^2\text{-h-°C} = 1.163 \text{ watt/m}^2\text{-K}$$

The following conversion factors may be used to convert the quantities in non S.I. units into S.I. units.

To convert	*To*	*Multiply by*
angstroms	m	10^{-10}
atmospheres	kg/m^2	10332
bars	kg/m^2	1.02×10^4
Btu	joules	1054.8
Btu	kWh	2.928×10^{-4}
circular mils	m^2	5.067×10^{-10}
cubic feet	m^3	0.02831
dynes	newtons	10^{-5}
ergs	joules	10^{-7}
ergs	kWh	0.2778×10^{-13}
feet	m	0.3048
foot-pounds	joules	1.356
foot-pounds	kg-m	0.1383
gauss	tesla	10^{-4}
grams (force)	newton	9.807×10^{-3}
horse power (metric)	watts	735.5
lines/sq. inch	tesla	1.55×10^{-5}

To convert	*To*	*Multiply by*
Maxwell	webers	10^{-8}
mho	Siemens	1
micron	metre	10^{-6}
miles	km	1.609
mils	cm	2.54×10^{-3}
poundals	newton	0.1383
pounds	kilogram	0.454
pounds (force)	newtons	0.448
pounds/sq. ft.	N/m^2	47.878
pounds/sq. inch	N/m^2	6894.43

E. Important Engineering Constants and Expressions in S.I. Units

S.No.	*Engineering constants and expressions*	*M.K.S. system*	*S-I units*
1.	Value of g_0	9.81 kg-m/kgf-sec^2	1 kg-m/N-sec^2
2.	Universal gas constant	848 kgf-m/kg mole-°K	848 × 9.81 = 8341 J/kg-mole-°K (∵ 1 kgf-m = 9.81 joules)
3.	Gas constant (R)	29.27 kgf m/kg-°K	$\frac{8314}{29}$ = 287 joules/kg-K
4.	Specific heat (for air)	for air c_v = 0.17 kcal/kg-°K c_v = 0.24 kcal/kg-°K	for air c_v = 0.17 × 4.814 = 0.71128 kJ/kg-K c_v = 0.24 × 4.184 = 1 kJ/kg-K
5.	Flow through nozzle- Exit velocity (C_2)	91.5 U where U is in kcal	$44.7\sqrt{U}$ where U is in kJ
6.	Refrigeration 1 ton	= 50 kcal/min	=210 kJ/min
7.	**Heat transfer** The Stefan Boltzmann Law is given by	$Q = \sigma T^4$ kcal/m^2-h where $\sigma = 4.9 \times 10^{-8}$ kcal/h-m^2-°K^4	$Q = \sigma T^4$ kJ/m^2-h where $\sigma = 5.67 \times 10^{-8}$ W/m^{2}K^4

F. Dimensions of Quantities

Different units can be represented dimensionally in terms of units of length L, mass M, and time T and current I. The dimensions can be derived as under:

1. Velocity = length/time = $L/T = LT^{-1}$
2. Acceleration = velocity/time = $LT^{-1}/T = LT^{-2}$
3. Force = mass × acceleration = MLT^{-2}
4. Charge (coulomb) = current × time = IT
5. Work or energy = force × distance = ML^2T^{-2}

6. EMF or potential = work/charge = $ML^2T^{-2}/IT = ML^2I^{-1}T^{-3}$
7. Power = work/time = $ML^2T^{-2}/T = ML^2T^{-3}$
8. Current density = current/area = $I/L^2 = IL^{-2}$
9. Resistance = emf/current = $ML^2I^{-1}T^{-3}/I = ML^2I^{-2}T^{-3}$
10 Electric flux density = electric flux or charge/area = $IT/L^2 = ITL^{-2}$
11. MMF = current × number of turns = I
12. Conductance = I/resistance = $I/ML^2I^{-2}T^{-3} = I^2T^3M^{-1}L^{-2}$
13. Electric field intensity = volt/metre = $ML^2I^{-1}T^{-3}/L = MLI^{-1}T^{-3}$
14. Resistivity = $\frac{\text{resistance} \times \text{area}}{\text{lenght}} = (ML^2I^{-2}T^{-3})(L^2)/L = ML^3I^{-2}T^{-3}$
15. Magnetic field intensity (H) = MMF/length = $I/L = IL^{-1}$
16. Magnetic flux = emf × time = $(ML^2I^{-1}T^{-3})(T) = ML^2I^{-1}T^{-2}$
17. Magnetic flux intensity = magnetic flux/area = $(ML^2I^{-1}T^{-2})/L^2 = MI^{-1}T^{-2}$
18. Impedance = emf/current = $ML^2I^{-2}T^{-3}$
19. Admittance = 1/impedance = $I^2T^3M^{-1}L^{-2}$
20. Inductance = magnetic flux/current = $ML^2T^{-1}I^{-1}/I = ML^2T^{-2}I^{-2}$
21. Capacitance = electric charge/potential = $IT/ML^2T^{-3}I^{-1} = M^{-1}L^{-2}T^4I^2$

1.5 MECHANICAL MEASUREMENTS

Mainly the following two types of measurements are involved in the mechanical engineering field:

1. Mechanics type (or self-operated type)
2. Power type

1. **Mechanics type of measurements:** This type of mechanical measurements is *commonly applied to experimental or developmental programmes.*

   The following are the three distinct methods or approaches essential to determine the complete solution to complex problems in mechanical design:

   (*i*) The empirical method
   (*ii*) The rational method
   (*iii*) The experimental method.

   (*i*) The empirical method:

   In this method use is made of the knowledge of satisfactory previous performance, either personally observed or generally recognized as "good practice". The results of such observations are available in the form of "*thumb rules*" in handbooks and codes.

   - This method, however, *cannot* be relied upon especially in case of complex systems' design, since this method of design is mainly based upon the judgement of the designer.

   (*ii*) The rational method:

   This method is strictly based upon well established scientific laws and relationships.

   - The rational method has a very limited scope in areas of mechanical engineering *other than* mechanics and thermodynamics, where the availability of laws and relationships is rather scarce.

(*iii*) The experimental method:

In the experimental design method, *the component is designed by trial and error based on existing knowledge and the designed component is put on the job.* The trials are carried out intelligently and the interpretation of the error is done wisely.

The experimental design, in fact, requires the greatest application of engineering ingenuity.

2. **Power type of Measurements:** These types of measurements are *generally used for monitoring of operational measurement in control system.*
   - *The entire area of automation and control is based upon power type of measurements.*

1.6 METHODS OF MEASUREMENT

The *broad classification* of methods of measurement is as follows:

1. Direct comparison methods
2. Indirect comparison methods.

1. Direct comparison method:

In this method the parameter to be measured is *directly compared with either a primary or a secondary standard.*

Direct comparison is quite commonly used for measurements of length. However, for measurement of mass the problem becomes much more intricate since it is just not possible for human beings to distinguish between wide margins of mass.

- Generally, this method is not always the most accurate or the best, it is not sensitive enough also.

2. Indirect comparison method:

In this method *the comparison is done with a standard through the use of a calibrated system.*

These methods for measurements are used in those cases where the desired parameter to be measured is *difficult to be measured directly, but it has got some correlation in the some other parameter which can be easily measured.*

Example: The elimination of bacteria in milk is directly dependent upon its temperature. Thus the bacteria elimination can be measured indirectly by *measuring the temperature of the milk.*

– In indirect measurements an *empirical relation is generally established between the measurement actually made and the results that are desired.*

  - When the primary purpose of making a measurement is to determine quality of a product, then quality should be measured *directly*. However, in case direct measurement is not possible, then, *indirect* measurement should be made.
  - *The measurement systems used in engineering application make use of indirect methods for measurement purposes.*

– A measurement system consists of a *transducing element* which converts the quantity to be measured with an *analogous signal.* This signal after being processed by some *intermediate means* is then fed to the end devices which provide the measurement results.

1.7 MODES OF MEASUREMENT

Following are the *three* modes of measurement:

1. Primary measurements
2. Secondary measurements
3. Tertiary measurements.

1. Primary measurements

- In this case the *sought value of a parameter is determined by comparing it directly with "reference standards"*
- There is *no conversion* of measurand in terms of length.

  Examples:

  (*i*) Measurement of time by counting the number of strokes of a clock.
  (*ii*) Matching of two lenghts when determining the length of an object with a ruler.
  (*iii*) Matching of two colours when judging the temperature of a red hot steel.

2. Secondary measurements

The indirect measurements involving 'one translation' are called ***secondary measurements.***

Examples:

(*i*) The pressure measurement by manometers.
(*ii*) The temperature measurement by mercury-in-glass thermometers.

3. Tertiary measurements

The indirect measurements involving 'two conversions' are called ***tertiary measurements.***

Examples:

(*i*) The measurement of temperature of an object by thermocouple (Fig. 1.3). The primary signal (temperature of object) is transmitted to a thermocouple which generates a *voltage* which is a function of the temperature (first translation/conversion). The voltage in turn is applied to a voltmeter through a pair of wires. The second conversion is then voltage into length. The tertiary signal is transmitted to brain of the observer.

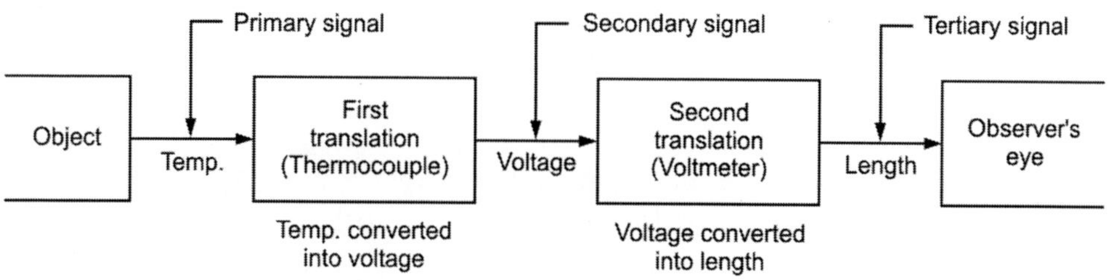

Fig. 1.3 Tertiary measurement: Measurement of temperature by a thermocouple.

(*ii*) The measurement of static pressure by a Bourdon tube pressure gauge,
(*iii*) The measurement of the speed of a rotating shaft by means of an electric tachometer. The unit of a measuring system where translation of measurand takes place is called the *transducer or translator.*

- Majority of measurement systems are tertiary systems and they include the whole range of mechanical, electrical, pneumatic electro-mechanical and electro-pneumatic instruments.

  Measurements may also be divided into following *two categories*:

  1. **Contact type.** In this case the sensor of the measuring device *contacts* the controlled medium.
  2. **Non-contact type.** Here the sensor *does not contact* the controlled medium.

- Non-contact measurements include *optical, radioactive and others.*

1.8 GENERALISED MEASUREMENT SYSTEM AND ITS FUNCTIONAL ELEMENTS

Figure 1.4. shows a measurement system (generalized) with different components (called *elements*).

The various elements are:

1. **Primary sensing element.** It is an element that is sensitive to the measured variable. The sensing elements *sense the condition, state or value of the process variable by extracting a small part of energy from the measurand, and then produce an output which reflects this condition, state or value of the measurand.*

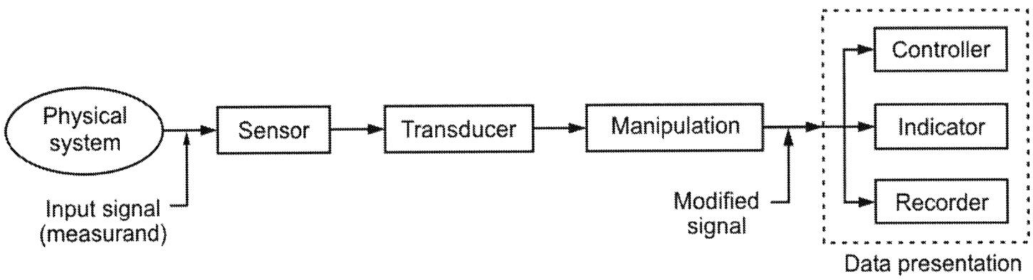

Fig. 1.4 Generalised measurement system.

2. **Variable conversion of transducer element.** This element *converts the signal from one physical form to another* without changing the information content of the signal.
3. **Manipulation element.** This element *operates* on the signal according to some *mathematical rule without changing the physical nature of the variable.*
4. **Data transmission element.** This element *transmits the signal from one location to another without changing its information contents.*
5. **Data presentation element.** This element *provides a display record or indication of the output from the manipulation elements.*

Stages of the general measurement system

The general measuring system consists of the following *three stages:*

1. **Stage-1:**
   - It senses desired input to exclusion of all others and *provides analogous output.*

     "*Types and Examples*":

     Mechanical: Contacting spindle, spring-mass, elastic devices (*e.g.*, Bourdon tube for pressure, proving ring for force), gyro.

     Electrical: Contacts, resistance, capacitance, inductance, piezoelectric crystals and polymers, thermocouple, semiconductor junction, etc.

     Optical: Photographic film, photoelectric diodes and transistors, photomultiplier tubes, holographic plates.

     Hydro-pneumatic: Buoyant float, orifice, venturi, vane, propeller.

2. **Stage-2:**
   - It *modifies transduced signal into form usable by final stage.*
   - It *usually increases amplitude and/or power* depending on requirement.

- It may also selectively filter *unwanted component or convert signal into pulsed form.*

  "*Types and Examples*":

  Mechanical: Gearing, cranks, slides, connecting links, cams, etc.

  Electrical: Amplifying or attenuating systems, bridges, filters, telemetering systems, various special purpose I.C. (Integrated circuit) devices.

  Optical: Mirrors, lenses, optical filters, optical fibers, spatial filters (pin hole, slit).

  Hydro-pneumatic: Piping, valving, dashpots, plenum chambers.

3. **Stage-3:**

- It *provides an indication or recording in* form that can be evaluated by an unaided human sense or by a controller.
- It *records data* digitally on a computer.

  "*Types and Examples*":

  Displacement types: Moving pointer and scale, moving scale and index, light beam and scale, electron beam and scale (oscilloscope), liquid column.

  Digital types: Direct alphanumeric readout.

  Recorder: Digital printing, inked pen and chart, direct photography, magnetic recording (hard disk or tape).

  Processor and computers: Various types of computing systems, either special purpose or general, used to feed read out/recording devices and/or controlling systems.

  Controllers. All types.

Illustration of measuring systems:

I. *Bourdon tube pressure gauge:*

Figure 1.5 shows the simple Bourdon tube pressure gauge — an example of a simple measurement system.

- In this case the Bourdon tube acts as the primary sensing element and a variable conversion element. It senses the input (pressure) quantity. On account of the pressure, the closed end of the Bourdon tube is displaced and thus the pressure is converted into a small displacement.

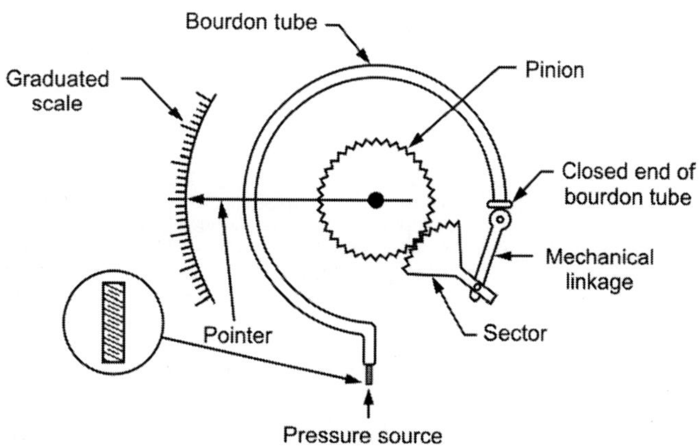

Fig. 1.5 Simple pressure Bourdon tube pressure gauge.

- The closed end of the Bourdon tube is connected to a gearing arrangement through mechanical linkage. The gearing arrangement *amplifies* the small displacement and consequently the pointer rotates through a large angle. Thus, the mechanical linkage acts as a data transmission element while the gearing arrangement acts as a data manipulation element.
- The final data presentation stage consists of a pointer and dial arrangement, which when calibrated with known, pressure inputs, gives an indication of the pressure signal applied to the Bourdon tube.

Figure 1.6 shows the schematic diagram of this measurement system.

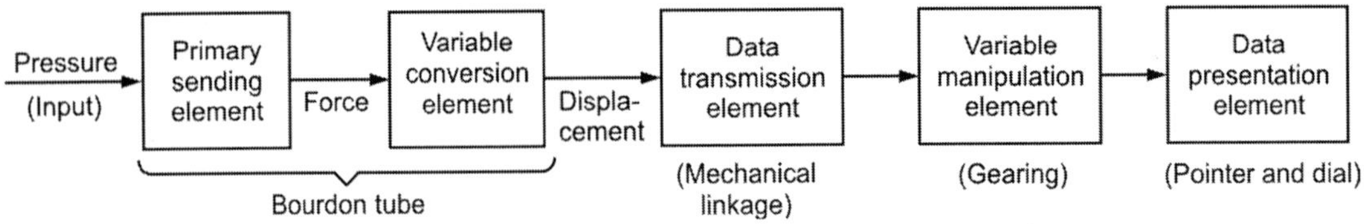

Fig. 1.6 Schematic diagram of a Bourdon tube pressure gauge.

II. Pressure actuated thermometer

Figure 1.7 shows a pressure-actuated thermometer.

- The liquid-filled bulb acts as the *primary sensor* and *variable conversion element* since a temperature change results in pressure build-up within the bulb, due to the constrained thermal expansion of filling liquid.
- The pressure tubing is employed to transmit the pressure to the Bourdon tube and thus functions as the *data transmission element.*

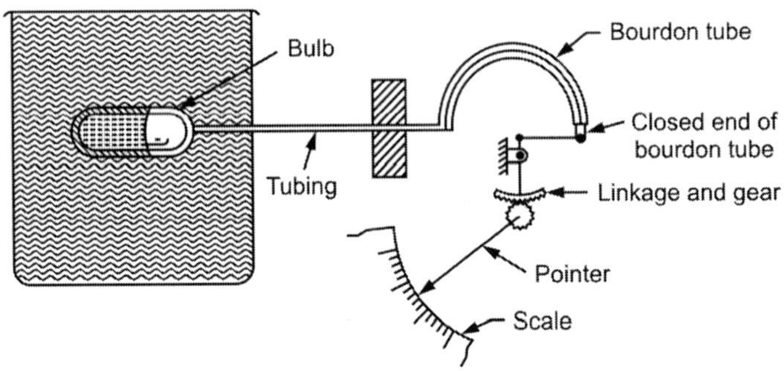

Fig. 1.7 Pressure-actuated thermometer.

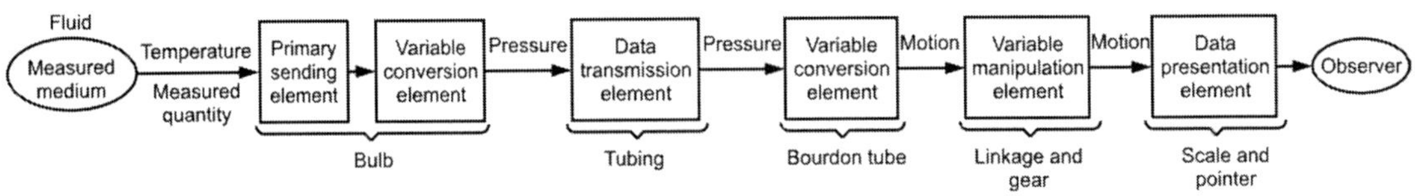

Fig. 1.8 Schematic diagram of a pressure-actuated thermometer.

- The Bourdon tube converts the fluid pressure into displacement of its tip, and such acts as the *variable conversion element.*
- The displacement is manipulated by the linkage and gear (*manipulation elements*) to give a larger pointer motion.
- The scale and the pointer serve as the data *presentation element.*

Figure 1.8 shows the schematic diagram of a pressure-actuated thermometer.

1.9 INSTRUMENTS

1.9.1 Introduction

The **instrument** *serves as an extension of human faculties and enables the man to determine the value of unknown quantity or variable which his unaided human faculties cannot measure.* A measuring instrument provides information about the physical value of some variable being measured.

- The instrument would sense a physical parameter (*e.g.,* velocity, pressure, temperature, etc., process and translate it into a format and range which can be interpreted by the observed.
- The instrument *must also provide the controls* by which the operator can obtain, manipulate and respond to the information.

  - An instrument, in *simple cases*, consists of a *single unit* which gives an output reading or sign according to the unknown variable (measurand) applied to it. Under *more complex measurements, situations*, a measuring instrument may consist of *transducing elements* which convert the measure and to analogous form. The *analogous signal* is then processed by some intermediate means and then fed to the end devices for presenting the measurement results for the purpose of display, record and control.
  - The instruments made by man not only retain their characteristics for extended periods of time but are also accurate and sensitive in their response.
  - *Measurements* involve the use of *instruments* as a physical means of determining quantities and variables.

1.9.2 Classification of Instruments

The instruments may be *classified* as follows:

1. Absolute and secondary instruments.
2. Analog and digital instruments.
3. Mechanical, electrical and electronic instruments.
4. Manual and automatic instruments.
5. Self-contained and remote indicating instruments.
6. Self-operated and power-operated instruments.
7. Deflection and null output instruments.

1. Absolute and secondary instruments

***(i) Absolute instruments*:** These *instruments give the magnitude of the quantity under measurement in terms of physical constants of the instrument.*

Examples: Tangent galvanometer and Rayleigh's current balance.

- These instruments are *seldom used except in 'standard institutions.*

***(ii) Secondary instruments*:** *These instruments are so constructed that the quantity being measured can only be measured by observing the output indicated by the instrument.*

These instruments are *calibrated* by comparison with an absolute instrument or another secondary instrument which has already been calibrated against an absolute instrument.

Examples: Voltmeter, glass thermometer, pressure gauge etc.

- The secondary instruments find *wide use in every sphere of measurement.*

2. Analog and digital instruments:

(*i*) *Analog instruments*: The signals of an analog unit vary in a continuous fashion and can take on infinite number of values in a given range.

Examples: Fuel gauge, ammeters and voltmeters, wrist watch, speedometer of an automobile, etc.

(*ii*) *Digital instruments*: Signals varying in *discrete steps* and taking on a finite number of different values in a given range are *digital signals* and the corresponding instruments are of *digital type.*

Examples: Odometer of an automobile, calibrated balance of a platform scale, timer on a scoreboard, etc.

- The digital devices have the following *advantages*:
  - High accuracy;
  - High speed;
  - Elimination of human operational errors.
- Owing to the application of digital computers for data handling, reduction and in automatic controls, the importance of *digital instrumentation* is increasing very fast. Thus, it becomes necessary to have both analog-to-digital (A/D) converters at input to the computers and digital-to-analog (D/A) converters at the output of the computers.

3. Mechanical, electrical and electronic instruments:

(*i*) *Mechanical instruments*:

- These instruments are very reliable for static and stable conditions.
- They are unable to respond rapidly to the measurements of dynamic and transient conditions due to the fact that they have moving parts that are rigid, heavy and bulky and consequently have a large mass. Mass presents inertia problems and hence these instruments cannot faithfully follow the rapid changes which are involved in dynamic instruments.
- Most of the mechanical instruments cause noise pollution.

(*ii*) *Electrical instruments*:

- The electrical methods of indicating the output of detectors *are more rapid than that of mechanical methods.*
- Unfortunately an electrical system normally depends upon a mechanical meter as an indicating device. This mechanical movement has some inertia due to which the *frequency response* of these instruments is poor.

(*iii*) *Electronic instruments*:

- Most of the scientific and industrial instruments require *very fast responses*. Such requirements cannot be met with by mechanical and electrical instruments.
- These instruments use *semiconductor devices. In electronic devices, since the only movement involved is that of electrons, the response time is extremely small owing to very small inertia of electrons.*
- With the use of electronic devices *very weak signals can be detected by using pre-amplifiers and amplifiers.*

Comparison between mechanical and electrical/electronic instruments:

Advantages and disadvantages of mechanical instruments:

Advantages:

(*i*) Simple in design and use.
(*ii*) More durable due to rugged construction.
(*iii*) Relatively cheaper in cost.
(*iv*) For operation of the instruments, no external power supplies are required.
(*v*) Quite reliable and accurate for the measurement of parameters which are stable and non-variant with time.
(*vi*) Long history of development and successful use.

Disadvantages:

(*i*) Poor frequency response to dynamic and transient measurements.
(*ii*) Large forces are required to overcome mechanical friction,
(*iii*) Incompatibility when remote indication or control is needed.
(*iv*) Cause noise pollution.

Advantages of electrical/electronic techniques of measurements:

(*i*) Less power consumption and less load on the system being measured.
(*ii*) Non-contact measurements are possible.
(*iii*) Light, compact and more reliable instrumentation.
(*iv*) Remote indication and recording feasible.
(*v*) The amplification produced is greater than that produced by a mechanical contrivance.
(*vi*) The mathematical processing of signals like summation, differentiating and integrating is possible.
(*vii*) The frequency and transient response is good.
(*viii*) Greater flexibility.
(*ix*) Lighter in construction.

- The electronic instruments are *widely used in detection of electro-magnetically produced signals such as radio, video and microwaves.*
- The use of these instruments has made it possible to *build analog and digital computers* which are quite instrumental in the modern developments in the fields of science and technology.

4. Manual and automatic instruments:

- In *case of manual* instruments services of an operator are required.
  Example: Measurement of temperature by a resistance thermometer incorporating a Wheatstone bridge in its circuit (an operator is required to indicate the temperature being measured).
- In an *automatic* type of instrument, an operator is not required.
  Example: Measurement of temperature by mercury-in-glass thermometer.

5. Self-contained and remote indicating instruments:

- A self-contained instrument *has all its different elements in one physical assembly.*
- In a remote indicating instrument, the primary sensing element *may be located at an adequate long distance* from the secondary indicating element. Such types of instruments are *finding wide use in the modern instrumentation technology.*

6. *Self-operated and power-operated instruments*

- A self-operated instrument does not require any outside power for its operation; the output energy is supplied wholly or almost wholly by the input signal.

  Examples: Mercury-in-glass thermometer, Dial indicator etc.

  The power operated instruments are those instruments which require some auxiliary source of power such as electricity, compressed air, hydraulic supply etc. for their operation. In such cases, the input signal supplies only an insignificant portion of the output power.

  Example: Electro-mechanical measurement system shown in Fig. 1.9.

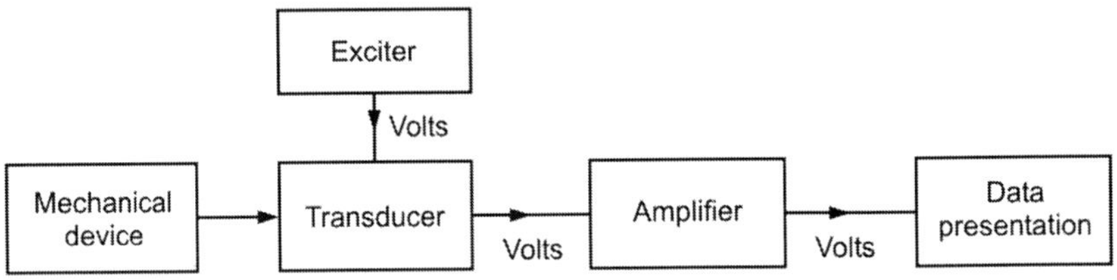

Fig. 1.9 Electro-mechanical measurement system.

7. *Deflection and null output instruments*

- In a "***deflection type***" *instrument*, the *deflection of the instrument provides a basis for determining the quantity under measurement*. The measured quantity produces some physical effect which deflects or produces a mechanical displacement of the moving system of the instrument. An opposing effect is built in the instrument which tries to oppose the deflection or the mechanical displacement of the moving system. The balance is achieved when opposing effect *equals* to cause producing the deflection or mechanical movement. The deflection or mechanical displacement at the point of balance then gives the value of measured quantity (which depends upon the calibration of the instrument).

  Examples: Permanent magnet moving coil (PMMC) ammeter etc.

  – These instruments are more suited for measurements under *dynamic conditions* than null type of instruments whose intrinsic response in slower.

- In a "*null type*" *of instrument, a zero or null indication leads to determination of the magnitude of measured quantity*. The null condition is dependent upon some other known conditions.

  Example: D.C. potentiometer.

  Advantages:

  (*i*) The "*null type*" instruments are *more accurate* than deflection type instruments.

  (*ii*) These instruments can be *highly sensitive* as compared with deflection type instruments.

1.9.3 Factors Relating to Selection of Instruments

The following are some important factors/points which are very essential to be considered for the proper selection of a correct instrument for any application:

1. The *accuracy* expected from the instrument.

   The following most significant and fundamental rule, which is perhaps the most important rule in measurement must always be kept in mind:

"Never demand an accuracy of measurement higher than that which is really needed, and never forget that each degree of accuracy, if demanded, is likely to have a disproportionate effect on the complexity and cost of measuring apparatus".

2. *When are the final data required* (*i.e.,* at the time of taking the measurement, or later one)?
3. The *cost criterion* (*i.e.,* the cost of the measuring process).
4. *In what form the data should be displayed* (*i.e.,* indicating, recording, integrating etc.)?
5. Whether quantity to be measured has *constant value or it is a time variant*?

   – In the measurement of time variant quantities the manner in which the quantity to be measured changes has a significant effect on the type of measuring instruments to be used. The *physical characteristics of the measuring device must always be taken into account since these characteristics determine the amount of interaction which takes place between the measurand quantity and the instrument.*

1.9.4 Functions of Instruments

Following are the *three* main functions of instruments:

1. Indicating function
2. Recording function
3. Controlling function.

1. Indicating function:

- The value of the quantity can be read by the movement of the needle on a calibrated scale provided on the instrument. Readings can be taken to any fraction within the limitations of the instrument and human eye.

  Examples: (*i*) The deflection of a pointer of a '*speedometer*' indicates the speed of the automobile at that moment,

  (*ii*) The use of a '*pressure gauge*' for indicating pressure.
- These days digital display (*i.e.,* the reading in numericals) is also commonly used. In fact where highly accurate and exact readings are desired, use of digital readouts is recommended.

2. Recording function:

- In several cases the instrument continuously records, with pen and ink, the value of measured quantity against some other variable or against time. A graph paper is provided whose speed can be adjusted. A recording instrument can be used to record more than one number of quantities also, in which case, either the recording of different parameters may be done by different pens or same pen may record one parameter after other.

  Example: A 'potentiometric type of recorder' used for monitoring temperature records the instantaneous values of temperatures on a strip chart recorder.

3. Controlling function:

- The controlling function is perhaps one of the most important functions particularly in the field of industrial control processes. Here, the information is used by the instrument/system to control the original measured quantity.

  Examples: (*i*) *Floats* for liquid level control.

  (*ii*) *Thermostats* for temperature control.

1.10 APPLICATIONS OF MEASUREMENT SYSTEMS

The instruments and measurement systems are employed for the following *applications*:

1. Monitoring of processes and operations.
2. Control of processes and operations.
3. Experimental engineering analysis.

1. Monitoring of processes and operations:

In these types of applications the measuring instruments *simply indicate the value or condition of parameter under study and do not serve any control function.*

Examples: (*i*) Water and electric energy meters installed in homes.

(*ii*) An ammeter or a voltmeter indicates the value of current or voltage being monitored at a particular instant.

2. Control of processes and operations:

There is a strong association between measurements and control. The instruments find a very useful application in automatic control systems.

- A common example is the typical refrigeration system which employs a thermostatic control. A temperature measuring device (often a bimetallic element) senses the room temperature, thus providing the information necessary for proper functioning of the control system.

3. Experimental engineering analysis:

The engineering problems can be solved by theoretical as well as experimental methods; several applications require the use of both the methods.

Following are the *uses* of experimental engineering analysis:

(*i*) To determine system parameters, variable and performance indices.

(*ii*) To formulate the generalised empirical relationships in cases where there is no proper theoretical backing.

(*iii*) To test the validity of theoretical predictions,

(*iv*) To solve mathematical relationships with the help of analogies.

1.11 STATIC AND DYNAMIC CHARACTERISTICS OF MEASURING INSTRUMENTS DEVICES

1.11.1 Introduction

The performance of an instrument is described by means of quantitative qualities, termed as ***characteristics***; these characteristics having been broken down into the sub areas of *static characteristics* and *dynamic characteristics.*

- The *static characteristics pertain to a system where the quantities to be measured are constant or vary slowly with time. Performance criteria based on dynamic relations (involving rapidly varying quantities) constitute dynamic characteristics.*

– The static characteristics, in a real sense, also influence the quality of measurement under dynamic conditions, but these characteristics (static) *show up as non-linear or statistical effects in otherwise linear differential equations giving the dynamic characteristics. These effects would make the differential equations analytically unmanageable and so the conventional approach is to treat the two aspects of the problem separately*. Thus, even though these effects influence the dynamic behaviour, the differential equations of dynamic performance generally *neglect* the effects of dry friction, backlash, hysteresis, statistical scatter etc.

- *The overall performance of an instrument is judged by a semiquantitative superimposition of the static and dynamic characteristics.*

1.11.2 Definitions Relating to Measuring Instruments

Some important terms used in connection with measuring instruments are discussed below:

1. **True or actual value.** *The actual magnitude of a signal input to a measuring system which can only be approached and never evaluated is termed as* ***true or actual value****.*
2. **Indicated value.** *It is the magnitude of a variable indicated by a measuring instrument.*
3. **Correction.** *The revision applied to the critical value so that the final result obtained improves the worth of the result is called* ***correction****.*
4. **Overall error.** *It is the difference of the scale reading and the true value.*
   – When the instrument is properly designed and correctly adjusted the consistent bias in error is very rare.
5. **Range.** *The region between the limits within which an instrument is designed to operate for measuring, indicating or recording a physical quantity is called the* ***range of the instrument****.*
6. **Sensitivity.** *The ratio of output response to a specified change in the input is called sensitivity.*
   - The minimum change in the measured variable which produces an effective response of the instrument is called "*Resolution sensitivity*". It is also called "*discrimination*".
   - The lowest level of measured variable which produces effective response of the instrument is called "*Threshold sensitivity*".
7. **Scale sensitivity.** *It is defined as the ratio of a change in scale reading to the corresponding change in pointer deflection.*
8. **Scale readability.** The scale readability (in analog instruments) *indicates the closeness with which the scale can be read.*
9. **Repeatability.** It is defined as the *variation of scale reading*; it is random in nature.
   - It is a measure of closeness with which a given input can be measured over and over again.
10. **Accuracy.** It may be defined as *conformity with or closeness to an accepted standard value (true value).*
    - Accuracy of an instrument is influenced by factors like static error, dynamic error, reproducibility, dead zone.
11. **Uncertainty.** Uncertainty denotes the *range of error*, *i.e.*, the region in which one guesses the error to be.

12. **Precision.** It refers to the *degree of agreement* within a group measurements.
    - It is usually expressed in *terms of the deviation in measurement.*
13. **Drift.** *An undesired gradual departure of the instrument output over a period of time that is unrelated to changes in input, operating conditions or lead is called* ***drift.***
14. **Linearity or non-linearity.** *Deviation of transducer output curve from a specified straight line.* The "*non-linearity*" may be: (*i*) *Terminal linearity* (deviation from a straight line through the end points); (*ii*) *Best-fit linearity* (deviation from the straight line which gives minimum errors, both plus and minus.
15. **Dead zone.** *It is the range within which variable can vary without being detected.*
16. **Dead time.** *It is the time before the instrument begins to respond after the measured quantity has been changed.*
17. **Speed of response.** *The quickness of an instrument to read the measured variable is called* ***speed of response.***
18. **Reproducibility.** *The degree of closeness with which the same value of a variable may be measured at different times is called* ***reproducibility.***
19. **Tolerance.** *It is the range of inaccuracy which can be tolerated in measurements.*
20. **Backlash.** It is defined as the *maximum distance or angle through which any part of a mechanical system may be moved in one direction without applying appreciable force or motion to the next part in a mechanical system.*
21. **Stiction (static friction).** *It is the force or torque that is necessary just to initiate motion from rest.*
22. **Noise.** It may be defined *extraneous disturbance generated in a measuring system which conveys no meaningful information w.r.t. desired signal.*

1.11.3 Static Characteristics

Measurements of applications in which parameter of interest is more or less constant; or varies very slowly with time are called **static measurements**. *A set of criteria (e.g., "accuracy", "error", "reproducibility", "drift", "sensitivity", "dead zone") that provide meaningful description of measurements under static conditions are called* **static characteristics**.

The main static characteristics may be summed up as follows:

(*i*) Accuracy
(*ii*) Sensitivity
(*iii*) Reproducibility
(*iv*) Drift
(*v*) Static error
(*vi*) Dead zone.

1.11.3.1 Accuracy, errors and correction

Accuracy:

It is the degree of correctness with which a measuring means yields the "true value" with reference to accepted engineering standards.

Accuracy is determined as the maximum amount by which the result differs from the true value (the average of an infinite number of measured values when the average deviation due to the various contributing factors tends to zero). It is *almost impossible to determine experimentally the true value. The true value is not indicated by any measurement system due to the loading effects, lags and mechanical problems (e.g., wear, hysteresis, noise pick-up etc.)*

Accuracy of the measured signal depends upon the following *factors*:

- Variation of the signal being measured;
- Intrinsic accuracy of the instrument itself;
- Accuracy of the observer;
- Whether or not the quantity is being truly impressed upon the instrument.

Errors:

The accuracy of any measurement system is measured in terms of its *errors*. Since it is impossible to measure the true value of a quantity, as such, an approximation of the "*true value*" obtained by sufficiently extended series of measurements and also taking into parameters and conditions to which corrections may be applied, we obtain, what is called the "*best-measured value*" of the *quantity*. *The difference between the best measured value and the true value of the quantity is known as* **static error**.

The absolute value of error does not indicate precisely the accuracy of measurements. It is the *relative static error* which provides the quality of measurement. The **relative static error** is *defined as the ratio of absolute static error to the true value of the quantity under measurement*.

Let, V_m = Measured value of quantity,

V_t = True value of quantity,

E_s = Static (or absolute) error of measurement, and

E_r = Relative static error.

Then, $$E_s = V_m - V_t \quad \text{...(1.1)}$$

and $$E_r = \frac{\text{Absolute error}}{\text{True value}} = \frac{E_s}{V_t} \quad \text{...(1.2)}$$

or, % age relative error, $\% E_r = E_r \times 100$ ...(1.3)

We have, $$V_t = V_m - E_s \quad \text{...[From eqn. (1.1)]}$$
$$= V_m - E_r \times V_t \quad \text{...[From eqn. (1.2)]}$$

or, $$V_t + E_r \times V_t = V_m$$

or, $$V_t = \frac{V_m}{1 + E_r} \quad \text{...(1.4)}$$

However, when E_r is small, $E_r << 1$,

Eqn. (1.4) may be written as:

$$V_t = V_m (1 - E_r) \quad \text{...(1.5)}$$

- The error may be either +ve or –ve. For +ve static errors the instrument reads *high* whereas it reads low for –ve static errors.

Correction

The difference between the true value and the measured value of a quantity is called **static correction** (C_s).

i.e., $$C_s = V_t - V_m (= -E_s) \quad \text{...(1.6)}$$

Equation (1.6) indicates that the static correction of the instrument reading is of the same magnitude as the error, but *opposite in sign* (*i.e.*, $C_s = -E_s$).

"**Accuracy**" may be expressed as:

(*i*) Point accuracy, or
(*ii*) Percent of scale range, or
(*iii*) Percent of true value.

(*i*) **Point accuracy.** In this case the accuracy of an instrument is stated for one or more points in a range.

Examples:

(*a*) A given thermometer may be stated to read ± 0.2°C between 50°C and 100°C.
(*b*) A scale of length may be read within ± 0.2 mm.

(*ii*) **Percent of scale range.** Here, the error is calculated on the basis of maximum value of the scale.

$$\% \text{ Error} = \left(\frac{V_m - V_t}{\text{Max. scale value}} \right) \times 100 \qquad ...(1.7)$$

Eqn. (1.7) refers to the intrinsic accuracy of the instrument itself and does not include procedural or personal performance.

(*iii*) **Percent of true value.** In this case the absolute error of measurement is expressed as percentage of true value of the unknown quantity.

i.e.,
$$\% \text{ Error} = \left(\frac{V_m - V_t}{V_t} \right) \times 100 \qquad ...(1.8)$$

– Here, with a drop in true value, the size or error decreases.
– In this type of expression of accuracy, the percentage error is the maximum for any point in the instrument range.

The percent accuracies on the basis of actual scale or full-scale reading are normally used to express equipment accuracies and do not include procedural or personnel performance which is actually included by the total error.

1.11.3.2 Static calibration

Static calibration is *a process by which all the static performance characteristics are obtained in one form or another*. In general, static calibration refers to a situation in which all inputs, whether desirable, interfering or modifying except one are kept at some constant values. Then the one input under study is varied over some range of constant values, which causes the output(s) to vary over some range of constant values. Thus an *output-input relationship is developed which comprise a static calibration valid under the stated constant conditions of all the other inputs*.

• The procedure of calibration involves a comparison of the particular instrument with either:

(*i*) *a primary standard,*
(*ii*) *a secondary standard with a higher accuracy than the instrument to be calibrated, or*
(*iii*) *an instrument of known accuracy.*

– All working instruments must be calibrated against some reference instruments which have a higher accuracy. The reference instruments in turn must be calibrated against instruments of still higher grade of accuracy or against primary standards or against other standards of known accuracy.

- The following **steps** are necessary in *performing a calibration*:

  1. Examine the construction of the instrument, and identify and list all the possible inputs,
  2. Decide as best as possible which of the inputs will be significant in the application for which instrument is to be calibrated.
  3. Procure apparatus that will allow to vary all the significant inputs over the ranges considered necessary. Procure standards to measure each input.
  4. By holding some inputs constants, varying others, and recording the output(s), develop the desired static input-output relations.

1.11.3.3 Range and span

Range. *The difference between the largest and the smallest readings of the instrument is called the* ***Range of an instrument****.* The range is expressed by stating the lower and upper values.

Span. *It represents the algebraic difference between the upper and lower range values of the instrument*. If the highest point of calibration is S_{max} units while the lowest is S_{min} units and that the calibration is continuous between the points, then we say that the instrument range is between S_{min} and S_{max}.

The *instrument span* is given by, $S_{max} - S_{min}$.

The above definitions apply both to analog as well as digital instruments.

Error calibration curve: Refer to Fig. 1.10

- It is an error curve which can be used for correcting instrument readings.
- The curve is plotted by calibrating the instrument against a suitable standard at a number of points on the scale.

  Examples: (*i*) Range: 2 kN/m^2 to 50 kN/m^2 ;

  Span : 50 – 2 = 48 kN/m^2

  (*ii*) Range : –5°C to 90°C ;

  Span : 90 – (–5) = 95°C.

- While determining the range of the instrument the frequency range must be taken into consideration. "*Frequency range*" is defined as frequencies over which measurements can be performed with a specified degree of accuracy.

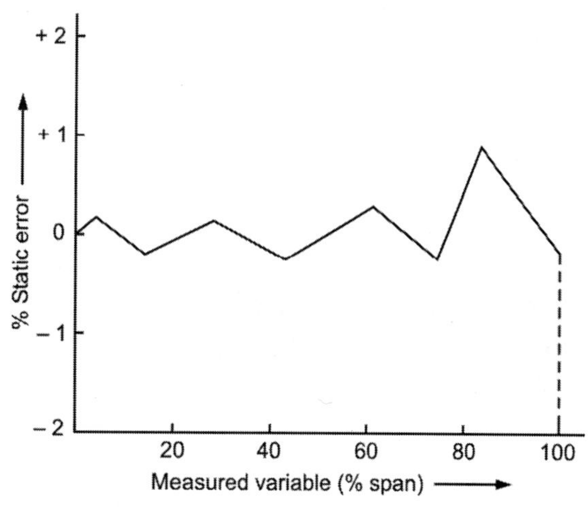

Fig. 1.10 Error calibration curve.

Example. A moving iron voltmeter may have a range of 0–120 V and 0–125 Hz frequency range with an accuracy of 0–5 or 1 per cent of full scale reading.

1.11.3.4 Scale readability

The term scale readability is frequently used in analog type of instruments. *It indicates the closeness with which the scale of an analog type of instrument can be read.*

The readability depends upon many *factors* such as:

(*i*) Number of graduations;
(*ii*) Spacing of graduations; .
(*iii*) Size of pointer;
(*iv*) Parallax effects;
(*v*) Discriminating power of the observer.

The readability, thus, is logically implied by the number of significant figures; the higher the number of significant figures, the better the readability of the instrument scale.

1.11.3.5 Repeatability and reproducibility

Although the meaning of the terms *repeatability* and *reproducibility* is same but they are applied in different contexts.

Repeatability *pertains to the closeness of output readings when the same input is applied repetitively over a short period of time with the same measurement conditions, same instrument and observer, same location and same conditions of use maintained throughout.*

Reproducibility *relates to the closeness of output readings for the same input when there are changes in the method of measurement, observer, measuring instrument, location, conditions of use and time of measurement.*

1.11.3.6 Drift

Drift *is an undesired gradual departure of the instrument output over a period of time that is unrelated to changes in input, operating conditions or load.*

An instrument is said to have no drift if it reproduces same readings at different times for same variation in measured variable.

The drift may be caused by the following *factors*:

(*i*) High mechanical stresses developed in some parts of instruments and systems.
(*ii*) Wear and tear.
(*iii*) Mechanical vibrations.
(*iv*) Temperature changes.
(*v*) Stray electric and magnetic fields.
(*vi*) Thermal e.m.fs.

Examples of drift:

- Drift occurs in "*resistance thermometers and thermocouples*" *due to the contamination of the metal and a change in its atomic or metallurgical structure.*
- Drift may occur in "*flow meters*" *due to wear and erosion of the orifice plate, nozzle or venturimeter.*

Classification of drift:

Drift may be *classified* as follows:

1. Zero drift; 2. Span drift; 3. Zonal drift.

1. **Zero drift:** Drift is called *zero drift if the whole of instrument calibration gradually shifts over by the same amount.*

   – It may be due to permanent set or slippage and can be corrected by *shifting pointer position.*

2. **Span drift:** *If the calibration from zero upwards changes proportionally, it is called "span (or sensitivity) drift".*
   - It may be due to change in spring gradient etc.
3. **Zonal drift:** *When the drift occurs only over a portion of span of an instrument, it is called "zonal drift".*
   - It industrial instruments, drift is an undesirable quantity since it is rarely apparent and cannot be easily compensated for.

Draft occurs very slowly and can be checked only by periodic inspection and maintenance.

1.11.3.7 Accuracy and precision

Accuracy. *The closeness with which an instrument reading approaches the true value of the quantity being measured is called* **accuracy**. Accuracy is determined as the maximum amount by which the result differs from the true value. Accuracy of an instrument is influenced by factors like static error, dynamic error, reproducibility, dead zone.

Precision. The term 'precise' means clearly or sharply defined. Precision is a *measure of reproducibility of measurements*. It is usually expressed in terms of the deviation in measurement.

- ***Significant figure*** *is an indication of the precision of measurement*. Significant figures convey actual information regarding the magnitude of precision.

Following ***examples*** will illustrate the difference in meaning of the two terms *accuracy* and *precision*:

(*i*) Let us consider the case of an ammeter having high precision which can take readings to 1/150 of an ampere, but has a wrong zero adjustment. Now we can take readings to 1/150 of an ampere which are consistent and clearly defined since the ammeter is *precise* as ever. However, the readings taken with this ammeter are not accurate, since they do not confirm to truth because of faulty adjustment of zero.

(*ii*) Take the case of a micrometer which is normal in every respect but its anvil is displaced from its true position. When the readings are taken by this instrument, these would be clearly defined and consistent and we would say that micrometer is as precise as ever. However, the readings do not conform to the truth as anvil is displaced from its true position. Thus, the readings taken by this micrometer are not accurate.

- It is worth noting that *precision is a necessary prerequisite to accuracy but it does not guarantee accuracy*.

1.11.3.8 Sensitivity

The ratio of the magnitude of output signal to the input signal or response of measuring system to the quantity being measured is called **sensitivity.**

It is represented by the slope of the calibration curve if the ordinates are expressed in actual units.

- When the calibration curve is linear [Fig. 1.11(*a*)], the sensitivity is constant. However, if the calibration curve is non-linear [Fig. 1.11(*b*)] the sensitivity is different at different points, being the slope of curve at various points.

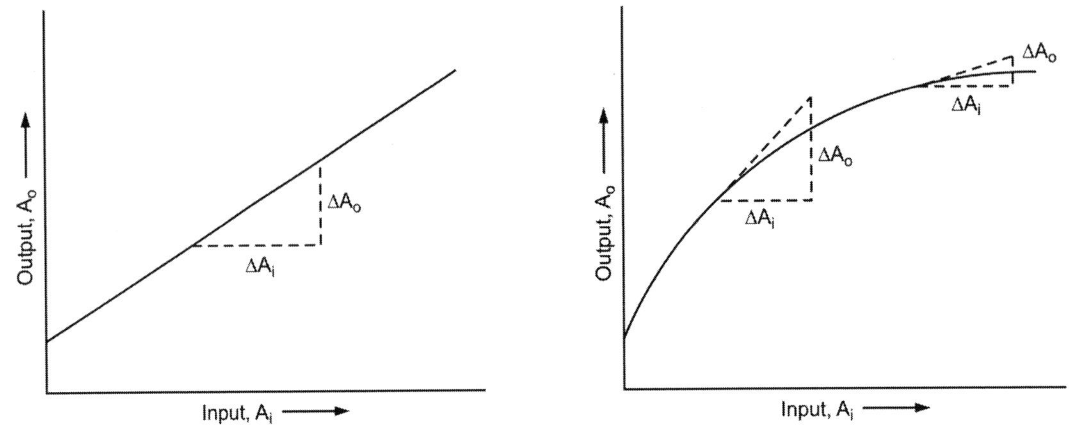

Fig. 1.11 Static sensitivity.

Static sensitivity, in general, is defined as:

$$\text{Static sensitivity(S)} = \frac{\text{Infinitesimal change in output}}{\text{Infinitesimal change in input}} = \frac{\Delta A_o}{\Delta A_i} \quad ...(1.9)$$

Similarly, *Inverse sensitivity* or *deflection factor* or *scale factor* $= \dfrac{\Delta A_i}{\Delta A_o}$ ...(1.10)

- The *sensitivity of an instrument should be high and thus the range should not be high in comparison to the value being measured.*
- Sensitivity has a wide range of units, and these depend upon the instrument or measurement system being investigated.

The *following points are worth noting*:

– In case the form of input to and output from the measurement system used with electrical/ electronic equipment is same, the term **gain** is used instead of sensitivity.

– The *increase* in displacement with the optical and mechanical instruments is described by the term **amplification.**

– When the input or output signal is changing with time, the term **transfer function** or **transfer operator** is used (in place of the terms sensitivity, gain or amplification—which apparently may the same and describe a relationship between output and input).

1.11.3.9 Linearity

The ability to reproduce the input characteristics symmetrically is called **Linearity.** It can be expressed by the straight line equation.

The linearity is simply a measure of maximum deviation of any of the calibration points from the straight line (drawn by using the method of least square from the given calibration data). Figure 1.12. shows the actual calibration curve and a straight line drawn from the origin using method of least squares.

Any departure from straight line relationship is ***non-linearity***. The non-linearity may be due to the following factors:

(*i*) Viscous flow or creep;
(*ii*) Non-linear elements in the measurement device;

(*iii*) Mechanical hysteresis;

(*iv*) The elastic after-effects in the mechanical system.

Two common types of non-linearity are:

1. **Terminal linearity.** It is the deviation from a straight line through the end points.
2. **Best fit linearity.** It is the deviation from the straight line which gives minimum errors, both plus and minus.

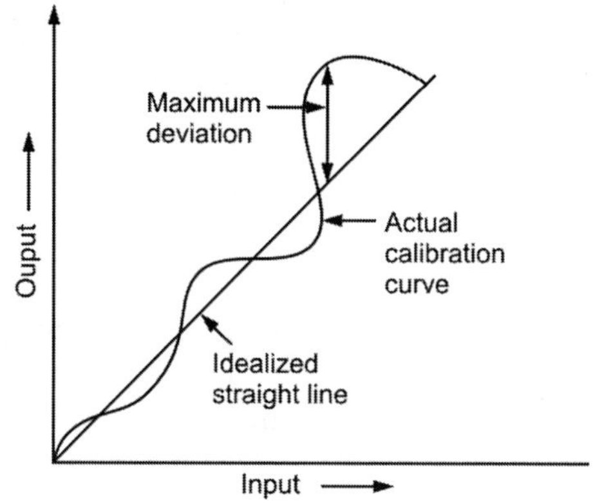

Fig. 1.12 Linearity exhibited as deviation between actual calibration curve and idealized straight line relationship.

- **Linearity** (*i.e.*, output is linearly proportional to input) is considered to be one of the best characteristics of an instrument or measurement system. This is so, because the conversion from a scale reading to the corresponding value of input quantity is very convenient. However, it may be understood that a non-linear behaviour does not essentially lead to inaccuracy; an instrument having non-linear calibration curve may be highly accurate as ever.
- Since accuracy and linearity are related closely to each other, it is better to keep the non-linearity as small as possible so as to result in small errors.

1.11.3.10 Hysteresis

The maximum differences in output at any measured value within the specified range when approaching the point first with increasing and then with decreasing input may be termed as **hysteresis.**

- It is a phenomenon which shows different output effects when loading and unloading. It is *non-coincidence of loading and unloading curves.*

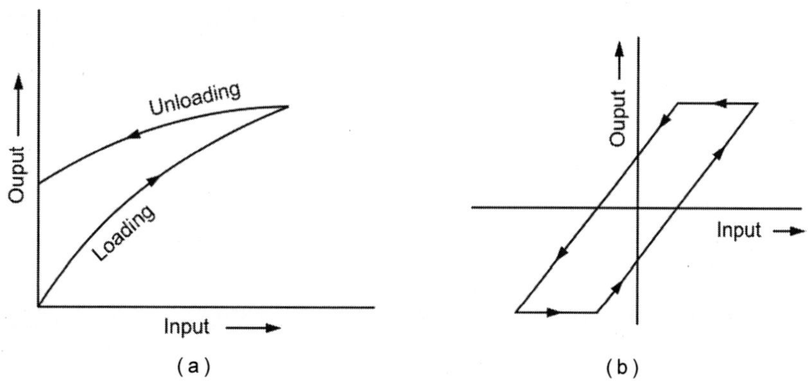

Fig. 1.13 Hysteresis effects.

Figure 1.13(*a*) shows output and input curves (loading and unloading) for an instrument which has no friction due to sliding parts. The non-coincidence of loading and unloading curves is on account of *internal friction or hysteresis damping.*

Figure 1.13(*b*) shows the input-output relations of instruments which do not have internal friction but have external sliding friction.

- The numerical value of hysteresis can be specified in terms of either output or input and is usually given as percentage *of full scale.*
- Hysteresis results from the *presence of irreversible phenomenon such as:*
  - Mechanical friction;
  - Slack motion in bearings;
  - Magnetic and thermal effects.

1.11.3.11 Threshold and Resolution

The minimum value below which no output change can be detected when the input of an instrument is increased gradually from zero is called the **threshold** *of the instrument.* Thus threshold defines that minimum value of input which is necessary to cause a detectable change from zero output.

- In a *digital system,* threshold is the input signal necessary to cause one least significant digit of the output reading to change.

  Threshold may be caused by *backlash or internal noise.*

Resolution or Discrimination:

When the input is slowly increased from some arbitrary (non-zero) input value, it is observed that the output does not change at all until a certain increment is exceeded; this increment is called **Resolution or discrimination** of the instrument. Thus resolution defines the smallest change of input for which there will be a change of output.

- In case of *analog instruments,* the resolution is determined by the observer's ability to judge the position of a pointer on a scale. Resolution is usually reckoned to be no better than ±0.2 of the smallest division of the scale.
- In case of digital instruments, resolution is determined by the number of neon tubes taken to show the measured value.
  - **Threshold** defines the *smallest measurable input* while the **resolution** defines the *smallest measurable input change.*
  - *"Threshold"* and *"resolution"* may be *expressed as an actual value or as a fraction or percentage of full scale value.*

1.11.3.12 Dead zone and dead time

Dead zone:

Refer to Fig. 1.14. *The largest change of input quantity for which there is no output of the instrument is termed as* ***dead zone.*** It may occur due to friction in the instrument which does not allow pointer to move till sufficient driving force is developed to overcome the friction loss.

"Dead zone" is caused by backlash and hysteresis in the instrument.

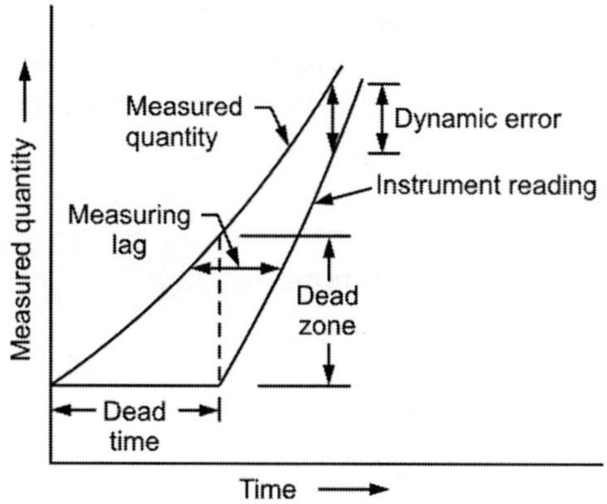

Fig. 1.14 Dead zone and dead time.

Dead time:

Refer to Fig. 1.14. *The time required by a measurement system to begin to respond to a change in the measurand is termed as* ***dead time***. It represents the time before the instrument begins to respond after the measured quantity has been changed.

1.11.4 Dynamic Characteristics

When an instrument is required to measure a time-varying process variable, one has to be concerned with *"dynamic characteristics"*, these *characteristics quantify the dynamic relation between the input and output.*

1.11.4.1 Dynamic response

The evaluation of the ability of a system to faithfully transmit and present all the pertinent information included in the input signal and to exclude all else, is called the **response**.

The behaviour of the system when inputs vary with time (i.e., inputs are dynamic in nature) and so does the output, is called **dynamic response.** Such a response in measurement systems is found to occur in industrial, aerospace and biological applications.

The dynamic inputs are of two types:

(*i*) Steady state periodic; (*ii*) Transient.

The *"steady state periodic"* quantity is one whose magnitude has a definite repeating time, whereas the time variation of a *"transient"* magnitude does not repeat.

- The response of a measurement system subjected to a time varying input can be divided into the following two parts:

  (*i*) Steady state response; (*ii*) Transient response.

"Steady state response" is simply the response when time reaches *infinity*.

"Transient response" in measurement systems, is defined as the part of response which goes to *zero* as time becomes *large*.

- **Periodically varying inputs may be:**
  - A *purely sinusoidal input* like the A.C. voltage with a frequency of 50 Hz.

    or

  - A complex time varying repetitive input which is a *combination of the fundamental* and *harmonics*.

When the systems are subjected to periodically varying inputs, they exhibit in their response a magnitude and phase relationship which is *different* from that of the input signal because of the *energy storage elements. The output is not a faithful representation of the input.* The harmonic contents of the input are not reproduced in the output both in magnitude and phase relationship as in the input. *The output is distorted both in terms of magnitude as well as phase relationships.*

1.11.4.2 Dynamic characteristics of a measurement system

The *dynamic characteristics* of a measurement system are:

1. Speed of response ... *Desirable*
2. Measuring lag ... *Undesirable*
3. Fidelity ... *Desirable*
4. Dynamic error ... *Undesirable*

1. **Speed of response.** The *"speed of response"* or *"responsiveness"* is defined as the *rapidity with which a measurement system responds to changes in the measured quantity.*
2. **Measuring lag.** It refers to *retardation or delay in the response of a measurement system to changes in measured quantity.*

   The lag *is caused by conditions such as capacitance, inertia, or resistance.*

   The measuring lags are of the following two types:

   (*i*) ***Retardation type lag.*** In this type of measuring lag the response of the measurement system *begins immediately after a change in measured quantity has occurred.*

   (*ii*) ***Time delay type lag.*** In this case the response of the measurement system *begins after a dead time after the application of the input.* If the measured quantity varies at a fast rate, the dead time has a severe adverse effect on the performance of the system. The measurement lags of this type are *very small* and are of the order of a fraction of a second and hence can be *ignored.*
3. **Fidelity.** It is defined as the *degree to which a measurement system indicates changes in the measured quantity without any dynamic error.* It refers to the ability of the system to reproduce the output in the same form as the input.
4. **Dynamic error.** The dynamic error, also called ***"measurement error"***, is the *difference between the true value of the quantity changing with time and the value indicated by the measurement system if no static error is assumed* (see Fig. 1.14.)
   - *The maximum amount by which the pointer moves beyond the steady state is called* **overshoot**.

Example 1.1. *On the application of a step input of 80 bar to a pressure gauge, the pointer swings to pressure of* 82.5 *bar and finally comes to rest at 81.3 bar. Determine:*

(*i*) *The overshoot of the gauge reading and express it as a percentage of the final reading.*

(*ii*) *The percentage error of the gauge.*

Solution: (*i*) Overshoot of the gauge reading = 82.5 – 81.3 = **1.2 bar (Ans.)**

$$\text{Percentage overshoot} = \frac{1.2}{81.3} \times 100 = \mathbf{1.476\%\ (Ans.)}$$

(*ii*) $$\text{Percentage error} = \frac{81.3 - 80}{80} \times 100 = \mathbf{1.625\ (Ans.)}$$

1.11.4.3 'Dynamic' analysis of measurement systems

The dynamic behaviour of measurement systems is studied in the following two domains:

1. Time domain analysis.
2. Frequency domain analysis.

1. Time domain analysis:

In this the input signal is applied to the measurement system and the behaviour of the system is studied as a *function of time.* The dynamic response of the system to different types of inputs, which are a function of time is analysed at different intervals of time after the application of the input signal. In most cases, the actual input signals vary in random fashion with respect to time and therefore cannot be mathematically defined. Consequently the performance of a system can be analysed (in the time domain analysis) by using the following standard *test signals/inputs:*

(*i*) Step input; (*ii*) Ramp input;
(*iii*) Parabolic input; (*iv*) Impulse input.

2. Frequency domain analysis:

This type of analysis of a system *pertains to the steady state response of the system to a sinusoidal input.* Here, the system is subjected to a sinusoidal input and the system response is studied with *frequency as the independent variable.*

- *Frequency response.* It is the maximum frequency of the measured variable that an instrument is capable of following without error. The usual requirement is that the frequency of measurand should not exceed 60 per cent of the natural frequency of the measuring instrument.

Standard test signals/inputs:

The most common standard inputs used for dynamic analysis are discussed below:

1. *Step function*:

Refer to Fig. 1.15(*a*). It is a *sudden change from one steady value to another.*
It is mathematically represented by the relationship:

$$x = 0 \quad \text{at} \quad t < 0$$
$$x = x_c \quad \text{at} \quad t \geq 0$$

where x_c is a constant value of the input signal x_i.

- The *"transient response" indicates the capacity of the system to cope with changes in the input signal.*

2. *Ramp or linear function*:

In this case, the *input varies linearly with time.*
This input is mathematically represented as:

$$x = 0 \quad \text{at} \quad t < 0$$
$$x = \Psi \quad \text{at} \quad t \geq 0$$

where, Ψ is the slope of the input versus time relationship.

- *The ramp-response becomes indicative of the steady state error in following the changes in the input signal.*

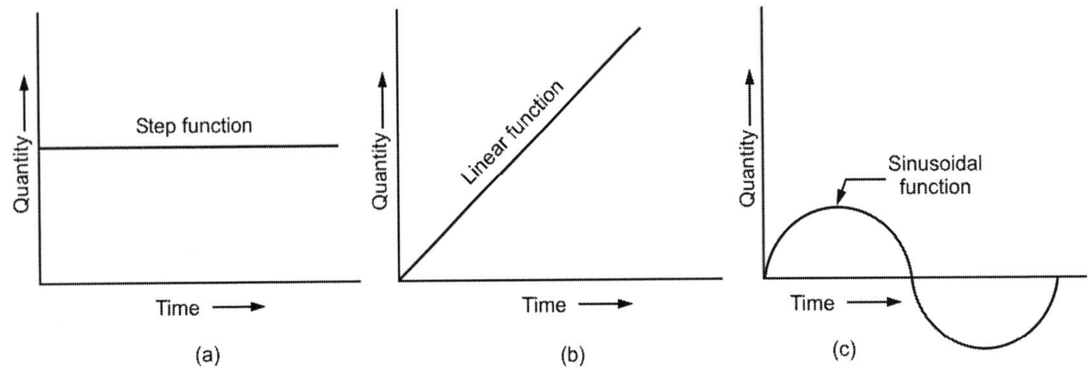

Fig. 1.15 Standard input function.

3. ***Sinusoidal function:***

In this case the *input varies sinusoidally with a constant maximum amplitude.* It is represented mathematically as follows:

$$I_i = A \sin \omega t$$

where, A = Amplitude, and

ω = Frequency in rad/s.

- *The frequency or harmonic response is a measure of the capability of the system to respond to inputs of cyclic nature.*

A general measurement system can be mathematically described by the following differential equation:

$$(A_n D^n + A_{n-1} D^{n-1} + ... + A_1 D + A_0) I_0 = (B_m D^m + B_{m-1} D^{m-1} + ... + B_1 D + S_0) I_i. \qquad ...(1.11)$$

where, A's and B's = Constants, depending upon the physical parameters of the system,

D^k = Operative derivative of the order k,

I_o = The information out of the measurement system, and

I_i = The input information.

The order of the measurement system is generally classified by the value of the power of n.

1. *Zero-order system:* $n = 0$ and $A_1, A_2, A_3 ..... A_n = 0$
2. *First-order system:* $n = 1$ and $A_2, A_3, A_4 ..... A_n = 0$
3. *Second-order system:* $n = 2$ and $A_3, A_4, A_5 ..... A_n = 0$

The above *method of classification is used for most of the instruments and systems.*

Examples of zero-order system:

(*i*) Mechanical levers;

(*ii*) Amplifiers;

(*iii*) Potentiometer (It gives and output voltage which is proportional to wiper's displacement) etc.

The zero-order system is characterised only by the static sensitivity (parameter), the value of which is obtained through the process of static calibration.

Examples of first-order system:

(*i*) Velocity of a true falling mass;
(*ii*) Air pressure buildup in bellows;
(*iii*) Measurement of temperature by mercury-in-glass thermometers,
(*iv*) Thermistors and thermocouples;
(*v*) Resistance-capacitance network.

Examples of second-order system:

(*i*) Piezoelectric pickeup;
(*ii*) Spring-mass system (used for acceleration and force measurements);
(*iii*) Pen control system on X-Y plotters;
(*iv*) U.V. galvanometer, etc.

Damping ratio:

In the design of instruments a term which is very frequently used is the ' *'damping ratio''* (γ) defined as the *ratio of the actual value of coefficient of viscous friction in movement and the value required to produce critical damping.*

i.e.,

$$\gamma = \frac{A_1}{2\sqrt{A_0 A_2}}$$

This dimensionless term is very useful because to determine its value, it is not necessary that the values of A_1, A_0 and A_2 may be known. In practice it is not easy to determine accurately the values of A_1 and A_2. Further, even if these values are known, they do not in themselves specify whether the instrument is under, over or critically damped, since a numerical calculation has to be performed with them first. Therefore, designers find *"damping ratio"* as a very convenient measure of the amount of damping present in the movement.

The terms *damping ratio* (γ) and *underdamped natural frequency* (ω_n) immediately conjure up a physical picture of the response of an instrument and both of the quantities are very easy to measure. Thus γ and ω_n easily do away with quantities A_2, A_1 and A_0.

1.12 TYPES OF ERRORS

In measurement systems, the *errors could originate from several sources*. Broadly these may be *classified* as follows:

1. *Gross errors*
2. *Systematic errors:*

   (*i*) Instrumental errors
   (*ii*) Environmental errors
   (*iii*) Observational errors

3. *Random errors.*

1.12.1 Gross Errors

These errors occur due to *human mistakes* in reading instruments and recording and calculating results of measurement. Although it is probably impossible to eliminate the gross errors completely, yet one should try to anticipate and correct them.

The mathematical analysis of gross errors is impossible since these may occur in different amounts. While some gross errors may be easily detected, others may go unnoticed.

These errors can be *avoided* by adopting *two means:*

(*i*) Immense care should be taken while taking the reading and recording the data.

(*ii*) Two, three, or even more readings should be taken for the quantity being measured.

- Designers of instruments to some extent can take care to avoid gross errors but much would depend upon the interest and competence of observer.

1.12.2 Systematic Errors

The systematic errors are repeated consistently with the repetition of the experiment and are caused by *such effect as:*

– sensitivity shift;

– zero off-set;

– known non-linearity.

- These errors *cannot* be determined by direct and repetitive observations of the measurand made each time with same technique.

These errors can be located only by having repeated measurements under different conditions or with different equipment and where possible by an entirely different method.

These errors may be *instrumental, environmental or observational errors.*

1. Instrumental errors:

These errors arise *due to:*

— Inherent shortcomings in the instruments.

— Misuse of the instruments.

— Loading effects of instruments.

Instrument errors are inherent in instruments due to construction, calibration or operation of the instruments. *These errors may cause the instrument to read too low or too high.*

Some *"examples"* of such errors are:

(*i*) Mechanical friction and wear, backlash, yielding of supports, pen or pointer drag and hysteresis of elastic members due to aging.

(*ii*) Constructional faults resulting from finite width of knife edges, excessive friction at the mating parts, lost motion due to necessary clearance in gear teeth etc.

(*iii*) Improper selection and poor maintenance of the instrument.

(*iv*) Unavoidable physical phenomenon due to friction, capillary attraction and imperfect rarefaction.

Assembly errors. These errors may be due to the following:

(*i*) Displaced scale, *i.e.*, incorrect fitting of the scale zero with respect to the actual zero position of the movement.

(*ii*) Non-uniform division of the scale.

(*iii*) Bent or distorted pointer.

The assembly errors do not change with time, so these can be easily discovered and rectified.

Misuse of instruments. A good instrument used in an unintelligent way may give erroneous results.

- *Loading effects.* In several measurement techniques, the detector/sensor utilizes the energy of the medium, thus altering slightly the parameter to be measured, resulting in errors.

2. Environment errors:

The environmental errors are due to conditions external to the measuring device (*e.g.,* effects of pressure, temperature, humidity, dust, vibrations, external magnetic or electrostatic fields). These errors are much more troublesome than assembly errors.

These errors can be *eliminated or reduced* by taking the following measures:

(*i*) Using the instrument in controlled conditions of pressure, temperature, humidity in which it was originally assembled and calibrated,

(*ii*) Measure deviations in the local conditions from the calibrated ones and then apply a suitable corrections to the instrument readings,

(*iii*) Automatic compensation for departures from the calibrated conditions by using sophisticated devices.

(*iv*) Make a complete new calibration under the local conditions,

(*v*) In case it is suspected that external magnetic or electrostatic fields can effect the readings of the instruments, magnetic or electrostatic shields may be provided.

(*vi*) Apply computed corrections. Normally, efforts are made to avoid the use of application of computed corrections, but where these corrections are needed and are necessary, they are incorporated for the computation of the results.

3. Observational errors:

These errors occur due to *carelessness of operators.* Even when an instrument has been properly selected, carefully installed and faithfully calibrated, shortcoming in the measurement occurs due to certain failings on the part of the observer.

The observational errors may be due to the following *reasons:*

(*i*) ***Parallax***—Since the parallax errors arise on account of pointer and the scale not being in the same plane, we can eliminate this error by having the pointer and the scale in the same plane.

(*ii*) Wrong scale reading, and wrong recording of data.

(*iii*) Inaccurate estimates of average reading, lack of ability to interpolate properly between graduations.

(*iv*) Incorrect conversion of units in between consecutive readings, and non-simultaneous observation of interdependent quantities.

(*v*) Tendency to read high or low, anticipate a signal and read too soon—Personal bias.

- Modern electrical instruments have *digital display of output* which completely eliminates the errors on account of human observational or sensing power as the output is *in the form of digits.*

1.12.3 Random Errors

- The random errors are *accidental, small and independent.*
- *They vary in an unpredictable manner.*
- The magnitude and direction of these errors *cannot* be predicted from a knowledge of measurement system; however these errors are *assumed to follow the law of probabilities.*

The most common causes of these errors are:

(*i*) Friction in instrument movement.
(*ii*) Backlash in the movement.
(*iii*) Parallax errors between pointer and scale.
(*iv*) Finite dimensions of the pointer and scale divisions.
(*v*) Hysteres in elastic members.
(*vi*) Mechanical vibrations.

1.12.4 Sources of Errors

Besides the errors so far enumerated (of the physical and mechanical nature) and the errors which necessarily result from the faulty calibration of a measuring systems, the errors arise from the *following sources also:*

(*i*) **Noise.** *It is defined as any signal that does not convey useful information.*

— Noise signals/errors can be reduced to a minimum level through filtering, careful selection of components, shielding and isolation of the entire measuring system.

(*ii*) **Response time.** *It is defined as the time taken by the instrument to show 63.2% change in a reading to a step input.* This factor contributes to the uncertainty of the measurement.

(*iii*) **Design limitations.** In the design of the instrument, certain inevitable factors such as friction and resolving power lead to uncertainty of measurements.

(*iv*) **Energy exchanged by interaction.** As earlier pointed out, whenever the energy required for operating the measuring system is extracted from the measurand, the value of the latter is altered to a greater or leaser extent. This alteration is dependent upon the capacity of the system.

(*v*) **Transmission.** During the transmission of information from the primary sensing element to the indicator, the signal may be *attenuated* due to any of the following *reasons:*

— It may suffer loss through leakage;
— It may be absorbed or otherwise consumed in the communication channel;
— It may be distorted by resonance, attenuation or delay phenomenon whose actions are selective on various signal component.

(*vi*) **Deterioration of measuring system.** The examples of some types of deterioration which occur in a measuring system and result in as a source of error in measurement are:

— Change in resistance of a circuit element through strain relief;
— Alteration of thermocouple characteristics through oxidizing or reducing atmosphere;
— Weakening of a permanent magnet.

(*vii*) **Ambient influences on measuring systems.**
(*viii*) **Errors of observation and interpretation.**

1.12.5 Statistical Analysis of Test Data

The *systematic and random errors* are evaluated and studied by *statistical procedures* which make it possible to state from a limited group of data the most probable value of a quantity, the probable uncertainty of a single observation, and the probable limits of uncertainty of the best value that can be derived from the data.

— The object of the statistical methods based on laws of chance which operate only on random errors and not on systematic errors, is to achieve consistency (precision) of value and not their accuracy.

The experimental data is obtained in the following *two forms of tests:*

(*i*) Single-sample test;

(*ii*) Multi-sample test.

(*i*) **Single-sample test.** *A single measurement (or succession of measurements) done under identical conditions excepting for time is known as "single-sample' test".*

(*ii*) **Multi-sample test.** In a multi-sample test, *repeated measurements of a given quantity are taken under different test conditions such as using different instruments, different ways of measurement and by employing different observers.* This type of test does not materialise result by simply making measurements with the same equipment, procedure, technique and same observer.

- For getting the exact value of the quantity under measurement, tests should he conducted using as many different procedures, techniques as experiments as practicable. It should be understood that the statistical means which help us to arrive at correct results are only valid for multi-sample tests.

1.12.5.1 Statistical averages

When several multi-sample observations are taken experimentally there is a scatter of the data about some central value, one of the methods, presenting the result in the form of a histogram (also called a *frequency distribution curve).* Thus, histogram (Fig. 1.16) is a pictorial representation of the test data where the ordinates indicate the number of observed readings of a particular value. With more and more data taken at smaller and smaller increments, the histogram finally changes into a *smooth curve.* The smooth curve is symmetrical with respect to the central value.

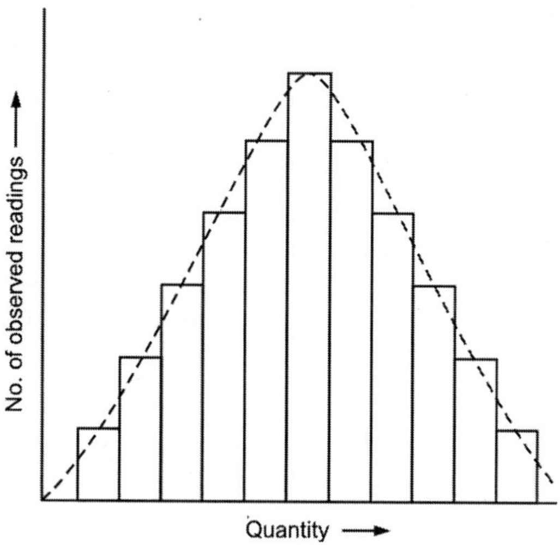

Fig. 1.16 Histogram.

(*i*) **Arithmetic mean (AM).** *The most probable value of measured variable (variate) is the arithmetic mean of the number of readings taken.* We arrive at the best approximation if the number of readings of the same quantity are very large. The mean is computed by summing all the values and dividing by the number of measurements. If $q_1, q_2, q_3 \dots q_n$ be the set of measured values of a *quantity,* then the arithmetic mean $\bar{q}$ will be given by,

$$\bar{q} = \frac{q_1 + q_2 + q_3 + \dots + q_n}{n} = \frac{1}{n}\sum_{i=1}^{n} q_i \qquad \dots(1.12)$$

(ii) Geometric mean (G.M). It is defined as the *n*th root of the product of *n* terms.

$$\text{Geometric mean (G.M)} = \sqrt[n]{q_1 \cdot q_2 \cdot q_3 \cdots q_n} \quad \text{...(1.13)}$$

(iii) Median. *The middle value of a set of an "odd" number of readings if variables are arranged in numerical order is called the* ***median***. For an *"even"* number of readings, the median is the *arithmetic average of the two central readings.*

(iv) Mode. *The value of the variable that occurs most frequently is termed as* ***mode***.

- For a symmetrical distribution curve, the three measures of centre value *viz. 'mean', 'median' and 'mode' are equal.*

1.12.5.2 Dispersion from mean

The property which denotes the extent to which the values are dispersed about the central value is termed as ***dispersion*** *or* ***stored*** *or* ***scatter***.

A large dispersion indicates that some factors involved in the measurement process are not under close control and therefore it becomes difficult to estimate the measured quantity with confidence and definiteness.

The simplest possible measure of dispersion is the **Range** *which is the difference between greatest and least values of data.*

(i) **Deviation.** *The departure of the observed reading from the arithmetic mean of the group of readings is termed as* **deviation.** Let the deviation of reading q_1 be d_1 and that of reading q_2 be d_2 etc.,

Then,

$$d_1 = q_1 - \bar{q}$$
$$d_2 = q_2 - \bar{q}$$
$$\cdots\cdots\cdots\cdots\cdots$$
$$d_n = q_n - \bar{q} \quad \text{...(1.14)}$$

and,

$$\bar{q} = \frac{\Sigma(q_n - d_n)}{n} \quad \text{...(1.15)}$$

$$\text{Algebraic sum of deviations} = d_1 + d_2 + d_3 + \ldots + d_n$$
$$= (q_1 - \bar{q}) + (q_2 - \bar{q}) + (q_3 - \bar{q}) + \ldots (q_n - \bar{q})$$
$$= (q_1 + q_2 + q_3 + \ldots + q_n) - n\bar{q} = 0 \; (\because q_1 + q_2 + q_3 + \ldots + q_n) = n\bar{q}$$

Thus, the *algebraic sum of deviations is zero.*

The above mathematical result implies that the deviations can be *positive as well as negative.*

(ii) **Average deviation.** *It is defined as the sum of the absolute values of deviations divided by the number of readings.*

Average deviation may be expressed as:

$$\bar{d} = \frac{|d_1| + |d_2| + |d_3| + \ldots + |d_n|}{n}$$

$$= \frac{\Sigma|d|}{n}, \text{ or, } \sum_{i=1}^{n} \frac{|q_i - \bar{q}|}{n} \quad \text{...(1.16)}$$

The average deviation is taken as a measure of accuracy of measurement.

- **The average deviation** is an *indication of the precision of the instruments* used in making the measurements. *Highly precise instruments yield a low average deviation between readings.*

(*iii*) **Standard deviation.** *"Standard deviation"* (or *"mean square deviation")* of an infinite number of data is defined *as the square root of the sum of the individual deviations squared divided by the number of readings.*

Thus, standard deviation,

$$\sigma = \sqrt{\frac{d_1^2 + d_2^2 + d_3^2 + ... + d_n^2}{n}}$$

$$= \sqrt{\frac{\Sigma d^2}{n}} = \sqrt{\sum_{i=1}^{n} \frac{(q_i - \overline{q})^2}{n}} \qquad ...(1.17)$$

- In practice, however the number of observations is *finite.* When the number of observations is *greater than 20,* standard deviation is denoted by the symbol σ while if it is *less than* 20 the symbol used is s. The standard devitation of a finite number of data is given by:

$$s = \sqrt{\frac{d_1^2 + d_2^2 + d_3^2 + ... d_n^2}{n-1}} = \sqrt{\frac{\Sigma d^2}{n-1}} \qquad ...(1.18)$$

Variance. *The square of the standard deviation is called* ***variance.***

Variance, $V = (\text{Standard deviation})^2$

$$= \sigma^2 = \frac{d_1^2 + d_2^2 + d_3^2 + ... d_n^2}{n-1} = \frac{\Sigma d^2}{n} \qquad ...(1.19)$$

But when the number of observation is less than 20,

Variance, $$V = s^2 = \frac{\Sigma d^2}{n-1} \qquad ...(1.20)$$

1.12.5.3 Best value from a sample of readings

Let, q_b = The best value (approximately the true value) from a set of readings/data,

Then, the deviation or individual component error,

$$d_i = q_i - q_b \qquad \text{where, } i = 1, 2, 3,$$

This deviation may be + ve, *zero,* or – ve.

(*i*) Sum of deviations:

$$\sum_{i=1}^{n} d_i = \sum_{i=1}^{n} (q_i - q_b) \qquad ...(1.21)$$

In case of a *best value,* the sum of deviations is equal to *zero.*

i.e., $$\sum_{i=1}^{n} (q_i - q_b) = 0 \qquad ...(1.22)$$

or, $$q_b = \frac{1}{n}\sum_{i=1}^{n} q_i, \text{ which is the arithmetic mean.}$$

Thus the arithmetic mean may be interpreted as the best value around which the chances of error being + ve or – ve likely to be equal.

(*ii*) $$d_i = q_i - q_b$$

Squaring both sides, we have:

$$d_i^2 = q_i^2 - 2q_i\, q_b + q_b^2$$

Sum of the square of deviations,

$$\sum_{i=1}^{n} d_i^2 = \sum_{i=1}^{n} q_i^2 - 2q_b \sum_{i=1}^{n} q_i + nq_b^2 \qquad ...(1.23)$$

The sum of the square of deviations, corresponding to the *best value* would be *minimum*. This condition can be obtained by setting the deviation w.r.t., q_b equal to zero;

i.e., $$\frac{d}{dq_b}\left[\sum_{i=1}^{n} d_i^2\right] = 0 \qquad ...(1.24)$$

or, $$\frac{d}{dq_b}\left[\sum_{i=1}^{n} q_i^2 - 2q_b \sum_{i=1}^{n} q_i + nq_b^2\right] = 0$$

As all the q_i^2 *s* are constant, the differentiation gives:

$$-2\sum_{i=1}^{n} q_i + 2nq_b = 0$$

or, $$q_b = \frac{1}{n}\sum_{i=1}^{n} q_i \qquad ...(1.25)$$

which is again the arithmetic mean. Obviously, *the arithmetic mean is the most probable value or the best value which one can obtain from a set of scattered measurement data.*

Example 1.2. *By using a micrometer screw the following readings were taken of a certain physical length:*

1.34, 1.38, 1.56, 1.47, 1.42, 1.44, 1.53, 1.48, 1.40, 1.59 mm

Assuming that only random errors are present, calculate the following:

(i) Arithmetic mean,
(ii) Average deviation,
(iii) Standard deviation, and
(iv) Variance.

Solution: *Given* : Reading of the micrometer screw = 1.34, 1.38, 1.56, 1.47, 1.42,1.44, 1.53, 1.48, 1.40, 1.59 mm.

(*i*) Arithmetic mean: We know that, arithmetic mean,

$$\bar{q} = \frac{1}{n}\sum_{i=1}^{n} q_i = \frac{q_1 + q_2 + q_3 + q_4 + q_5 + q_6 + q_7 + q_8 + q_9 + q_{10}}{10}$$

$$= \frac{1.34 + 1.38 + 1.56 + 1.47 + 1.42 + 1.44 + 1.53 + 1.48 + 1.40 + 1.59}{10}$$

= 1.461 mm (Ans.)

(*ii*) Average deviation: From eqn. (1.14), the *deviations* are :

$$d_1 = q_1 - \bar{q} = 1.34 - 1.461 = -0.121$$

$$d_2 = q_2 - \bar{q} = 1.38 - 1.461 = -0.081$$

$$d_3 = q_3 - \bar{q} = 1.56 - 1.461 = +\,0.099$$

$$d_4 = q_4 - \bar{q} = 1.47 - 1.461 = +\,0.009$$

$$d_5 = q_5 - \bar{q} = 1.42 - 1.461 = -\,0.041$$

$$d_6 = q_6 - \bar{q} = 1.44 - 1.461 = -\,0.021$$

$$d_7 = q_7 - \bar{q} = 1.53 - 1.461 = +\,0.069$$

$$d_8 = q_8 - \bar{q} = 1.48 - 1.461 = +\,0.019$$

$$d_9 = q_9 - \bar{q} = 1.40 - 1.461 = -\,0.061$$

$$d_{10} = q_{10} - \bar{q} = 1.59 - 1.461 = 0.129$$

∴ ***Average deviation,***

$$\bar{d} = \frac{\Sigma |d|}{n}\left(\text{or } \sum_{i=1}^{n} \frac{|q_i - \bar{q}|}{n}\right)$$

$$= \frac{0.121 + 0.081 + 0.099 + 0.009 + 0.041 + 0.021 + 0.069 + 0.019 + 0.061 + 0.129}{10}$$

$$= \mathbf{0.65\ (Ans.)}$$

It may be noted that for calculation of average deviations, the *signs of deviations are not considered.*

(*iii*) Standard deviation:

Since the number of readings is 10, which is less than 20, therefore, the following relation is used for calculation of standard deviation.

Standard deviation, $s = \sqrt{\dfrac{\Sigma d^2}{n-1}}$ ...[Eqn. (1.18)]

or,
$$s = \sqrt{\frac{(-0.121)^2 + (-0.081)^2 + (+0.099)^2 + (0.009)^2 + (-0.041)^2 + (-0.021)^2 + (+0.069)^2 + (0.019)^2 + (-0.061)^2 + (0.129)^2}{(10-1)}}$$

$$= \sqrt{\frac{0.05869}{9}} = \mathbf{0.0807\ mm\ (Ans.)}$$

(*iv*) Variance: *Variance,* $V = s^2 = (0.0807)^2 = \mathbf{0.00651\ mm^2}$ **(Ans.)**

1.12.5.4 Normal or gaussian curve of errors

In experimental work the random effects are normally studied with reference to the Gaussian or normal error distribution. Here, the following *assumptions* are made:

(*i*) All observations include a large number of small random disturbing effects, and these small random errors may be both +ve and –ve. Thus, there is a strong central tendency, *i.e., there is a greater probability of small errors and large errors of both* +ve *and* –ve *values.*

(*ii*) There is equal probability of +ve and –ve disturbing effects. As such the probability curve of errors plotted against the magnitude of error will be *symmetrical about the zero value*. The *law of probability states that the normal occurrence of deviations from average value of an infinite number of measurements or observations can be expressed as:*

$$P = \frac{1}{\sigma\sqrt{2\pi}} e^{-x^2/2\sigma^2} \qquad ...(1.26)$$

where P = Number of readings at any deviation '*d*', *i.e.*, probability of occurrence of deviation *d*,

σ = Standard deviation, and

x = Magnitude of deviation from mean.

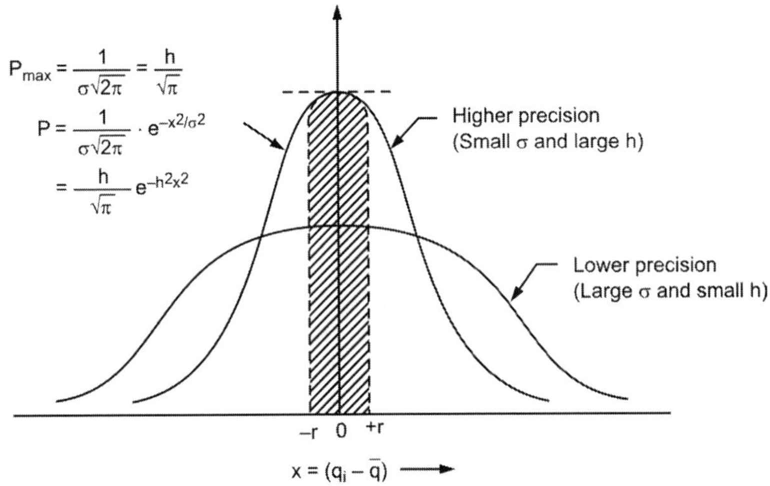

Fig. 1.17 Gaussian distribution curve.

Figure 1.17. shows a Gaussian distribution curve in which P has been plotted as a function of x. The curve is symmetrical about the mean value and may be regarded as the limiting form of a histogram. The two values of standard deviation σ_1 and σ_2 ($\sigma_1 > \sigma_2$) have been illustrated in the given figure. *Larger the value of σ, the flatter the curve and hence larger the expected error of all measurements.*

HIGHLIGHTS

1. *Measurement* is the act, or the result of a quantitative comparison between a predetermined standard and an unknown magnitude. Measurements provide us with a means of describing various phenomena in a quantitative terms.
2. The physical quantity or the characteristic condition which is the object of measurement in an instrumentation system is variously termed as *"measurand", "measurement variable", "instrumentation variable" or "process variable".*
3. The technology of using instruments to measure and control the physical and chemical properties of materials is called *"instrumentation".*
4. The *"measurement standard"* is the physical embodiment of the unit of measurement as well as that of its submultiple value.
5. The highest standard of either a base unit or a derived unit is called a *"primary standard".* The "secondary standards" are the reference calibrated standards designated and calibrated from the primary standards. *"Working standards"* are the normal standards which are used by the workers and technicians who actually carry out the measurements.

6. The two types of measurements involved in the mechanical engineering field are:
   (*i*) Mechanics (or self-operated type), and
   (*ii*) Power type.
7. Methods of measurements are: (*i*) Direct method, (*ii*) Indirect method.
   – In *direct comparison method* the parameter to be measured is directly compared with either a primary or a secondary standard.
   – In *indirect comparison method* the comparison is done with a standard through the use of a calibrated system.
8. The three modes of measurement are:
   (*i*) Primary measurements,
   (*ii*) Secondary measurements, and
   (*iii*) Tertiary measurements.
9. A generalised measurement system consists of the following components/elements:
   (*i*) Primary sensing element
   (*ii*) Variable conversion or transducer element
   (*iii*) Manipulation element
   (*iv*) Data transmission element
   (*v*) Data presentation element.
10. The *"instrument"* serves as an extension of human faculties and enables the man to determine the value of unknown quantity or variable which his unaided human facilities cannot measure.
11. The instruments may be *classified* as follows:
    (*i*) Absolute and secondary instruments,
    (*ii*) Analog and digital instruments,
    (*iii*) Mechanical, electrical and electronic instruments,
    (*iv*) Manual and automatic instruments,
    (*v*) Self-contained and power-operated instruments,
    (*vi*) Self-operated and power-operated instruments and
    (*vii*) Deflection and null output instruments.
12. Three main functions of instruments are:
    (*i*) Indicating function,
    (*ii*) Recording function, and
    (*iii*) Controlling function.
13. The *static characteristics* pertain to a system where the quantities to be measured are constant or vary slowly with time. The main static characteristics are: *Accuracy, sensitivity, reproducibility, drift, static error and dead zone.*
14. The *relative static error* is defined as the ratio of absolute static error to the true value of the quantity under measurement.
15. The difference between the true value and the measured value of a quantity is called *static correction.*
16. *Static calibration* is a process by which all the static performance characteristics are obtained in one form or another.
17. The difference between the largest and the smallest reading of the instrument is called the *range* of the instrument.
18. *Span* represents the algebraic difference between the upper and lower range values of the instrument.
19. An error curve which can be used for correcting instrument readings is called *"Error calibration curve".*
20. *Scale readability* indicates the closeness with which the scale of an analog type of instrument can be read.
21. *Drift* is an undesired gradual departure of the instrument output over a period of time that is unrelated to the changes in input, operating conditions or load.

22. The closeness with which an instrument reading approaches the true value of the quantity being measured is called *accuracy*.
23. *Precision* is a measure of reproducibility of measurements.
24. The ratio of the magnitude of the output signal to the input signal or response of measuring system to the quantity being measured is called *sensitivity*.
25. The ability to reproduce the input characteristics symmetrically is called *linearity*.
26. The maximum differences in output at any measured value within the specified range when approaching the point first increasing and then with decreasing input may be termed as *hysteresis.*
27. The minimum value below which no output change can be detected when the input of an instrument is increased gradually from zero is called the *threshold* of an instrument.
28. *Threshold defines* the smallest measurable input while the *resolution* defines the smallest measurable input change.
29. The largest change of input quantity for which there is no output of the instrument is termed as *dead zone.*
30. The time required by a measurement system to begin to respond to a change in the measurement is called *dead time.*
31. The incapability of the system to faithfully measure the input signal in undistorted form is called the *loading effect.*
32. *Noise* is a spurious current or voltage extraneous to the current or voltage of interest in an electrical or electronic circuit.
33. *"Dynamic characteristics"* quantify the dynamic relation between the input and output.
34. The evaluation of the ability of a system to faithfully transmit and present all the pertinent information included in the input signal and to exclude all else, is called *response.*
35. The behaviour of the system when inputs vary with time *(i.e.,* inputs are dynamic in nature) and so does the output is called *dynamic response.*
36. Response of a measurement system may be: (*i*) Transient response; *(ii)* Steady state response.
37. The rapidity with which a measurement system responds to changes in the measured quantity is called the *speed of response* or *"responsiveness"*.
38. *Measuring lag* refers to retardation or delay in the response of a measurement system to changes in measured quantity.
39. *Fidelity* is defined as the degree to which the measurement system indicates changes in the measured quantity without any dynamic error.
40. The *dynamic error* or *measurement error* is the difference between the true value of the quantity changing with time and the value indicated by the measurement system if no static error is assumed.
41. "Examples" of the systems:

    (*i*) *Zero-order systems:* Mechanical levers, amplifiers etc.

    (*ii*) *First-order systems:* Velocity of a free falling mass; air pressure build-up in bellows, thermistors and thermocouples etc.

    (*iii*) *Second-order systems:* Piezoelectric pick-ups, spring-mass system, pen control system on X-Y plotters etc.

42. *Damping ratio* (γ) is defined as the ratio of the actual value of coefficient of viscous friction in movement and the value required to produce critical damping.
43. The limits of the deviations from the specified value are defined as *limiting errors* or *guarantee errors.*
44. The *relative error* is defined as the ratio of the error to the specified (nominal) magnitude of a quantity.
45. The errors are classified as:

    1. Gross errors
    2. Systematic errors

        (*i*) Instrumental errors

        (*ii*) Environmental errors

        (*iii*) Observational errors.

    3. Random errors.

46. *Sources of errors are:*
    Noise, Response time; Design limitations; Energy exchanged by interaction; Transmission; Determination of measuring system; Ambient influences on measuring systems, Errors of observation and interpretation.
47. *Arithmetic mean* of the number of readings taken is the most probable value of measured variable.
48. *Geometric mean* is defined as the *n*th root of the product of *n* terms.
49. *Median* is the middle value of a set of an "odd" number of readings, if variables are arranged in numerical order. For an "even" number of readings, the median is the arithmetic average of the two central readings.
50. *Mode* is the value of the variable that occurs most frequently.
51. *Dispersion* (or *spread or scatter)* is the property which denotes the extent to which the values are dispersed about the central value.
52. The departure of the observed readings from the arithmetic mean of the group of readings is termed as *deviation.*
53. *Standard deviation* of an infinite number of data is defined as the square root of the sum of individual deviations squared divided by the number of readings.
54. The square of the standard deviation is called *variance.*
55. The *'law of probability'* states that the normal occurrence of deviations from the average value of an infinite number of measurements or observations can be expressed as:

    $$P = \frac{1}{\sigma\sqrt{2\pi}} e^{-x^2/2\sigma^2}$$

    where, P = Number of readings at any deviation '*d*', *i.e.,* probability of occurrence of deviations *'d',*
    σ = Standard deviation, and
    x = Magnitude of deviation from mean.

56. *Chauvenet's criterion* states as follows :
    "An observed reading may be rejected if the probability of obtaining the particular deviation from the mean is less than $\frac{1}{2n}$". The quantity *n* refers to the number of observations which are large enough to presume that the data follows the Gaussian error distribution.

THEORETICAL QUESTIONS

General Concepts

1. What do you understand by the term "measurement"?
2. Describe the significance of measurements.
3. Explain briefly the term "measurand".
4. State the basic requirements for getting the meaningful result of the measurement.
5. Explain the difference between an analog signal and a digital signal.
6. Define the term "instrumentation".
7. How are standards of measurement classified?
8. Explain clearly the difference between primary, secondary and working standards.
9. Explain briefly the following methods to complete solutions to complex problem in mechanical design:
    (*i*) The empirical method
    (*ii*) The rational method
    (*iii*) The experimental methods.
10. Distinguish between direct and indirect methods of measurement.
11. Describe the mechanics and power types of measurements. Explain with suitable examples.
12. What are primary, secondary and tertiary measurements? Explain with examples.

13. Explain the difference between the following:
    (*i*) Measurand and measurement
    (*ii*) Measurement and instrument
    (*iii*) Input and output.
14. Draw a block diagram representation of a generalised measurement system. Identify the various elements and point out the functions performed by each element/component.
15. Explain briefly various stages of the general measurement system.
16. Describe briefly the following examples of measurement systems:
    (*i*) Bourdon tube pressure gauge.
    (*ii*) Pressure-actuated thermometer.
17. How are instruments classified?
18. Explain briefly any three types of the instruments:
    (*i*) Absolute and secondary instruments,
    (*ii*) Analog and digital instruments,
    (*iii*) Mechanical, electrical and electronic instruments
    (*iv*) Self-operated and power-operated instruments,
    (*v*) Deflection and null-point instruments.
19. State the advantages and disadvantages of mechanical instruments over electrical/electronic instruments.
20. State the advantages of electrical/electronic techniques of measurements.
21. Discuss the factor relating selection of instruments.
22. Describe briefly the following main functions of the instruments:
    (*i*) Indicating function
    (*ii*) Recording function
    (*iii*) Controlling function.

Static and dynamic characteristics

23. What do you mean by characteristics of a measuring instrument?
24. What are static characteristics ? How do they differ from dynamic characteristics?
25. Define the following terms as applied to the measuring instrument:
    (*i*) True value (*ii*) Correction
    (*iii*) Range (*iv*) Sensitivity
    (*v*) Accuracy (*vi*) Linearity
    (*vii*) Dead zone (*viii*) Backlash.

Static characteristics

26. Enumenate the main static characteristics of measuring instruments.
27. Explain briefly the terms: Accuracy, errors and correction.
28. Explain the following terms:
    (*i*) Static error (*ii*) Static correction
    (*iii*) Relative error (*iv*) Percentage relative error.
29. How can "accuracy" be expressed ? Explain briefly.
30. What do you mean by "Static calibration"? Give the steps which are necessary in performing a calibration.
31. Differentiate between "Scale range" and "Scale span" giving suitable examples.
32. What is an "Error calibration curve" ?

33. What do you mean by "Scale readability" ?'
34. Differentiate between the terms "Repeatability" and "Reproducibility".
35. Explain the term "Drift" using suitable examples.
36. How is "drift" classified ?
37. Explain briefly the following:

    (*i*) Zero drift (*ii*) Span drift
    (*iii*) Zonal drift.

38. Differentiate between the terms "Accuracy" and "Precision" with suitable examples.
39. What is "sensitivity" ?
40. Define the term "Deflection factor".
41. What is "linearity" ? Explain briefly.
42. What do you mean by the term "Non-linearity" ? Explain briefly "Terminal linearity" and "Best fit linearity"
43. Describe briefly the term "Hystereis".
44. Explain clearly the difference between "Threshold" and "Resolution".
45. Explain briefly the following:

    (*a*) Dead zone (*b*) Dead time.

46. What do you mean by the term "loading effect" as applied to measuring instruments ? Explain briefly.
47. Explain briefly the terms "Static stiffness" and "Static compliance".
48. What is "noise"?
49. What is signal to noise ratio?
50. Explain briefly the following:

    (*i*) Generated noise; (*ii*) Conducted noise;
    (*iii*) Radiated noise.

Dynamic characteristics

51. Define the following terms:

    (*i*) Response; (*ii*) Dynamic response of a measuring system.

52. Explain briefly the following types of dynamic inputs:

    (*i*) Transients; (*ii*) Steady state periodic.

53. Define the following terms:

    (*i*) Transient response; (*ii*) Steady state response.

54. Define the dynamic response of a system and distinguish between steady state response and transient response.
55. Define the following terms:

    (*i*) Speed of response; (*ii*) Measuring lag;
    (*iii*) Fidelity; (*iv*) Dynamic error;
    (*v*) Overshoot.

56. What are the different standard inputs for studying the dynamic response of a system? Explain them briefly.

Error in measurement

57. What are limiting errors? Explain briefly.
58. Explain briefly the term *"relative error."*
59. How are errors classified?
60. What are "Gross errors"? How can these be avoided?

61. What are "Systematic errors" Explain briefly the following systematic errors with examples:
    (*i*) Instrumental errors
    (*ii*) Environmental errors
    (*iii*) Observational errors.
62. How can the systematic errors be located?
63. What are assembly errors? Can they be easily discovered and rectified? If yes, why?
64. State the factors which give rise to observational errors?
65. What are random errors? State the common causes of their occurrence.
66. Enumerate the various sources of errors.
67. Explain briefly the following sources of errors:
    (*i*) Noise
    (*ii*) Response time
    (*iii*) Design limitations
    (*iv*) Energy exchanged by interaction
    (*v*) Transmission
    (*vi*) Deterioration of measuring system.
68. What are single-sample and multi-sample tests ? Explain briefly.
69. Explain briefly the following terms:
    (*i*) Arithmetic mean
    (*ii*) Geometric mean
    (*iii*) Median
    (*iv*) Mode.
70. Describe briefly the following terms:
    (*i*) Deviation
    (*ii*) Average deviation
    (*iii*) Standard deviation
    (*iv*) Variance.

❑❑❑

Sensors and Transducers

Chapter 2

2.1 INTRODUCTION

The primary sensing element (sensor) is the first and foremost requirement for measurement and automatic controls. The **sensors** *sense the condition, state or value of the process variable and produce an output which reflects this condition, state or value. The* **transducers** *transform the energy of the process variable to an output of some other type of energy which is able to operate some control device.* Sometimes a *secondary transducer* may be employed to transform the output of the primary sensor to still another type of energy.

Examples:

- In the ordinary *dial indicator* the indicating spindle acts as a sensor/detector for displacement. It simply performs the function of sensor/detector and nothing else.
- The function of a *Bourdon tube* of a pressure gauge is *twofold:* Firstly to sense the pressure and secondly to give the resulting effect or output in the form of displacement. Here the tube acts a sensor/detector transducer.
- In a *compressive load cell,* the platform detects the force and gives an output in the form of deflection. This deflection may be further converted into an electrical output by strain gauges (called *secondary transducer).*

For the measurement of particular quantity, different types of sensors and transducers are available and the choice of a suitable unit depends upon the static and dynamic performance characteristics.

2.2 MECHANICAL DETECTOR-TRANSDUCER ELEMENTS

The various mechanical detector-transducer elements may be enumerated and discussed as follows:

1. Elastic members/elements
2. "Mass" sensing elements
3. Thermal detectors
4. Hydro-pneumatic elements.

1. Elastic members/elements:

These elements *work on the principle of direct tension or compression, bending and torsion.* These are invariably used to change force into displacement. The following elastic members/elements are commonly used:

(*i*) *Proving ring (stress ring).* It is a ring of known physical dimension and mechanical properties. An external tensile or compressive force applied across the ring diameter causes distortion which is proportional to that force. The distortion is measured by means of a dial gauge, a sensitive micrometer, or a strain gauge.

The proving rings have been used as *standards for calibrating tensile testing machines and for accurate measurement of large plastic loads.*

(*ii*) *Elastic torsion member.* Several times torque meters make use of elastic torsion members which twist in proportion to applied torque and deformation is used as a measure of torque.

(*iii*) *Springs.* In a spring type indicating scale, unknown weight applied to the free end of spring causes displacement which is indicated by the pointer.

(*iv*) *Bourdon tube, bellows, diaphragm.* Most pressure measuring devices use either a Bourdon tube, bellows or diaphragms. The action of these devices is based on the elastic deformation brought about by the force resulting from pressure summation.

2. "Mass" sensing elements:

- The *inertia of a concentrated mass* provides another basic mechanical detector-transducer element, which is used in the *accelerometers* and *vibration pick-ups* and serves to measure the characteristics of dynamic motion (*e.g.*, displacement, velocity, acceleration, frequency, etc.) through application of Newton's second law of motion.
- Any simple mechanically vibrating member *(e.g., a pendulum)* would sever as a time or frequency transducer, chopping the passage of time into discrete bits.
- Further the *manometer,* used for pressure measurement, also works on the *principle of mass displacement.*

3. Thermal detectors:

These are the devices employed to measure the *temperature* of solids, liquids and gases. They sense the temperature by employing one of the following *primary effects:*

(*i*) Change in physical state;
(*ii*) Change in chemical state;
(*iii*) Change in electrical properties;
(*iv*) Change in radiating ability.

The following *thermal detectors* are most commonly used

(*i*) Glass thermometers

(*ii*) Pressure gauge thermometers
(*iii*) Bimetallic thermometers
(*iv*) Resistance thermometers
(*v*) Thermistors
(*vi*) Pyrometers
(*vii*) Thermocouples.

4. Hydro-pneumatic sensors:

Following are the *common examples* of the hydro-pneumatic sensors:

(*a*) **Applied to static conditions:**

(*i*) *Simple float.* A simple float converts the fluid level into displacement; it makes no allowance for change in the density of the supporting liquid.

(*ii*) *Hydrometer.* It senses specific gravity and converts it into displacement. It uses the immersion depth as a means for detecting variation in specific gravity of the supplying liquid.

(*b*) **Applied to dynamic conditions:**

(*i*) *Orifices and venturies.* These are used for flow measurements in pipes and provide information in the form of pressure change as a result of transformation of energy.

(*ii*) *Pitot tube.* It measures the pressure resulting from total-flow rate rather than the change of rate.

(*iii*) *Vanes in the form of air foils or turbine wheels.* These are also used to sense fluid flow.

2.3 DEFINITION OF TRANSDUCER

A broad definition of a transducer is as follows:

"A **transducer** *is a device which converts the energy from one form to another".*

Most of the transducers either convert electrical energy into mechanical displacement and /or convert some non-electrical physical quantity (*e.g.,* force, sound, temperature etc.) to an electrical signal.

A transducer performs the following *functions* in an electronic instrumentation system:

1. *Detects or senses the presence, magnitude and changes in physical quantity being measured.*
2. *Provides a proportional electrical output signal (see Fig.* 2.1.)
   - A transducer can be broadly defined as a *device which converts an non-electrical quantity into an electrical quantity.*

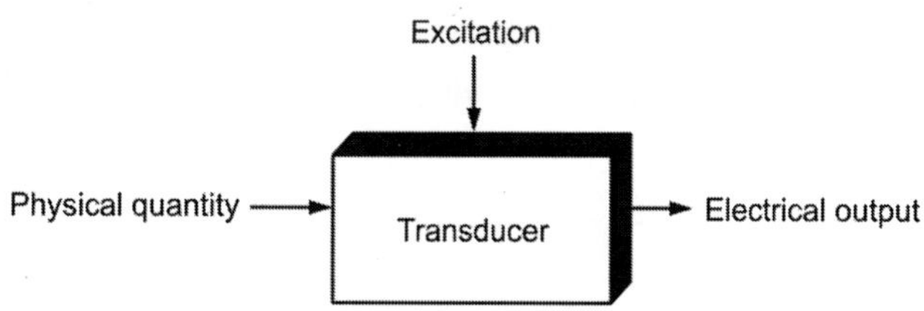

Fig. 2.1 Transducer.

An **inverse transducer** *is defined as a device which converts an electrical quantity into a non-electrical quantity.* It is a precision actuator which has an electrical input and a low power non-electrical output. A *piezoelectric* crystal acts as an inverse transducer because when a voltage is applied across its surfaces, it changes its dimensions causing a mechanical displacement.

2.4 CLASSIFICATION OF TRANSDUCERS

A. Broad classification

Transducers are broadly classified into two groups as follows:

1. ***Active transducers.*** They are also known as *self-generating type transducers.* These transducers *develop their own voltage or current. The energy required for production of an output signal is obtained from the physical phenomenon being measured.*

   Examples. *Thermocouples and thermopiles, piezoelectric pick-up, photovoltaic cell.*

2. ***Passive transducers.*** *They are known as externally-powered transducer.* These transducers *derive the power required for the energy conversion from an external power source.* However, they *may absorb some energy from the physical phenomenon under study.*

   Examples. *Resistance thermometers and thermistors, potentiometric devices, differential transformer, photoemission cell etc.*

B. Classification based on the type of output:

1. ***Analogue transducers.*** These transducers convert the input physical phenomenon into an analogous output which is a continuous function of time.

   Examples. *Strain gauge, a thermocouple, a thermistor or an* LVDT (linear voltage differential transformer).

2. ***Digital transducers.*** These transducers convert the input physical phenomenon into an electrical output which may be in form of pulse.

C. Classification based on electrical principle involved :

1. ***Variable-resistance type:***
   (*i*) Strain and pressure gauges (*ii*) Thermistors, resistance thermometers
   (*iii*) Photoconductive cell etc.

2. ***Variable-inductance type:***
   (*i*) Linear variable transformer (LVDT) (*ii*) Reluctance pick-up
   (*iii*) Eddy current gauge.

3. ***Variable-capacitance type:***
   (*i*) Capacitor microphone (*ii*) Pressure gauge
   (*iii*) Dielectric gauge.

4. ***Voltage-generating type:***
   (*i*) Thermocouple (*ii*) Photovoltaic cell
   (*iii*) Rotational motion tachometer
   (*iv*) Piezoelectric pick-up.

5. ***Voltage-divider type:***
   (*i*) Potentiometer position censor (*ii*) Pressure-actuated voltage divider.

Table 2.1 shows the measurements versus transduction methods.

- While *describing a particular transducer* the information must be available about the following aspects:

(*i*) The measurand.
(*ii*) The sensing element which responds directly to the measurand.
(*iii*) The principle of operation of the transducer and where the output of the transducer originates...
(*iv*) The useful range.

TABLE 2.1 Measurements versus Transduction Methods

S. No.	Quantity to be measured	Type of transducer
1.	*Displacement*	— Resistive — Inductive — Capacitive — Piezoelectric — Magnetoelectric — Radioactive — Electron tube
2.	*Thickness*	— Inductive — Capacitive — Piezoelectric — Photoelectric — Radioactive
3.	*Velocity*	— Resistive — Inductive — Capacitive — Piezoelectric — Photoelectric — Magnetoelectric — Radioactive — Electron tube
4.	*Acceleration*	— Resistive — Inductive — Capacitive — Piezoelectric — Magnetoelectric — Magnetostrictive — Electron tube
5.	*Mass*	— Inductive — Piezoelectric — Magnetoelectric — Radioactive
6.	*Force*	— Resistive — Inductive — Piezoelectric — Radioactive
7.	*Pressure*	— Resistive — Inductive — Capacitive — Piezoelectric — Thermoelectric — Magnetoelectric — Magnetostrictive — Radioactive — Electron tube
8.	*Flow*	— Resistive — Inductive — Capacitive — Piezoelectric — Magnetoelectric — Radioactive
9.	*Level*	— Resistive — Capacitive — Piezoelectric — Photoelectric — Radioactive
10.	*Temperature*	— Resistive — Photoelectric — Thermoelectric — Radioactive
11.	*Humidity*	— Resistive — Capacitive.
12.	*Viscosity*	— Resistive — Capacitive — Piezoelectric — Magnetostrictive

2.4.1 Transducer Sensitivity

The relationship between the measurand and the transducer output signal is referred to as ***Transducer sensitivity***.

i.e.,
$$\text{Transducer sensitivity} = \frac{\text{Output signal increment}}{\text{Measuand increment}}$$

Sensitivity of a transducer should be *usually as high as possible* since then it becomes easier to take the measurements.

2.4.2 Specifications for Transducers

While selecting the proper transducer for any applications, or ordering the transducers, the following *specifications* should be thoroughly considered :

(*i*) Ranges available.
(*ii*) Squaring system.
(*iii*) Sensitivity.
(*iv*) Maximum working temperature.
(*v*) Method of cooling employed.
(*vi*) Mounting details,
(*vii*) Maximum depth.
(*viii*) Linearity and hysteresis.
(*ix*) Output for zero input.
(*x*) Temperature coefficient of zero drift.
(*xi*) Natural frequency.

2.5 TRANSDUCERS ACTUATING MECHANISMS

Transducers are also known as *gauges, pick-ups and signal generators.* Most of the pick-ups have following *two basic elements:*

(*i*) Activating device.
(*ii*) Transducing element.

Figure 2.2 shows some typical actuating mechanisms

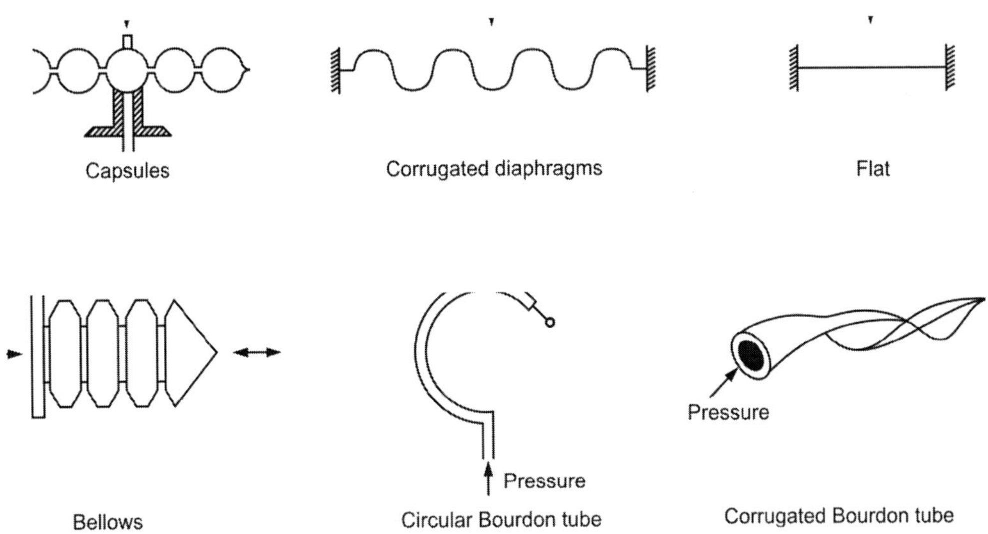

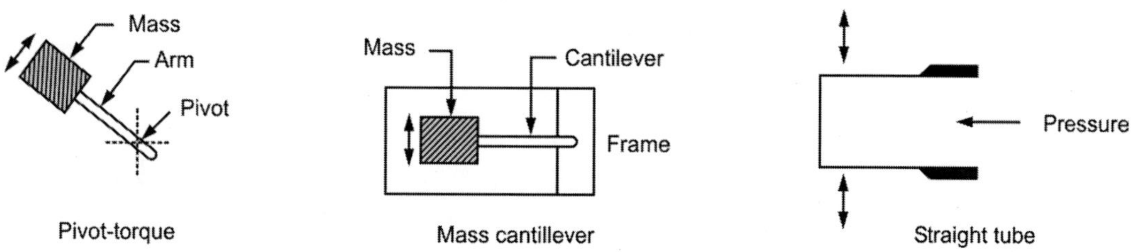

Fig. 2.2 Transducer actuating mechanisms.

2.6 RESISTANCE TRANSDUCERS

The resistance of a metal conductor is expressed by a simple equation that involves a few physical quantities.

The relationship is given by $R = \frac{\rho L}{A}$;

where, R = Resistance, Ω

ρ = Resistivity of conductor materials, Ω-*m*

L = Length of conductor, *m,* and

A = Cross-sectional area of the conductor, m^2.

Any method of varying one of the quantities involved in the above relationship can be the designed basis of an *electrical resistance transducer.* These are a number of ways in which resistance can be changed by a physical phenomenon.

The *translational and rotational "potentiometers"* which work on the basis of change in the value of resistance with change in length of the conductor can be used for measurement of translational or rotary displacements.

Strain gauges *work on the principle that the resistance of a conductor or a semiconductor changes when strained.* This property can be used for measurement of displacement, force and pressure.

The resistivity of materials changes with the change of temperature thus causing a change of resistance. This property may be used for measurement of *"temperature".*

In a *resistance transducer* an indication of measured physical quantity is given-by *a change in the resistance.* It may be *classified* (as discussed above) as follows :

1. Mechanically varied resistance — *Potentiometer*
2. Thermal resistance change — *Resistance thermometers*
3. Resistivity change — *Resistance strain gauge.*

2.6.1 Linear and Angular Motion Potentiometers

Such potentiometers *convert the linear motion or the angular motion of a rotating shaft into changes in resistance.* The device is a variable resistor whose resistance is varied by the movement of a slider over a resistance element.

— Translatory devices have strokes from 2.5 mm to 5 mm.

— Rotational devices have full scale ranging from 10° to 60° full turn.

The potentiometer shown in Fig. 2.3 and 2.4 form a part of the bridge circuit whose output voltage is changed by the slider position.

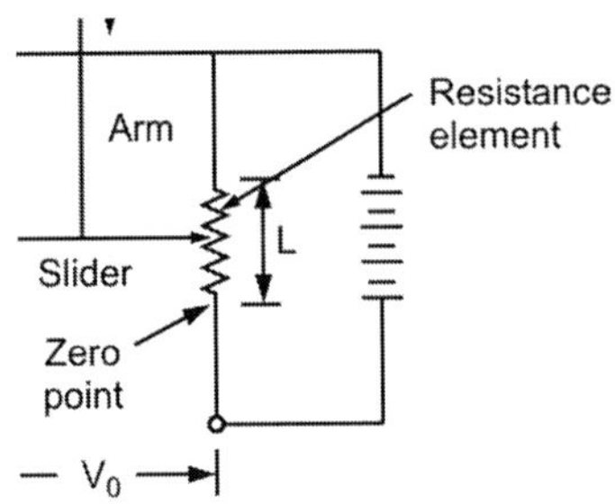

Fig. 2.3 Linear motion potentiometer.

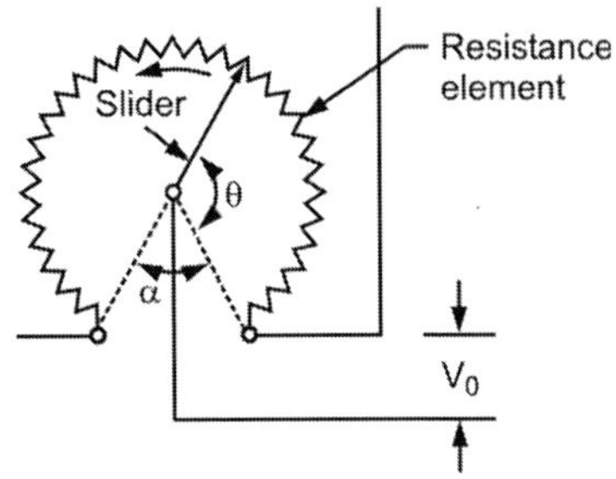

Fig. 2.4 Rotary motion potentiometer.

- — The slider is powered by the mechanical part on which the linear displacement of angular measurements are to be made.
- — Due to arm movement, the slider moves over the resistance element and thus shorts out a portion of the resistance. *The change in resistance in the potentiometer is then an indication of the amount of motion and the direction of movement is indicated by whether the resistance is increasing or decreasing.* The unbalanced voltage is measured directly or fed into an amplifier and recorded.

The potentiometers are used in many transducers designed to measure:

(*i*) Pressure
(*ii*) Force
(*iii*) Acceleration
(*iv*) Liquid level.

The potentiometers have the following *advantages* and *disadvantages:*

Advantages:

(*i*) High output.
(*ii*) Less expensive.
(*iii*) Available in different sizes, shapes and ranges.
(*iv*) Simple to operate.
(*v*) Their electrical efficiency is very high.
(*vi*) Rugged construction.
(*vii*) Insensitivity towards vibration and temperature.

Disadvantages:

(*i*) Limited life due to early wear of the sliding ram.
(*ii*) The output tends to noisy and erratic in high speed operation or when in high vibration environment.
(*iii*) In wire wound potentiometers the resolution is limited while in cermit and metal film potentiometers, the resolution is infinite.

Power rating of potentiometers:

The potentiometers are designed with a definite power rating which is related directly to their heat dissipating capacity. The manufacturer normally designs a series of potentiometers of single turn with a diameter of 50 mm with a wide range of ohmic values ranging from *100Ω to 10kΩ* in *steps of 100Ω.* These potentiometers are essentially of the *same size* and of the *same mechanical configuration.* They have the *same heat transfer capabilities.* Their rating is typically *5 W at an ambient temperature of* 21°C. This limits their input excitation voltage.

2.6.2 Thermistors and Resistance Thermometers

These transducers are thermally sensitive variable resistors made of certain conducting and ceramic-like semiconducting materials. They are used as temperature detecting elements used to sense temperature for the purpose of measurements and control.

Thermistors are essentially semiconductors which behave as resistors with a *high negative temperature coefficient of resistance.* The high sensitivity to temperature changes make the thermistors extremely useful for precision temperature (– 60°C to + 15°C) measurements, control and compensation. Their resistance ranges from 0.5Ω to 0.75 MΩ.

Thermistors are composed of sintered mixture of metallic oxides such as manganese, nickel, cobalt, copper, iron and uranium.

Figure 2.5 shows the commercial forms of thermistors.

Applications of thermistors:

1. *Measurements of temperature*.................. (major application).
2. Temperature compensation in complex electronic equipment, magnetic amplifiers and instrumentation equipment.
3. Measurement of power at high frequencies
4. Vacuum measurements.
5. Measurements of level, flow and pressure of liquids.
6. Measurement of thermal conductivity.

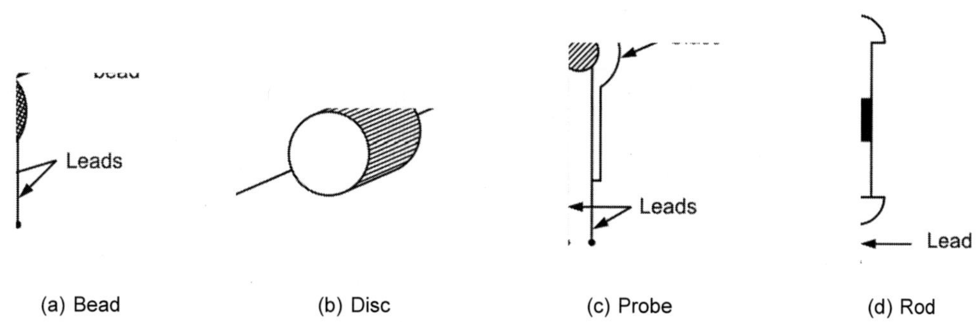

Fig. 2.5 Commercial forms of thermistors.

2.7 VARIABLE INDUCTANCE TRANSDUCERS

These are based on a change in the magnetic characteristic of an electrical circuit in response to a measureand which may be displacement, velocity, acceleration etc.

Variable inductive transducers may be *classified* as follows:

1. **Self-generating type.** *In this type voltage is generated because of the relative motion between a conductor and a magnetic field.*
   These may be further *classified* as follows:
   (*i*) Electromagnetic type.
   (*ii*) Electrodynamic type.
   (*iii*) Eddy current type.
2. **Passive type.** In this type the *motion of an object results in changes in the inductance of the coils of the transducer.*
   These may be further *classified* as follows:
   (*i*) Variable reluctance.
   (*ii*) Mutual inductance.
   (*iii*) Differential transfer type.

2.7.1 Self-generating Type

2.7.1.1 Electromagnetic Type

Figure 2.6 shows an electromagnetic type of self-generating variable inductance transducer.

— It consists of a permanent magnet core on which a coil is directly wound.
— When a plate of iron or other ferromagnetic material is moved with respect to the magnet, the flux field expands or collapses and a voltage is induced in the coil.
— This device is used for *indication of angular speed.* The measurements of speed can be made with great accuracy when the pick-up is placed near the teeth of a rotating gear.

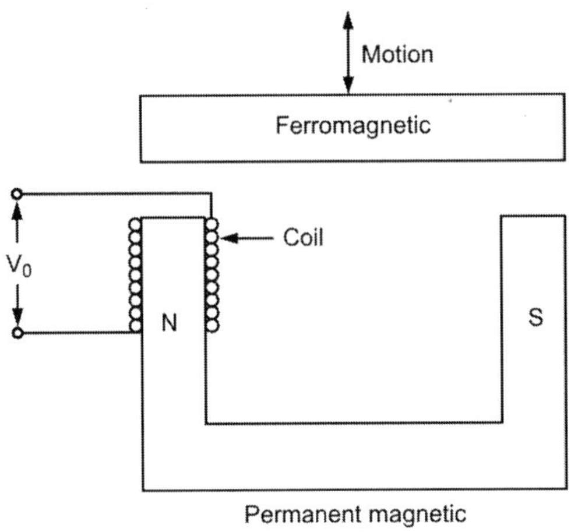

Fig. 2.6 Self-generating variable inductance transducer—Electromagnetic type.

2.7.1.2 Electrodynamic Type

This type of transducer (linear and rotational is shown in Fig. 2.7.

— In this type, coil moves within the field of the magnet. The turns of the coil are perpendicular to the intersecting lines of force.

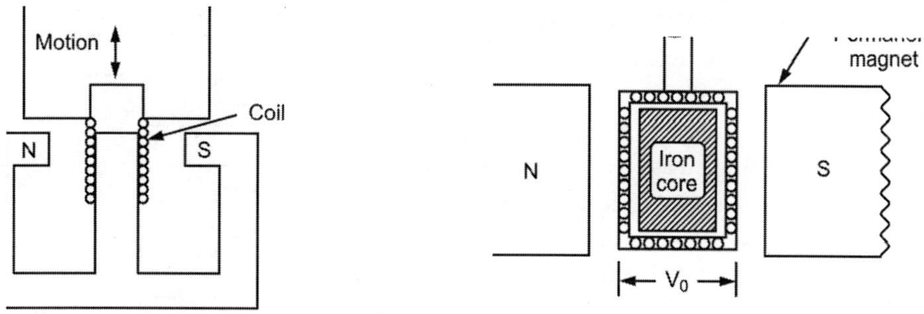

Permanent magnet
Electrodynamic (linear)
(*a*)

Electrodynamic (rotational)
(*b*)

Fig. 2.7 Self-generating variable inductance transducer—Electrodynamic type.

— When the coil moves it induces a voltage which at any moment is *proportional to the velocity of the coil.*

The principle of these transducer is used in the *magnetic flow meters.*

2.7.1.3 Eddy Current Type

Figure 2.8 shows an eddy current type of self-generating variable inductance transducer.

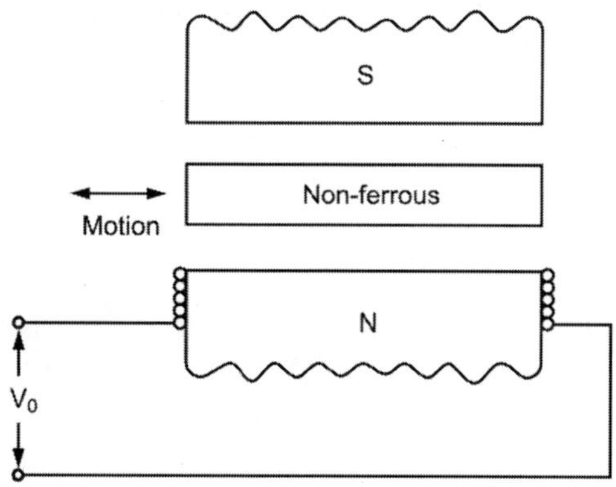

Fig. 2.8 Self-generating variable inductance transducer—Eddy current type.

2.7.2 Passive Type

2.7.2.1 Variable Reluctance Transducer

In these transducers (comprising of a magnetic field and core with a gap between the core and the fixed coils) a change in the reluctance of the magnetic circuit by a mechanical input results in a similar change both in the inductance and inductive reactance of the coils. The change in inductance is then measured by suitable circuitry and related to the value of mechanical input.

The magnetic circuit reactance may be changed by affecting a change:

(*i*) in the *air gap* or

(*ii*) in the *amount/type of core material.*

— Transducers which make use of *air gap change* are referred as *reluctance type.*

— Transducers which utilize a *variable core* are referred as *permeance type.*

A variable *reluctance transducer* is shown in Fig. 2.9. Here the inductance of a single coil is changed through the variable air gap. The change in inductance may be calibrated in terms of movement of the armature.

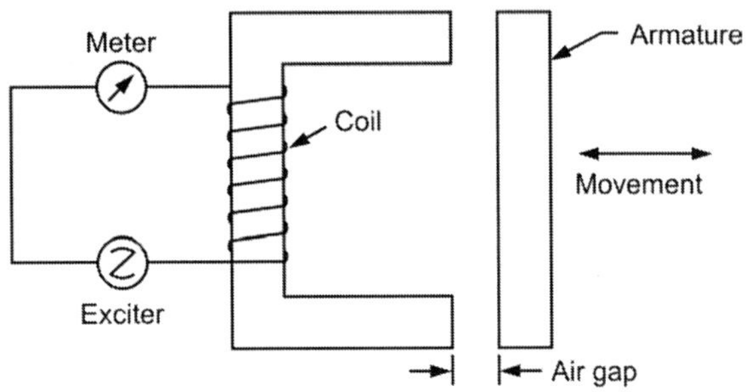

Fig. 2.9 Variable reluctance transducer.

This principle of variable reluctance is used for the measurement of dynamic quantities such as:

(*i*) Pressure
(*ii*) Force
(*iii*) Displacement
(*iv*) Acceleration
(*v*) Angular position etc.

Variable Permeance Transducer:

Figure 2.10 show a *variable permeance transducer in* which the inductance of coil is changed by *varying the core material.*

— The transducer consists of a coil of many turns of wire wound on a tube of insulating material with a moveable core of magnetic material.

— When the coil is energized and the core enters the solenoid cell, the *inductance of the coil increases in proportion to the amount of metal within the coil.* It is primarily used for the measurement of: (*i*) Displacement; (*ii*) Strain; (*iii*) Force.

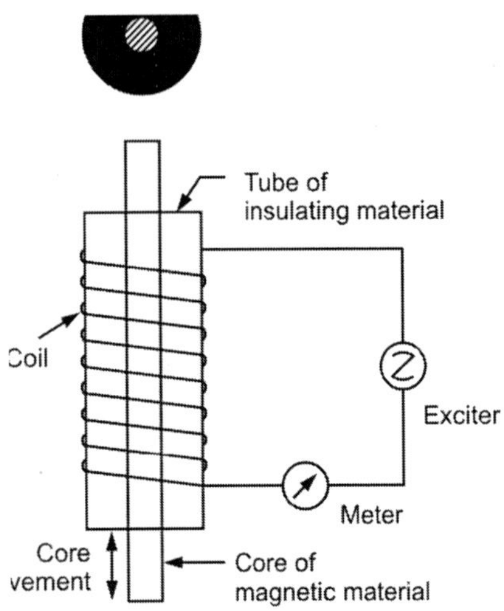

Fig. 2.10 Variable permeance transducer (self inductance arrangements).

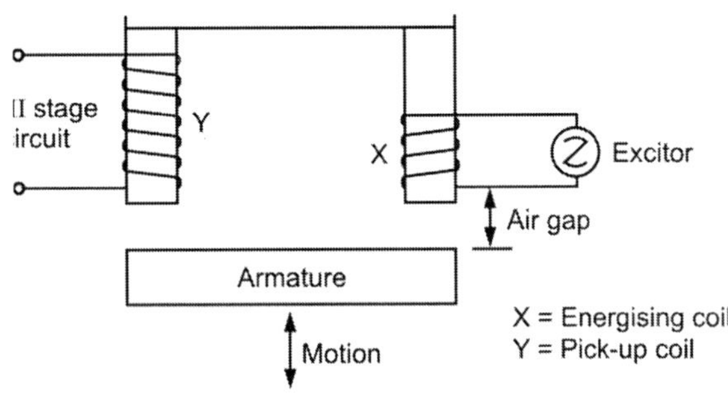

Fig. 2.11 Mutual inductance transducer.

2.7.2.2 Mutual Inductance Transducer

A two-coil mutual inductance transducer is illustrated in Fig. 2.11. It consists of an energising coil X and a pick-up coil *Y*. A change in the position of the armature by a mechanical input changes the air gap. This causes a change in the output from coil *Y*, which may be used as measure of the displacement of the armature, *i.e.*, the input mechanical input.

2.7.2.3 Linear-variable-differential Transformer (LVDT)

LVDT *is a passive inductive transducer and is commonly employed to measure force (or weight, pressure and acceleration etc. which depend on force) in terms of the amount and direction of displacement of an object.*

Construction. Refer to Fig. 2.12 (*a*).

— It consists of one primary winding (*P*) and two secondary windings (S_1 and S_2) which are placed on either side of the primary mounted on the same magnetic core. The magnetic core is free to move axially inside the coil assembly and the motion being measured is mechanically coupled to it.

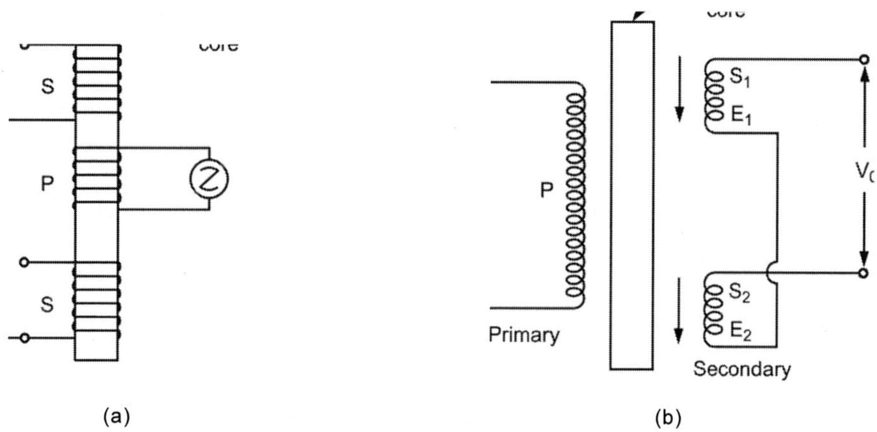

Fig. 2.12 Linear-variable-differential transducer (LVDT).

— The two secondaries S_1 and S_2 have equal number of turns but are connected in series opposition so that e.m.fs, (E_1 and E_2) induced in them are 180° out of phase with each other and hence, cancel each other out. [See Fig. 2.12 (*b*)]

— The primary is energised from a suitable A.C. source.

Working:

— When the core is in the centre (*called reference position)* the induced voltages E_1 and E_2 are *equal and opposite.* Hence they cancel out and the output voltages V_0 is zero.

— When the external applied force moves the core towards coil S_2, E_2 is increased but E_1 is *decreased* in magnitude though they are still antiphase with each other. The net voltage available is ($E_2 - E_1$) and is in *phase with* E_2.

Similarly, when the magnitude core moves towards coil S_1, $E_1 > E_2$ and $V_0 = E_1 - E_2$ and is in *phase with* E_1.

Thus, from above discussion, we find that the magnitude of V_0 is a *function of the distance moved by the core* and its *polarity* or *phase* indicates as to in which direction it has moved.

If core is attached to moving object, the *magnitude* of V_0 *gives the position of that object.*

Advantages:

1. It gives a high output and therefore many a times there is no need for intermediate amplification devices.
2. The transducer possesses a high sensitivity as high as 40 V/mm.
3. It shows a low hysteresis and hence repeatability is excellent under all conditions.
4. Most of the LVDTs consume a power of less than 1 W.
5. Less friction and less noise (due to absence of sliding contacts).
6. These transducers can usually tolerate a high degree of shock and vibration without any adverse effects.
7. It can operate over a temperature range from – 265°C to 600°C.
8. It is available in radiation-resistant design for operation in nuclear reactors.

Disadvantages:

1. These transducers are sensitive to stray magnetic fields but shielding is possible. This is done by providing magnetic shields with longitudinal slots.
2. Relatively large displacements are required for appreciable differential output.
3. The receiving instrument must be selected to operate on A.C. signals or demodulator network must be used if a D.C. output is required.
4. Several times, the transducer performances is affected by vibrations.
5. The dynamic response is limited mechanically by the mass of core and electrically by the frequency of applied voltage. The frequency of the carrier should be atleast ten times the highest frequency component to be measured.

Applications:

1. Measurement of material thickness in hot strip or slab steel mills.
2. In accelerometers.
3. Jet engine controls in close proximity to exhaust gases.

Note. LVDT *is not suited for fast dynamic measurements on account of mass of the core.*

2.8 CAPACITIVE TRANSDUCERS

The principle of operation of capacitive transducers is based upon the familiar equation for capacitance of a parallel plate capacitor:

$$\text{Capacitance, } C = \frac{\in A}{d} = \frac{\in_r \in_o A}{d} \qquad ...(2.1)$$

where,

$\in = \in_r \in_o$ = Pemittivity of medium, F/m,

$\in_r$ = Relative permittivity, (for air $\in_r = 1$),

$\in_o$ = Permittivity of free space = 8.85×10^{-12} Fm,

A = Overlapping area of plates, and

d = Distance between the two plates

Any physical quantity which can cause a change in $\in$, A or d can be measured by the *capacitance gauge.*

The *displacement is measured by measuring the change in capacitance brought about by :* (i) Change in area, or (ii) Change in distance between the plates.

The change in capacitance on account of change in dielectric is used to measure change in liquid and gas levels.

2.8.1 Capacitive Transducers—Using Change in Area of Plates

Figure 2.13 (*a*), (*b*) shows the elementary diagrams of the arrangements of a capacitive transducer where capacitance change occurs because of change in the area of plates.

Since capacitance is directly proportional to the effective area of the plates, response of such a system is linear.

Figure 2.13 (*c*) shows variation of the capacitance.

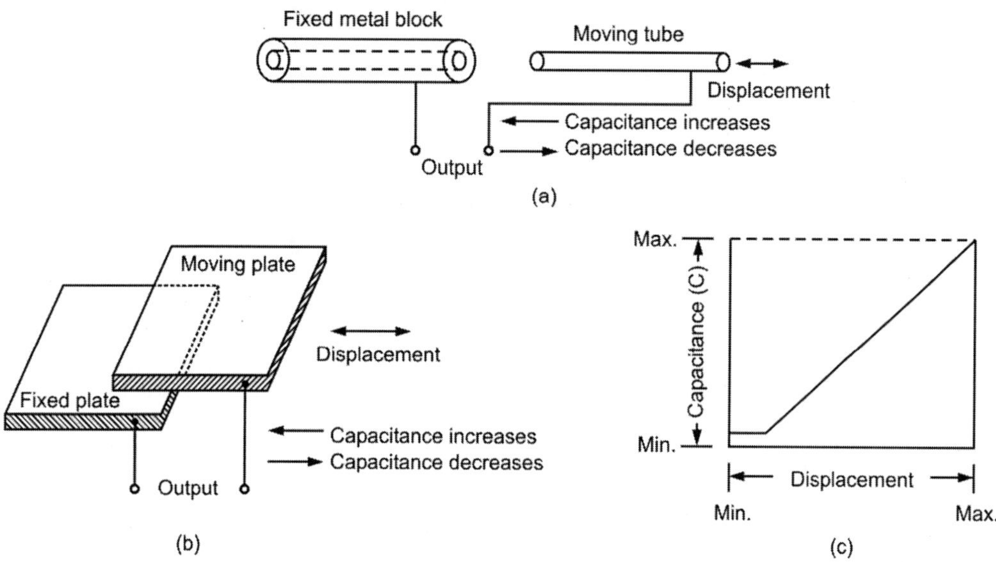

Fig. 2.13 Capacitive transducers working on the principle of change of capacitance with change of area.

2.8.2 Capacitive Transducer—Using Change in Distance Between the Plates

Figure 2.14 shows the basic form of a capacitive transducer utilizing the effect *of change of capacitance with change in distance between the plates.*

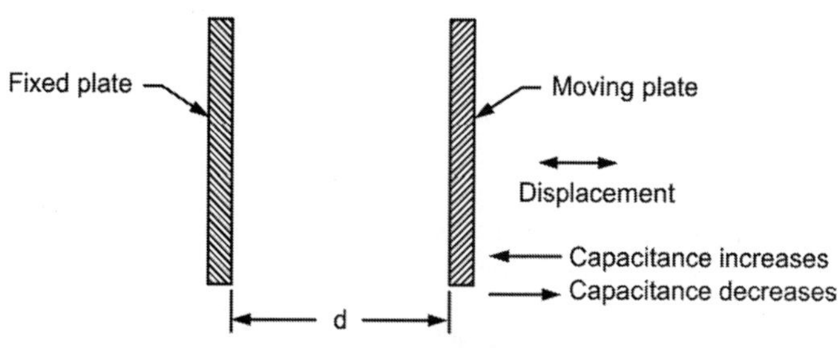

Fig. 2.14 Capacitive transducer.

One is a *fixed plate* and the displacement to be measured is applied to the other plate which is *movable.* Since, the capacitance, C varies inversely as the distance between the plates the response of this transducer is *not linear.*

Differential capacitor system

In a differential capacitor system, let the normal position of the central plate be represented by a solid line as shown in Fig. 2.15. The capacitance C_1 and C_2 are then identical,

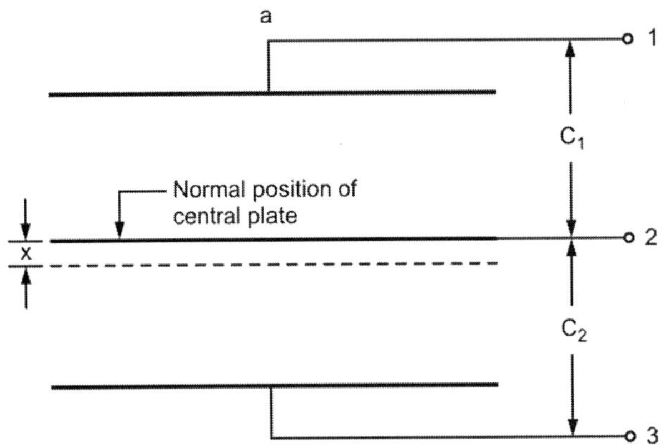

Fig. 2.15 Differential capacitor system.

i.e.,

$$C_1 = C_2 = C = \frac{\in A}{d} \qquad \text{...(2.2)}$$

When the central plate is displaced parallel to itself through a distance x, the capacitances are

$$C_1 = \frac{\in A}{d + x}, C_2 = \frac{\in A}{d - x} \qquad \text{...(2.3)}$$

For an alternating voltage E applied between the terminals 1 and 2, the voltages across C_1 and C_2 are given by

$$E_1 = \frac{EC_2}{C_1 + C_2} = E\frac{d + x}{2d}$$

and,

$$E_2 = \frac{EC_1}{C_1 + C_2} = E\frac{d - x}{2d} \qquad \text{...(2.4)}$$

When the differential measurement circuit is fed with output from the terminal pairs 1 and 3, and 2 and 3, the difference voltage would be recorded.

$$E_1 - E_2 = E\frac{x}{d} \qquad \text{...(2.5)}$$

The difference of voltage is a linear function of the displacement of the linear plate. The differential method can be used for displacement of 10^{-8} mm to 10 mm with an accuracy of 0.1%.

Advantages and disadvantages of capacitive transducers:

Advantages:

The major *advantages* of capacitive transducers are:

1. Require extremely small force for operation (hence very useful for use in small systems).
2. Extremely sensitive.
3. Require small power for operation.
4. High input impedance; therefore, loading effects are minimum.
5. Frequency response is good.
6. A resolution of the order of 2.5×10^{-3} mm can be obtained.
7. Can be used for applications where stray magnetic fields render the inductive transducers useless.

Disadvantages:

The principal *disadvantages* of capacitive transducers are:

1. The metallic parts must be insulated from each other. The frames must be earthed to reduce the effects of stray capacitances.
2. They show non-linear behaviour several times on account of *edge effects; guard rings* must be used to eliminate this effect.
3. The cable connecting the transducer to the measuring point is also a source of error. The cable may be a source of loading resulting in loss of sensitivity. Also loading makes the low frequency response poor.

Uses of the capacitive transducers. The capacitive transducers are used for the following *purposes*:

1. To measure *both linear and angular displacements.*
2. *To measure force and pressure.*
3. Used as *pressure transducers* in all those cases where the dielectric constant of a medium changes with pressure.
4. To measure *humidity* in gases.
5. Used in conjunction with mechanical modifiers for measurement of *volume, density, weight, input level* etc.

2.9 PIEZOELECTRIC TRANSDUCERS

2.9.1 Piezoelectric Materials

A "piezoelectric material" is one in which an electric potential appears across certain surfaces of a crystal if the dimensions of the crystals are changed by the application of a mechanical force. This potential is produced by the displacement of external charges. The effect is reversible, *i.e.,* conversely, if a varying potential is applied to the proper axis of the crystal, it will change the dimensions of the crystal thereby deforming it. This effect is known as **piezoelectric effect.**

Elements exhibiting piezoelectric qualities are sometimes known as *electro-resistive elements.* Common piezoelectric materials are : *Ammonium dihydrogen phosphate, Rochelle salts, lithium sulphate, dipotassium tartrate, potassium dihydrogen phosphate, quartz and ceramics A and B.*

There are two main groups of piezoelectric crystals:

1. *Natural crystals.....* such as *quartz* and *tourmaline.*
2. *Synthetic crystals......* such as *Rochelle salt, lithium sulphate, dipotassium tartrate* etc.

2.9.2 Desirable Properties of Piezoelectric Materials

The *desirable properties* of piezoelectric materials are:

(*i*) Stability,
(*ii*) High output,
(*iii*) Insensitivity to temperature and humidity,
(*iv*) The ability to be formed into most desirable shape.

Natural crystals entail the following advantages:

(*i*) Higher mechanical and thermal stability,
(*ii*) Ability to withstand higher stresses,
(*iii*) Low leakage,
(*iv*) Good frequency response.

Synthetic materials, in general, have a *higher voltage sensitivity.*

- *"Quartz"* is the most stable piezoelectric material. However, its output is quite small.
- *"Rochelle"* salt provides the highest output but it can be worked over a limited humidity range and has to be protected against moisture. The highest temperature is limited to 45°C.
- *"Barium titanate "* has the advantage that it can be formed into a variety of shapes and sizes since it is polycrystalline. It has also a higher dielectric constant.

2.9.3 Working of a Piezoelectric Device

A typical mode of operation of a piezoelectric device employed for measuring varying force applied to a simple plate is shown in Fig. 2.16. *The magnitude and polarity of the induced charge on the crystal surface is proportional to the magnitude and direction of the applied force.* The charge at the electrode gives rise to voltage (E), given by:

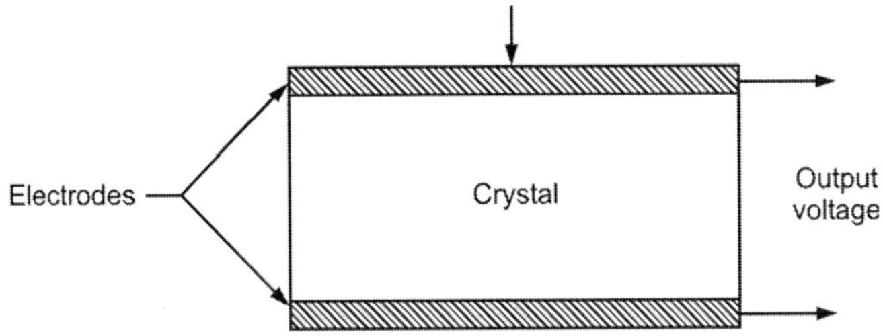

Fig, 2.16 Piezoelectric transducer.

$$E = \frac{gt\,F}{A} = gtp$$

where, g = Voltage sensitivity in Vm/N, $\left[g = \frac{K}{t},\ K = \text{Piezoelectric constant},\ t = \text{Thickness of the crystal.} \right]$

F = Force in N (newton),

A = Area of the crystal in m^2, and

p = Pressure $\left(= \frac{F}{A} \right)$ in N/m^2.

2.9.4 Advantages and Disadvantages of Piezoelectric Transducers

Advantages:

1. High frequency response.
2. Small size.
3. High output.
4. Rugged construction.
5. Negliible phase shift.

Disadvantages:

1. Output affected by changes in temperature.
2. Cannot measure static conditions.

Applications:

These transducers find the following *fields of application:*

1. Acceleraters.
2. Pressure cells.
3. Force cells.
4. Ceramic microphones.
5. Phonograph pick-up.
6. Cartridges.
7. Industrial cleansing apparatus.
8. Under-water detection system.

HIGHLIGHTS

1. The technology of using instruments to measure and control the physical and chemical properties of materials is called *instrumentation.*
2. Modes of measurements are:

   (*i*) Primary measurements (*ii*) Secondary measurements
   (*iii*) Tertiary measurements.
3. A *transducer* is a device which converts the energy from one from to another.

4. Transducers may be *classified* as follows:
   A. (*i*) Active transducers (*ii*) Passive transducers
   B. (*i*) Variable-resistance type (*ii*) Digital transducers
   C. (*i*) Variable-resistance type (*ii*) Variable inductance type
   (*iii*) Variable-capacitance type (*iv*) Voltage-generating type
   (*v*) Voltage-divider type.

THEORETICAL QUESTIONS

1. Define the term "instrumentation".
2. List the various modes of measurement.
3. Enumerate the elements of a measurement system.
4. What is transducer?
5. What are the functions of a transducer in an electronic instrumentation system?
6. How are transducers classified?
7. What are the advantages of electromechanical transducers?
8. Explain briefly with diagrams important transducer actuating mechanisms.
9. Describe briefly the following:
   (*i*) Thermistors and resistance thermometers.
   (*ii*) Wire resistance strain gauges.
10. Give the classification of variable inductance transducers.
11. Explain briefly any two of the following transducers:
    (*i*) Self-generating variable inductance transducer—Electromagnetic type
    (*ii*) Variable reluctance transducer
    (*iii*) Mutual inductance transducer
    (*iv*) Linear-variable-differential transformer (LVDT).
12. What is the principle on which a capacitive transducer works?
13. What are the advantages and disadvantages of capacitive transducers?
14. Give the applications of capacitive transducers.
15. What is a piezo-electric transducer? List the advantages and disadvantages of piezo-electric transducers.

❑❑❑

Signal Conditioning, Transmission and Processing

Chapter 3

3.1 INTRODUCTION

The *"first stage" of the instrumentation or measurement system which detects the measurand (which is basically a physical quantity) is termed as* **detector-transducer stage.** In this stage, in most of the cases, the quantity is detected and is transduced into an electrical form.

The output from the first stage *needs certain modifications* before it becomes compatible with the data presentation stage. The necessary modification is carried out in the *"intermediate stage",* more commonly referred to as the **signal conditioning stage.**

The *"last stage"* of the measurement system may consist of indicating, recording, displaying, data processing elements or may consist of control elements.

- *Measurement of dynamic mechanical quantities* places special requirements on the elements in the signal conditioning stage.

Large amplifications, as well good transient response, are often desired, both of which are difficult to obtain by mechanical hydraulic, or pneumatic methods. Consequently, *electrical or electronic elements are usually required.*

(*i*) When amplification is required *frictional forces are also amplified, resulting in considerable undesirable signal loading*. These effects, coupled with backlash and elastic deformations, result in poor response.

(*ii*) Initial loading results in reduced frequency response and in certain cases, depending on the particular configuration of the system, phase response is also a problem.

Advantages of electrical signal conditioning

In several detector-transducer combinations which provide an output in electrical form, it is convenient to perform further signal conditioning electrically.

- Such conditioning may typically *include:*
  - Converting resistance changes to voltage changes;
  - Subtracting offset voltages;
  - Increasing signal voltages;
  - Removing unwanted frequency components.
- Electrical methods are also preferred for their ease of *power amplification.*
  - *Additional power may be fed into the system to provide a greater output power than input by the users of "power amplifiers",* which have no important mechanical counter part in most instrumentation. (It is true that hydraulic and pneumatic systems may be set up to increase signal power; however, their use is limited to relatively slow-acting control applications, primarily in the fields of chemical processing and electric power generation). This technology is of a particular value when recording procedures employ stylus-type recorders, mirror galvanometers, or magnetic-disc methods.

3.2 FUNCTIONS OF SIGNAL CONDITIONING EQUIPMENT

The signal conditioning equipment may be required to perform the *following functions on the transduced signal:*

1. Amplification
2. Modification or modulation
3. Impedance matching
4. Data processing
5. Data transmission.

1. **Amplification.** It means *enhancement of the signal level* which is often in the low level range. The amplification system must bring the level of transducer signal to a value adequate enough to make it useful for *conversion, processing, indicating* and *recording.*
2. **Modification or modulation.** It means to *change the form of signal.* The signal may be *smoothened, linearised, filtered or converted into digital form.*
3. **Impedance matching.** The signal conditioning equipment arranges the input, and output impedances of the matching device so as to prevent loading of the transducer and to maintain a high signal level at the recorder.
4. **Data processing.** To carry out mathematical operations (*e.g.,* addition, subtraction, differentiation, integration etc.) before indication or recording of data.
5. **Data transmission.** To transmit signal from one location to another without changing the contents of the information.

The whole task of signal conditioning requires the following:

- Ingenuity;
- Proper selection of components;
- Faithful reproduction of signal.
  - The elements of signal conditioning are designed in such a fashion as to be *insensitive to all extraneous inputs. The accuracy, range and dynamic response are all designed to be compatible with the detector transducer.*

- In several situations the *"signal conditioning"* or *"data acquisition equipment"* is an *excitation* and *amplification system for passive transducers.* It may be an *amplification system for active transducers.* In both the applications, the transducer output is brought upto adequate level to make it useful for *conversion, processing, indicating* and *recording.*
  - In case of *"passive transducers"* (*e.g.,* strain gauges, potentiometer resistance thermometers inductive and capacitive transducers) excitation is needed because these transducers do not generate their own voltage or current; the excitation is provided from external sources.
  - The "active transducers" (*e.g.,* technogenerators, thermocouples, inductive pick-ups and piezoelectric crystals) do not require excitation from an external source since they produce their own electrical output. However, these signals have a low voltage level and as such they need to be amplified.

The excitation sources may be:

- D.C. voltage source.
- A.C. voltage source.

Figures 3.1 and 3.2 show D.C. and A.C. signal conditioning systems respectively.

D.C. signal conditioning system:

Refer to Fig. 3.1. The resistance transducers like strain gauges constitute one or more than one arm of a Wheatstone bridge which is excited by an isolated D.C. source. The bridge can be balanced by a potentiometer and can also be calibrated for unbalanced conditions.

Characteristics of a D.C. amplifier:

(*i*) It should have extremely good thermal and long term stability.

(*ii*) It may require balanced differential inputs giving high mode rejection ratio (CMRR); CMRR is a *measure of ratio of desired signal to undesired signal, this value should be as high as possible.*

Advantages:

(*i*) D.C. amplifier is easy to calibrate at low frequencies.

(*ii*) It is able to recover from an overload condition unlike its A.C. counterpart.

Disadvantages:

- The major disadvantage of a D.C. amplifier is that it suffers from the *problem of drift.* As a result, the *low frequency spurious signals come out as data information.* This problem is overcome by the use of the drift amplifiers.
- The D.C. amplifier is followed by a lowpass filter which eliminates high frequency components or noise from the data signal.

Uses. D.C. systems are generally used for common resistance transducers such *as potentiometers and resistance strain gauges.*

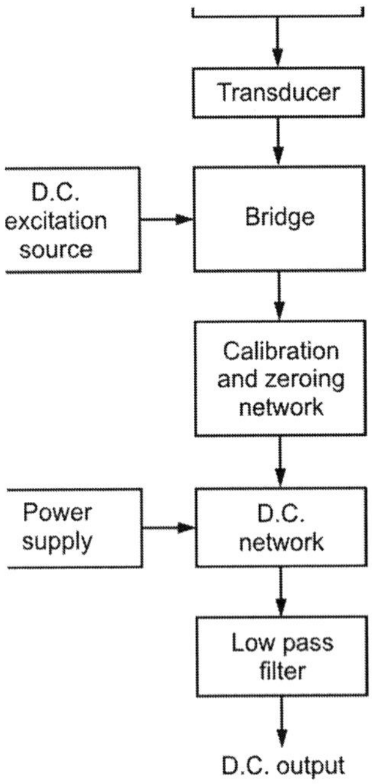

Fig. 3.1 D.C. signal conditioning system.

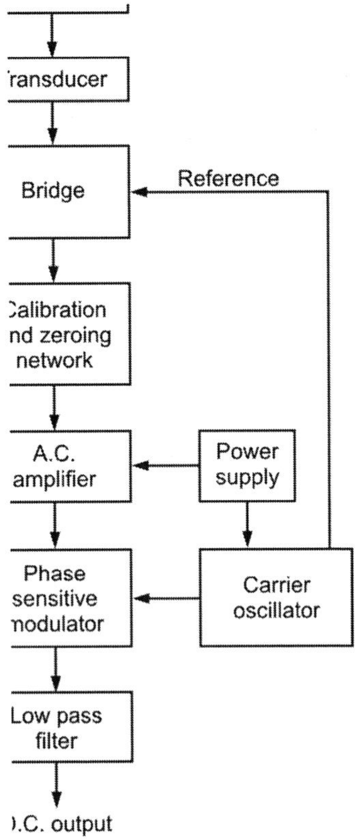

Fig. 3.2 A.C. signal conditioning system.

A.C. signal conditioning system:

Refer to Fig. 3.2. The problems which are encountered in D.C. systems are overcome through *carrier type A.C. signal conditioning system.*

The transducer parameter variations amplitude modulate the carrier frequencies at the bridge output and the waveform is *amplified* and *demodulated.* The demodulation is phase sensitive so that polarity of D.C. output indicates the direction of the parameter change in the bridge output.

In carrier systems, it is *very easy to obtain very high rejection of mains frequency pick-up.*

- Active filters can be used to reject this frequency and prevent overloading of A.C. amplifier.
- The carrier frequency components of the data signal are filtered out by the phase-sensitive demodulators.

Uses. A.C. systems are used for variable reactance transducers and for systems where signals have to be transmitted long via cables to connect the transducers to the signal conditioning equipment.

- The physical quantities like pressure, temperature, acceleration, strain etc. after having being *transduced* into their analogous electrical form and *amplified* to sufficient current or voltage levels (say 1 to 10 V) are futher *processed by electronic circuits.*
  - The signal, is some applications, does not need any further processing and the amplified signal may be directly applied to indicating or recording or control instruments.

– Several applications, however, involve further processing of signals which involve linear and non-linear operations.

3.3 AMPLIFICATION

An amplifier *is a device which is used to increase or augment the weak signal. It may operate on mechanical (levers, gears etc.) optical, pneumatic and hydraulic, or electrical and electronic principles.* The *ratio of output signal (I_o) to input signal (I_i) for an amplifier is termed as* **gain, amplification** *or* **magnification.** The gain or amplification (G) is expressed as:

$$G = \frac{I_o}{I_i} \quad ...(3.1)$$

Since $\frac{I_o}{I_i}$ are in the same units, the gain G is a *dimensionless quantity.*

Invariably, in order to get greater magnification, two or more amplifiers are arranged in series/ cascades. The overall gain of the arrangement (assuming that no loading occurs) is given by the product of individual gains of the amplifying units,

3.4 TYPES OF AMPLIFIERS

The amplifiers, on the basis of principle of working, may be categorised as follows:

1. Mechanical amplifiers.
2. Fluid amplifiers.
3. Optical amplifiers.
4. Electrical and electronic amplifiers.

3.5 MECHANICAL AMPLIFIERS

The mechanical amplifiers may be further *classified* as follows:

(*i*) Simple and compound levers: The compound lever has two or more levers linked together so that output from one lever provides the input to the other.

Example. The *Huggenberger extensometer* is one of the most popular and *accurate mechanical amplifier.* It uses a system of compound levers to give *very high magnification* to the order of 2000 *or even more.*

(*ii*) Simple and compound gears: The simple and compound gear trains are used quite frequently *to provide mechanical amplification of either angular displacement or rotary speed.*

A *"compound gear train"* gives *greater modification with the additional advantage of no change in the direction of input signal.*

Examples. The gear trains are used for the magnification of displacement in the *Bourdon tube pressure gauge* and in the *dial-test indicator* where linear movement is translated into rotation by means of rack and pinion.

Limitations of mechanical amplification:

The mechanical amplification, as earlier stated, usually suffers from the errors caused by the following *factors.*

- Internal loading;
- Friction at the mating parts;
- Elastic deformation;
- Backlash.

3.6 FLUID AMPLIFIERS

Fluid amplifiers may be *classified* as follows:

(*i*) **Hydraulic amplifier:** When a small displacement is applied to a piston operating inside a cylinder containing some liquid, there occurs a large displacement of the liquid in the output tube which has a small diameter.

Examples. This principle is employed in the *mercury-in-glass thermometer* and the *single-column manometers.*

(*ii*) **Pneumatic amplifier:** Pneumatic methods are extensively used and can be applied to any type of measurement.

3.7 OPTICAL AMPLIFIERS

In optical amplification, a ray of light strikes a mirror with an angle of incidence i and gets reflected with angle of reflection equal to the angle of incidence. When the mirror rotates through an angle θ, the angle of incidence change to $(i + \theta)$. Before rotation of the minor, the angle between the incident ray and reflected ray is $2i$ and after rotation it is $2(i + \theta)$. Obviously there is angular magnification of 2θ between the incident and reflected rays. In order to get a greater magnification, more number of mirrors surfaces may be used.

Example. This principle to amplify the input signals is used in the following cases;

– Optical levers;
– U.V galvanometers;
– Mechanical-pointer galvanometers.

3.8 ELECTRICAL AND ELECTRONIC AMPLIFIERS

The electrical amplifiers are used to *increase the magnitude of weak voltage or current signals resulting front electromechanical transducers.*

3.8.1 Desirable Characteristics of Electronic Amplifiers

The following are the desirable characteristics of electronic amplifiers:

(*i*) *High input impedance* so that its loading effect on the transducer in minimum.
(*ii*) *Low output impedance* so that the amplifier is *not unduly loaded by the display or recording* device.
(*iii*) *Frequency response* should be as *good as that of the transducer.*

3.8.2 A.C. and D.C. Amplifiers

The instrumentation systems usually employ the following two types of electronic amplifiers,

(*i*) A.C. coupled amplifiers,
(*ii*) D.C. coupled amplifiers.

- For an "A.C. *amplifier"* bandwidth is the range of frequencies between which gain or amplitude ratio is constant to within –3dB (3dB down points). This corresponds to the frequencies at which the voltage output amplitude falls by 29.3% to 70.7% of the maximum value.
  – The "A.C. *amplifiers"* are *only capable of dealing with rapid, repetitive signals but are usually simpler and cheaper* when compared with their D.C. counterparts.

- In "A.C. amplifier systems" the amplifier drift and spurious noise are *not significant;* the mains frequency pick-up *rejection is very high.*

- The *"D.C. amplifiers"* are capable *of amplifying static, slowly changing or rapid-repetitive input signals.*
  - The "DC. amplifire systems" are *easy to calibrate at low frequencies, and have the ability to recover rapidly from overload conditions.*

3.8.3 Modulated and Unmodulated Signals

The measurands may be *"pure"* in the sense that analog electrical signal contains *nothing more than the real time variation of the measurand information itself.*

On the other hand, the signal may be *"mixed"* with a **carrier** which consists of a *voltage oscillation at some frequency higher than that of the signal.* A common rule of thumb is that *the frequency ratio should be at least 10:1.*

The measurand affects the *carrier* by varying either its *amplitude* or its *frequency:*

- In the *former* case the *carrier frequency is held constant and its amplitude is varied by the measurand. This process is known as* **Amplitude modulation (or AM).**
- In the *latter* case the *carrier amplitude is held constant and its frequency is varied by the measurand.* This is known as Frequency modulation (or FM).

The most familiar use of AM and FM transfer of signals is in AM and FM *radio broadcasting.*

When *"modulation"* is used in instrumentation "amplitude modulation" (AM) is the *more common form.*

- Nearly any mechanical signal from a passive pick-up can be transduced into an analogous AM form. Sensors based on either *inductance (e.g.,* differential transformer) or *capacitance* (*e.g.,* capacitance pickup for liquid level) require an A.C. excitation.

In addition, however, resistance-type sensors may also use an A.C. excitation, as with some strain gauge circuits.

It is required to extract signal information from the modulated carrier.

- This operation, *when AM is used,* may take several forms;
  - The simplest is merely to display the entire signal using an *oscilloscope or oscillograph,* and then to *"read"* the result from the envelope of the carrier.
  - More commonly. The *mixed signal and carrier are "***demodulated***"* by *"rectification* and *filtering".*
- FM demodulation is more complex operation and may be accomplished through the use of
  - Frequency discrimination,
  - ratio detection, or
  - 1C phase-locked loops.

3.8.4 Integrated Circuits (ICs)

The integrated circuits (ICs), as the name implies, are groups of circuit elements combined to perform specific purposes. For the most part the elements consist of *transistors, diodes, resistors,* and, to lesser extent, *capacitors,* all connected and packaged in convenient plug-in units.

ICs from the building blocks are used to construct more complex circuits such as:

- Differential amplifiers;
- Mixers (for combining signals);
- Timers:
- Filters:
- Audio preamps,
- Auto-power amplifiers,
- Voltage references;
- Regulators and comparators,
- Several digital devices.

3.8.5 Operational Amplifiers (Op-Amp)

An ***operational amplifier (Op-Amp)*** *is a linear integrated circuit (IC) that has a very high voltage gain, a high input impedance and a low output impedance.*

It is so called because it can be employed to carry out mathematical operations like *"addition", "subtraction", "multiplication", "division" "integration", "differentiation"* etc.

- Operational amplifiers are linear integrated circuits that work on *relatively low supply voltage.*
- They are *reliable and inexpensive.*
- An *ideal operational amplifier* is device of infinite voltage gain, infinite bandwidth, infinite input impedance (open) and zero output impedance.
- An Op-amp may contain two dozen transistors, a dozen resistors and one or two capacitors. ***Examples:*** μA 709; LM 108-LM 208, CA 741 CT and CA741T.

3.8.6 Attenuators

An ***attenuator*** *is a two-port resistive network and is used to reduce the signal level by a given amount.*

In a number of applications, it is necessary to introduce a specified loss between the source and a matched load without altering the impedance relationship. Attenuators may be used for this purpose.

Attenuator may be *symmetrical* or *asymmetrical,* and can be either *fixed or variable. A fixed attenuator with constant attenuation* is called a **pad.**

- Variable attenuators are used as control volumes in radio broadcasting sections.
- Attenuators are also used in *laboratory to obtain small value of voltage or current for testing circuits.*

The attenuation is expressed in **decibels (dB)** or, in **naper**. The attenuation offered by a network in decibels is given by:

$$\text{Attenuation in } dB = 10 \log_{10}\left(\frac{P_i}{P_o}\right) \quad ...(3.23)$$

where, P_i is the input power and P_o is the output power.

The attenuators may be of the following types:

1. Resistance attenuator.
2. Symmetrical T-attenuator.
3. L-Type attenuator.
4. π-type attenuator.

3.8.7 Filters

Filtering *is the process of attenuating unwanted components of a measurement while permitting the desired component to pass.*

The filter *is an electronic circuit which can pass or stop a particular band of frequencies through it.* The filters were first designed by G.A. Campbell and D.Z. Zobel at Bell laboratories.

The band of frequencies which will pass through filter is called the **pass band** *and the band of all remaining frequencies is called* **attenuation band.** In case of **ideal filter,** all frequencies of pass band will pass without suffering from any attenuation while the band of all remaining frequencies of attenuation band will be suppressed completely.

Classification of filters:

The filters may be *classified* as follows:

A. *On the basis of passing and attenuating of frequencies:*

1. **Low pass filters:**
   - These are those filters which pass only low frequencies through them and which reject all high frequencies above the cut-off frequencies.
   - A low pass filter is also called *"lag network"* because it causes a phase lag in the output signal.
   - This type of filter is also called *"integrating network"*.
2. **High pass filters:**
   - These are those filters which pass only high frequencies through them and which reject all low frequencies below the cut-off frequency.
   - The high pass filter is a differentiating network and is also called as *"lead network"* because it will cause a phase lead in the output signal.
3. **Band pass filters:**
   - These are those filters which pass a band of frequencies through them and which reject all other frequencies to pass through them.
4. **Band stop filters:**
   - These filters, which are also known a s *"band elimination filters"*, are those which reject a band of frequencies to pass through them and which allow the other frequencies to pass through them.

B. On the basis of relation between series and shunt Impedances:

1. **Constant filters (or prototype filters)**. In this filter the series impedance z_1 and stunt impedance z_2 are interrelated by the relation:

   $$z_1 z_2 = k^2,$$

   where, k is a constant independent of frequency.
2. ***m*-derived filters.** These filters do not have the product of series and shunt impedances equal to k^2, but have the same characteristic impedance as the corresponding k section, with sharper attenumation characteristic.

3.9 SIGNAL TRANSMISSION

The terms *"measuring devices"* and *"transmitters"* generally go side-by-side and it is very difficult to make any distinction between them.

A measuring device converts a primary indication into some form of energy that can easily be displayed on a scale; some transmitters also do the same things. In the *stricter sense* **"transmitters"** *could be considered as devices which transmit the value of the primary variable at a considerable distance from the primary element.* If transmission is to be carried over *very long distances,* then devices are known as **"telemeters".**

The terms **data transmission** *and* **"telemetry"** *refer to the process by which the measurand is transferred to a remote location for the purpose of being processed, recorded and displayed.*

For transmission purposes, the measured variable is converted into a transmittable signal (either pneumatic or electrical), so that it can be received by a remote indicating, recording, or controlling device. The *selection of transmission device* depends upon the *nature of the variable and the distance the signal is required to be sent.*

For data transmission various methods have been developed; the choice of a particular method depends upon:

(*i*) The physical variable;

(*ii*) The distance involved.

- **The hydraulic and pneumatic methods** are employed for transmission over a *short distance.*
  - The pneumatic type transmission devices are generally suitable for transmission upto maximum distance of 200 m.
- **The electrical/electronic methods** are suitable *equally for short as well as long distance transmission.*
  - Generally short transmission is carried out on own communication connections between sending and receiving devices.
  - The *telemeters* which are designed for *long-distance transmission* may be designed to transmit over their own wire or over phone wires or by microwave.

3.9.1 Mechanical Transmission

The *"rack and pinion arrangement"* and the *"gear trains"* as used in *Bourdon-tube pressure* gauge and *dial indicator gauge* constitute mechanical transmission. They *amplify the displacement* and also *transmit the signal to a pointer which moves across a calibrated dial.*

3.9.2 Hydraulic Transmission

Figure 3.3 shows the hydraulic method of transmission, which is commonly used. Here four bellows are employed, two at the transmission end and two at the receiving end. The four bellows are connected by an impulse pipeline and the whole system is filled with liquid. When the actuating link, on the transmission end, is operated by the measurand, then one bellow is expanded and other is contracted. This expansion and contraction is communicated to receiving end, which moves the receiving pointer an equal amount. The purpose of using two bellows on either side is to compensate for changes in ambient temperature.

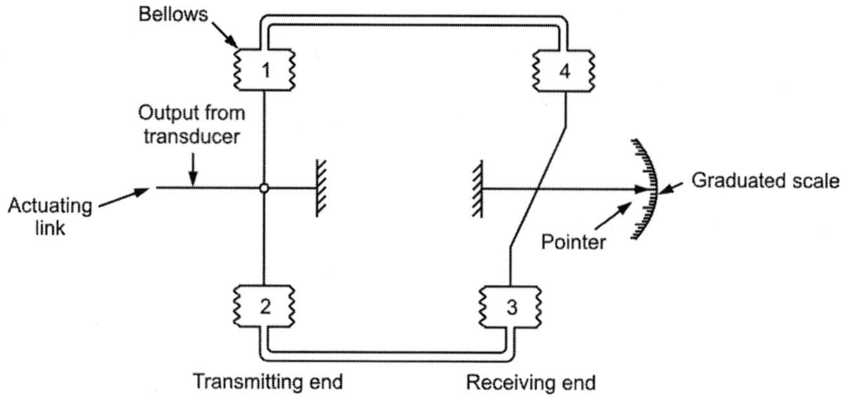

Fig. 3.3 Hydraulic method of transmission.

3.9.3 Pneumatic Transmission

Figure 3.4 shows the one of the pneumatic methods of transmission ***(Flapper nozzle mechanism).***

It consists of an open nozzle which is supplied with air through a restriction/orifice (its diameter being *smaller* than nozzle diameter for proper functioning). In front of the nozzle there is a flapper which is positioned by the measuring element. The force on the flapper is produced by a transducer which converts the measurand into linear displacement. The flapper is pivoted about a point and the other end, it contains some balancing counter weight.

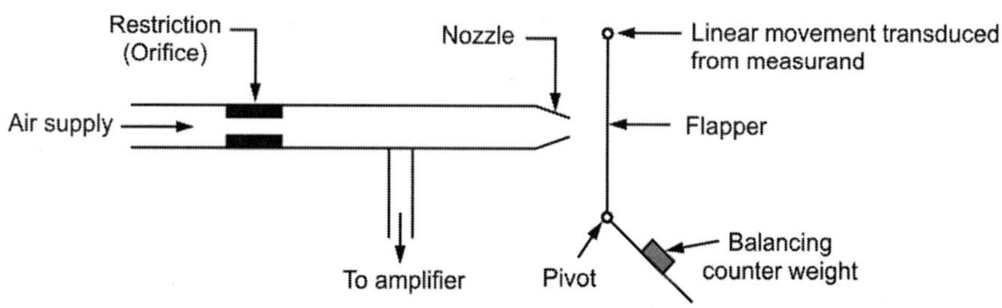

Fig. 3.4 Schematics of pneumatic transmission–Flapper nozzle mechanism.

When the flapper is moved against, the nozzle, the air cannot escape and maximum air passes to the amplifier, and when flapper is moved away from the nozzle, minimum air passes to the amplifier as most of the air escapes to atmosphere. Thus, the movement of flapper from one extreme position to another serves to control the amplifier, which produces an air pressure proportional to the measurand of adequate strength for transmission over the required distance.

3.9.4 Magnetic Transmission

Figure 3.5 shows the schematics of magnetic transmission. In this arrangement/device, an armature is attached at the end of the mechanical moving part whose movement is to be transmitted outside the armature moving inside a non-magnetic tube. A magnet is placed around the armature outside the tube. The magnet follows the movement of the armature and repositions a pneumatic transmitter. The magnet movement could also be utilised to operate an electronic transmitter.

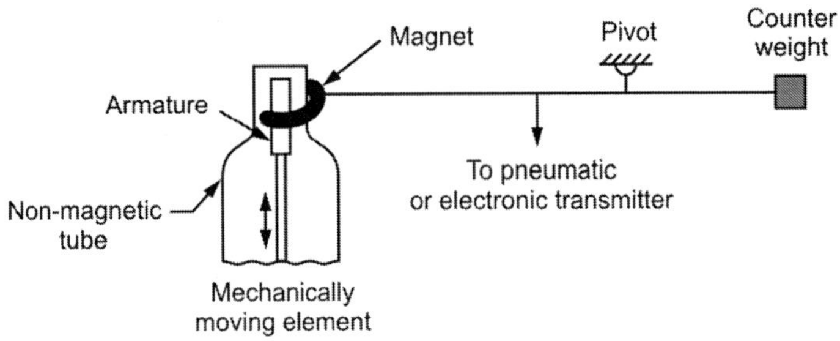

Fig. 3.5 Schematic Transmission.

3.9.5 Electric Type of Transmitters

Most of the electric type of transmitters employ A.C. bridge circuits in which degree of coupling between inductances is varied by changing the amount of iron core within a coil. The common *examples* are:

1. Wheatstone bridge transmitter
2. Inductance bridge.
3. Impedance bridge.
4. Differential transformer.
5. Self synchronous motor (Selsyn)
6. Resistance manometers.

3.10 SINGLE DISPLAY AND RECORDING DEVICES

3.10.1 Introduction

The main purpose of any measurement system is to provide information concerning the state and condition of the physical phenomenon being investigated. The measuring systems may be activated either directly from the measuring means *(e.g.,* bellows, pressure spiral etc., to which indicating pointer is attached directly through lever and leverage system) or by means of a servo-operated system (null-balance system which incorporates a feedback circuit in a closed loop). The last stage of a measurement system is the *data presentation stage;* if the results of the system are meaningful they must be displayed for *instant observation by a display device or for storage for observation at a later stage by a recorder* (The data presentation devices may be called as *"output devices").*

The following *factors* decide about the *choice between the display devices and recorders:*

(*i*) The information content of the output.
(*ii*) The expected use of the output.

The output devices may be *categorized* as follows:

1. Single number output devices.
2. Time domain output devices.

1. Single number output devices:

Such devices indicate the value of some particular quantity under condition such that the value to be measured can be regarded as time variant over the time interval during which measurements are made; thus a *single number will represent measurement.*

"Indicating instruments" and *digital display units* belong to this class.

2. Time domain output devices:

The indicating instruments or the digital display units (suitable only when the *output varies at a very slow rate)* do not serve the purpose *when the values of the quantity are to be taken as a function of time.*

For fast changing outputs (where signal waveform or shape is the desired information) ..... *"Cathode ray oscilloscope (CRO)"* is used.

For keeping a *permanent record of the variation of the output with time ..... "Cathode ray tube photographs", direct writing recorders" strips chart recorders", magnetic tape recorders* etc. are used.

The machine interpretable outputs can be had from:

- Magnetic tapes;
- Punched paper tapes;
- Punched cards;
- Pulsed signals.

- The information available from an instrument may take the following forms:
  (*i*) *Quantitative information* (*e.g.*, angular spread in r.p.m. force in newtons).
  (*ii*) *Qualitative information* (*e.g.*, the approximate value or direction of change of some variable, a check reading)
  (*iii*) Status informations (*e.g.*, On/Off, in/out),
  (*iv*) Alphanumeric and symbolic information (*e.g.*, the labels and instructions; letters *A* to Z, the numerals 0 to 9, punctuation marks and various other simple symbols can be generated and interpreted).
- A *good display, functionally, is one which permits the best combination of speed accuracy and sensitivity when transferring* the *necessary information from the instrument to the operator.*

3.10.2 Electrical Indicating Instruments

Qualitative the electrical indicating instruments are widely used for measurement of current, voltage, resistance and power. These instruments can be *classified* as follows:

1. Analog instruments.
2. Digital instruments.

Analog type of meters use scale and needle (pointer) type arrangement to display the value of the measured parameter. *The measured parameter is converted into electrical signal which further actuates some electro-mechanical device to move the pointer continuously as the parameter changes.*

Digital type meters *indicate the reading in exact numerals.*

3.10.3 Digital Instruments

The **digital instruments** *indicate the value of the measured in the form of decimal number* (whereas the analog instruments display the quantity to be measured in terms of deflection of a pointer, *i.e.*, an analog displacement or an angle corresponding to the electrical quantity).

The **digital meters** *work on the principal of "qantization".* The analog quantity to be measured is first subdivided or quantized into a number of small intervals upto many decimal places. The objective of the digital instrument is then to determine in which portion of the subdivision the measured can thus be identified as an integral multiple of the smallest unit called the quantum, chosen for subdivision. The measuring procedure thus reduces to one of counting the number of quanta present in the measurand.

A digital instrument *can be considered as a counter which counts the pulses in a predetermined time.* Digital transducers whose output is in the form of pulses are used to monitor the desired parameter. *Accuracy of digital instrument is dependent on the number of pulses generated by transducer because the fraction of pulse cannot be generated and in counting there can be ambiguity of only one pulse at start/stop.* Hence, *more are pulses corresponding to a measure, less the possibility of error corresponding to one pulse* and *more the accuracy.*

The *information* in the electronic digital read-out (display) devices is *presented as a series of digits on tubes, screen or printed on a piece of paper.* The relevant characters (letters of alphabet from A to Z, numerals from 0 to 9, punctuation mark and other symbols in common use) can be generated by:

(*i*) Semiconductor light emitting diodes (LED).
(*ii*) Liquid crystal displays (LCD).
(*iii*) Numerical indicators tubes (NIT).
(*iv*) Hot filament or bar tubes.

3.10.3.1 Light Emitting Diode (LED)

A P-N junction can absorb light energy and produce electric current. The opposite process is also possible, that is a junction diode can emit light. The emission of light occurs under forward bias condition due to recombination of electrons and holes. The emitted light may be visible or invisible.

A P-N *junction diode, which emits light when forward biased is known as a* **light emitting diode (LED).**

The amount of light output is *directly proportional to the forward current.* Thus, higher the forward current, higher is the light output.

Figure 3.6 (*a*) shows the schematic symbol of a light emitting diode. The arrows pointing away from the diode symbol represent the *light,* which is being transmitted away from the junction.

Figure 3.6 shows the basic structure of a light emitting diode.

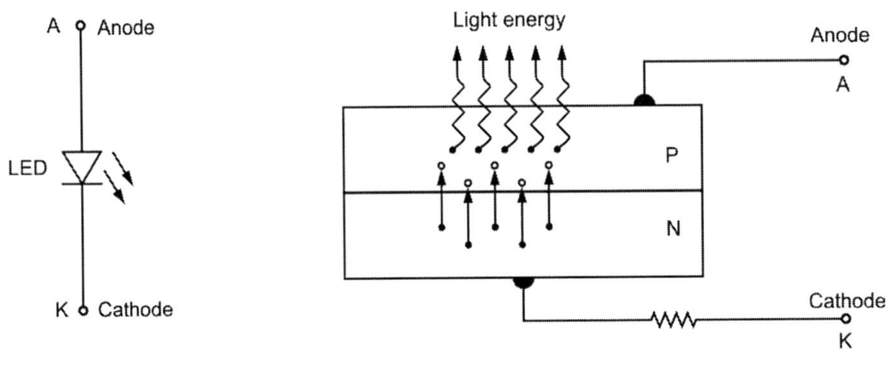

(a) Symbol (b) Basic structure

Fig. 3.6 Light emitting diode (LED).

In a forward based P-N junction, free electron from N-type material diffuse into P-region. Once in P-region these free electrons encounter holes and eventually recombine. Thus the free conduction electron fills a vacancy in covalent structure and thus becomes a valence electron. In doing so the electron loses a certain amount of energy as it jumps form conduction band to the valence band. In Si or Ge *diode, the energy that recombining electrons lose is dissipated in the form of heat. But if other semiconductor material such as gallium arsenide and gallium phosphide are used to form P-N diode, the energy lost by recombining electrons is given off in the form of* **light energy.**

- Diodes made of gallium arsenide (GaAs) emit infrared radiation invisible to eyes and such diodes are referred to as IRED—Infrared emitting diodes.

  Figure 3.7 shows two curves used to determine LED **operating characteristics.**
- Figure 3.7 (*a*) is forward bias V-I curve for a typical IRED, the type used in burglar alarms. Forward bias of around 1 V is required to produce significant forward current.
- Figure 3.7 (*b*) is a plot of radiant output power *vs* forward current. The radiant output power is rather small (μW) and *indicates a very low efficiency of electrical to radiant energy conversion.*

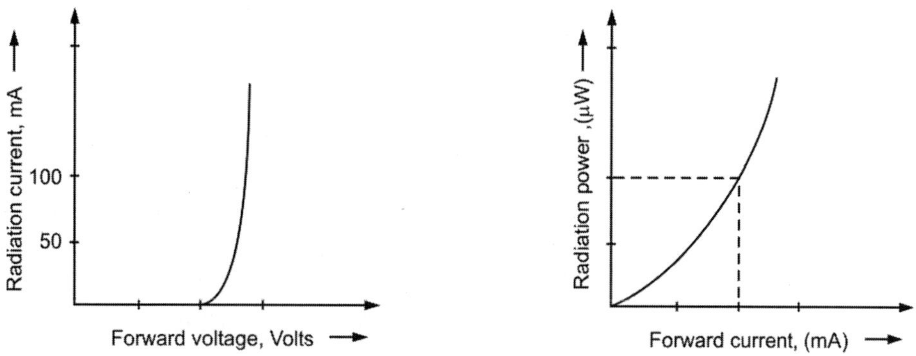

Fig. 3.7 Operating characteristics—LED.

Advantages:

1. LEDs are miniature in size and they can be stacked together to form numeric and alphanumeric display in high density matrix.
2. The light output from an LED is a function of current flowing thought it. Therefore, intensity of light emitted from LEDs can be smoothly controlled.
3. LEDs have a high efficiency as emitted of electromagnetic radiation. They require moderate power for their operation. They are useful where miniaturization in the low D.C. power are important.
4. They are very useful where dynamic operation of large number of arrays is involved.
5. Economical and high degree of reliability.
6. Being rugged, they can withstand shocks and vibrations.

Disadvantage: LEDs are not suited for large areas display, primarily because of their high cost.

Applications:

Since LEDs operate at voltage levels 1.5 V to 3.3 V, they are highly compatible with solid state circuitry. Their uses include the following:

(*i*) Panel indicator
(*ii*) Digital watches
(*iii*) Multimeters
(*iv*) Calculators
(*v*) Itercoms
(*vi*) Switch boards
(*vii*) Burglar-alarm systems
(*viii*) Optical fibre communication systems etc.

3.10.3.2 Liquid Crystal Displays (LCD)

- A liquid crystal is a material, usually an organic compound, which flows like a liquid at room temperature; its molecular structure has some properties normally associated with solids (*e.g.*, chloesteryl nonanoate and p-azoxyanisole).
- When light is incident on an activated layer of a liquid crystal, it is *either absorbed or else is scattered by the disoriented molecules.*
- A liquid crystal 'cell' (Fig. 3.8) consists of a thin layer (about 10 μm) of a liquid crystal sandwiched between two glass sheets with transparent electrodes deposited on their inside faces.
  - When both glass sheets are transparent, the cell is known as *transmittive type cell.*

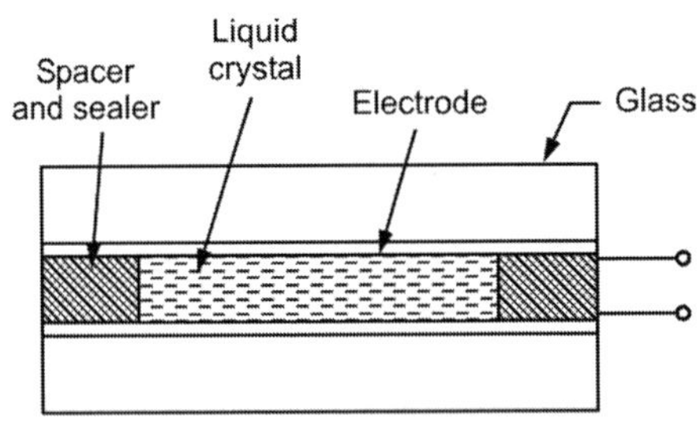

Fig. 3.8 A liquid crystal cell.

  - When one glass is transparent and the other has a reflective coating, the cell is called *reflective type.*
- Liquid crystal display produces no illumination of its own; it depends entirely on illumination falling on it from an external source for its visual effect.

Advantages:

1. Extremely low power requirement.
2. Long life time-about 50 000 hours.

Uses:

1. Cellular phone display.
2. Desk top LCD monitors.
3. Note book computer display.
4. Watches and portable instruments.
5. Pocket T.V. receiver.

3.10.3.3 Numerical Indicator Tubes (NIT)

- A numerical indicator tube (NIT) consists of a gas filled glass tube having ten cathodes in the form of numbers and an anode.
- On the application of adequately high voltage (breakdown voltage between 60 to 150 V) between the anode and cathode, the gas becomes ionized and a current flows through the gas between the electrodes. The gas molecules are excited to a higher energy state and upon relaxation to their original state, they emit light and a glow appears around the cathode. Due to this glow a particular number is displayed.

3.10.3.4 Hot Film or Bar Tubes

A hot filament indicator consists of small bars of light emitting tubes; these tubes may be the tiny florescent tubes similar to those used for lighting. The bars make up the characters of display and get illuminated when an electric current flows through them.

The hot filament indicators have a slow response time, to the time of about 20 ms.

Segmental and dot matrix displays. The different characters appearing on the digital read out systems are required to be legible and this aspect depends upon *stroke width, form of character, contrast* and *illumination.* Although the desirable size of numerals depends upon the viewing distance, the visual displays are read most accurately when (*i*) ratio of stroke width to height is about 1 : 6 to 1 : 8 for black on white and somewhat higher (upto 1:10) for white on black and (*ii*) The width is at least two-thirds the height.

It has been observed through research that the characters formed by 7 × 5 dot matrix are superior to those generated by the seven-segment displays.

3.10.3.5 Digclampter

- A 'Digclampter' Fig. 3.9 is a combined meter for measurement of *Voltage, Current, Resistance* and *Temperature.*
- It gives the reading in *digital form.*
- It can *measure even under strong magnetic field.*
- It has LED *display which provides low power consumption and also has automatic zero adjustment in display.*
- It can *measure D.C. volts, A.C. volts, A.C. amps., Resistance, Temperature.*
- It works with good accuracy and overload circuit.

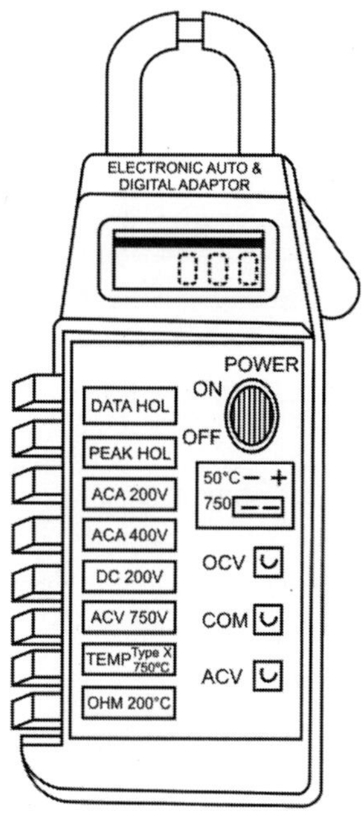

Fig. 3.9 Digclampter.

Ranges

D.C. Voltage.................... 200 V
A.C. Voltage.................... 750 V
Resistance........................ 2000 kΩ
A.C. Current.................... 200 A, 400 A, 500 A
Temperature..................... 40°C to 1000°C

3.10.3.6 Digital Tachometer for r.p.m. Measurements

A digital tachometer for r.p.m. measurements is shown in Fig. 3.10. It is a precision instrument. It measures the r.p.m. of rotating objects. It uses specially developed LSI device for computing r.p.m. It is 4 digital (0.3 red LED) instrument which can measure 5 to 9999 r.p.m. with ± 1 r.p.m. accuracy upto 999 r.p.m. and ± 10 r.p.m. over 1000 r.p.m.

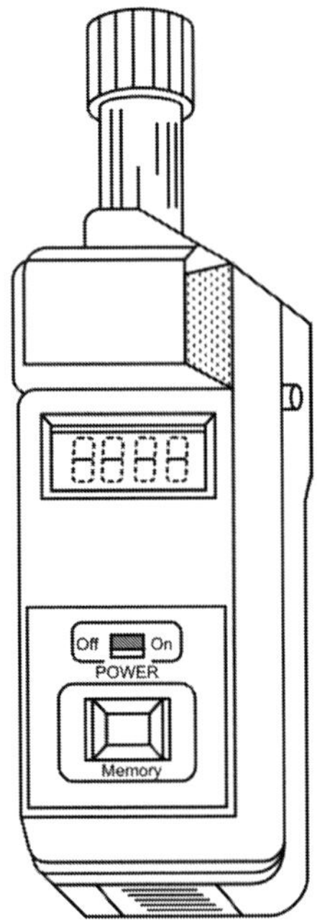

Fig. 3.10 Digital tachometer.

Working:

- Apply a deflective marker to the object being measured.
- Place the tachometer "Power Switch" in "ON" position.
- Press the "Measure" button and align the visible light beam (Distance 5 to 15 cm) with the applied target. Verify the Monitor lamp lights each time the target passes through the light beam. Release the "Measure Switch" when the reading stabilizes.

Such instrument has Memory facility. The last reading taken is automatically latched in the memory. This reading may be displayed simply by pressing the Memory Recall Button.

3.10.4 Recording Devices

A **recorder** *records electrical and non-electrical quantities as a function of time.* The record may show how one variable varies with respect to another, or how the input signal varies with time.

The record serves the following *objectives:*

(*i*) It preserves the details of measurement at a particular time.

(*ii*) It provides at a glance the overall picture of the performance of unit.

(*iii*) It provides immediate reflection on the action taken by the operator.

3.10.4.1 Types of Recorders

In an instrumentation system, one of the important considerations is the method by which the data required in recorded. The recording method should be consistent with the type of system. If we are dealing with a wholly analog system, then *analog recording techniques* should be used. While, on the other hand, if the system has a digital output, *digital recording devices* are employed. Two types of recorders are:

1. Analog recorders:

(*i*) Graphic recorders

(*a*) Strip chart recorders

- Galvanometer type
- Null type

– Potentiometric recorders

– Bridge recorders

– LVDT recorders.

(*b*) X-Y recorders

(*ii*) Oscillographic recorders.

(*iii*) Magnetic tape recorders.

2. Digital recorders

The above recorders are discussed briefly below:

3.10.4.2 Strip Chart Recorders

Figure 3.11 shows the basic constructional features of a strip chart recorder.

A strip chart consists of the following:

- A long wall of graph paper moving *vertically*.
- A system for driving the paper at some selected speed.
- A stylus for marking paper on the moving graph paper (Most recorders use a pointer attached to the stylus, which (pointer) moves over a calibrated scale thus showing instantaneous value of the quantity being measured).

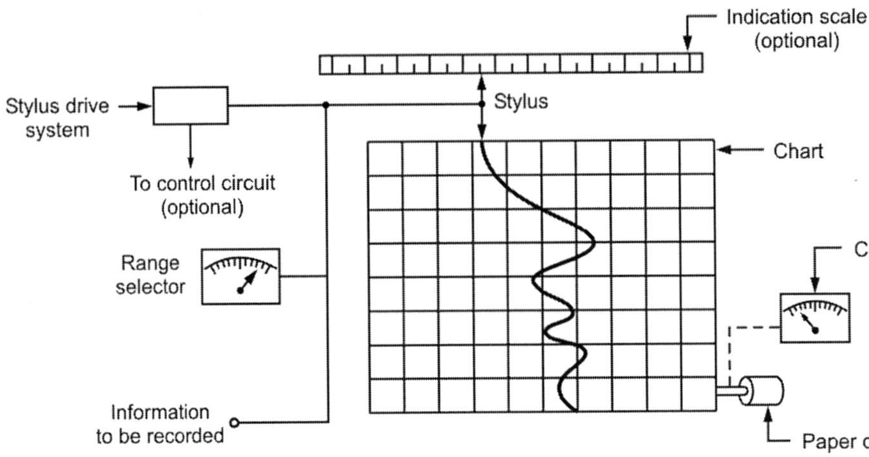

Fig. 3.11 Strip chart recorder.

- A stylus driving system which moves the stylus in nearby exact replica or analog of the quantity being measured (A spring wound mechanism may be used but in most of the recorders a *synchronous motor* is used for driving the paper).

Marking mechanisms. The most commonly used mechanisms employed for marking marks on the paper are:

(*i*) Marking with ink filled stylus.
(*ii*) Marking with heated stylus.
(*iii*) Chopper bar.
(*iv*) Electric stylus marking.
(*v*) Electrostatic stylus.
(*vi*) Optical marking method.

***Tracing** systems.* For producing graphic representations, the following two types of tracing systems are used:

(*i*) Curvilinear system.
(*ii*) Rectilinear system.

Galvanometer type strip chart recorders:

- This type of recorder operates on the *"deflection principle".*
- The deflection is produced by a galvanometer, (d'Arsonval) which produces a torque on account of a current passing through its coil. This current is proportional to the quantity being measured.
- These recorders can work on ranges for a few mA to several mA or from a few mV to several mV.
- The moving galvanometer type recorder is *comparatively inexpensive instrument,* having a narrow bandwidth of 0 to 10 Hz. It has a sensitivity of 0.4 mV/mm or from a chart of 100 mm width a full scale deflection of 40 m V is obtained. Linear amplifiers are used for measurement of smaller voltages.
- This type of recorder is not useful for recording fast variations in either current or voltage or power.

Null type strip chart recorder:

- This type of recorder operates on *"comparison basis"*. The null type strip chart recorders are of the following types:

1. Potentiometric recorders.
2. Bridge recorders.
3. LVDT recorders.

The most common application of **potentiometric recorder** is *for recording and control of process temperatures.* Self-balancing potentiometers are unduly used in industry because of the following *reasons:*

(*i*) Their action is automatic and thus eliminates the constant operation of an operator.
(*ii*) They draw a curve of the quantity of being measured with the help of recording mechanism.
(*iii*) They can be mounted on switchboard or panel and thus act as mounting devices for the quantity under measurement.

Single-point and multi-point recorders:

- *Instruments that record changes of only one measured variable are called* **single-point recorders.**
- **A multi-point recorder** may have as many as 24 inputs, with traces displaced in 6 colours.

3.10.4.3 X-Y Recorders

A X-Y recorder is an instrument *which gives a graphic record of the relationship between two variables.* This system has a pen which can be positioned along the two axes with the writing paper remaining stationary. There are *two* amplifier units, one amplifier actuates the pen in the Y-direction as the input signal is applied, while the second amplifier actuates the pen in X-direction. The movements of the pen in X- and Y-directions are automatically controlled by means of a motor, pulleys and a linear potentiometer. Obviously, trace of the marking pen will be due to the combined effects of two signals applied simultaneously. In these recorders, an e.m.f. is plotted as a function of another e.m.f. There are many variations of X-Y recorders. With the help of these *recorders and approximate transducers a physical quantity may be plotted against another physical quantity.*

A few examples in which *use of X-Y recorders is made* are as under:

(*i*) Plotting of stress-strain curves, hysteresis curves and vibrations amplitude against swept frequency.
(*ii*) Pressure -volume diagrams for I.C. engines.
(*iii*) Pressure-flow studies for lungs,
(*iv*) Lift drag wind tunnel tests,
(*v*) Electrical characteristics of materials such as resistance versus temperature and plotting the output from electronic calculators and computers.
(*vi*) Speed-torque characteristics of motors.
(*vii*) Regulation curves of power supplies.
(*viii*) Plotting of characteristics of vacuum tubes, zener diodes, rectifiers and transistors etc.

3.10.4.4 Ultraviolet (U.V) Recorders

These recorders are basically electro-mechanical oscillographic recorders and modified version of Duddel's oscillographs.

An ultraviolet recorder consists of a number of galvanometer (moving coil) elements mounted in a single magnet block as shown in Fig. 3.12. A paper sensitive to ultraviolet light is used for producing a trace for the purpose of recording. The u.v. light is projected on the paper with the help of mirrors attached to the moving coils.

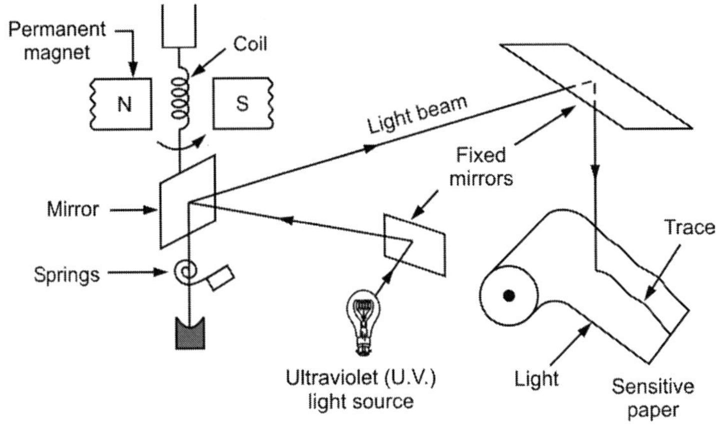

Fig. 3.12 Ultraviolet (u.v.) recorder.

Working. When a current is passed through the moving (galvanometer) coil, it deflects under the influence of the magnetic field of the permanent magnet. The ultraviolet light falling on the mirrors is deflected and projected on to the u.v. light sensitive paper through a lens and mirror system. The paper is driven past the moving high spot and thus a trace of variation of current with respect to time is produced.

The recorder, in addition to the input currents, may have the following additional traces

(*i*) Grid lines.
(*ii*) Timing lines.
(*iii*) Trace identification.

- The ultraviolet (u.v.) recorders, compared to the mechanical and pen recorders, *have better frequency and response characteristics;* the typical values are :

  Frequency response = 0 to 300 Hz; 0 to – 12 kHz (maximum)

  Response time = 16 μs

  – The maximum frequency that may be recorded depends upon the frequency response of the galvanometer used. When high frequency signals are to be recorded, the marking paper is moved with sufficient speed so as to spread out the trace along the time-axis. *The u.v. recorders have an additional advantage of multi-trace recording.*

- *"Typical applications"* of U.V. recorders are in recording:

  (*i*) Regulation transients of generators;
  (*ii*) Output of transducers;
  (*iii*) Control system performance.

  – These recorders are also used for *recording the magnitude of low frequency signals which cannot be measured with analog (pointers) type instruments.*

3.10.4.5 Magnetic Tape Recorders

These recorders have response characteristics which enable them to be used at *higher frequencies;* hence they find an extensive use in Instrumentation systems.

A magnetic tape recorder consists of the following *basic components:*

1. Recording head.

2. Magnetic tape.
3. Reproducing head.
4. Tape transport mechanism,
5. Conditioning devices.

Advantages:

(*i*) Low distortion.
(*ii*) Wide frequency range from D.C. to several MHz.
(*iii*) Wide dynamic range which exceeds 50 dB.
(*iv*) The recorded signal is immediately available with no time lost in processing. The recorded signal can be played back, or reproduced as many times as desired without loss of signal,
(*v*) Multi-channel recording possible.

3.10.4.6 Cathode Ray Oscilloscope (CRO)

A **cathode-ray oscilloscope** *is an instrument which presents signal waveforms visually.* It is also useful for *comparing two signals in phase, frequency or amplitude.*

A CRO can operate upto 50 MHz, can allow viewing of signals within a time span of a few nanoseconds and can provide a number of waveform displays simultaneously on the screen. It also has the ability to hold the displays for a short or long time for many hours) so that original signal may be compared with one coming on later.

A block diagram of cathode-ray oscilloscope is shown in Fig. 3.13.

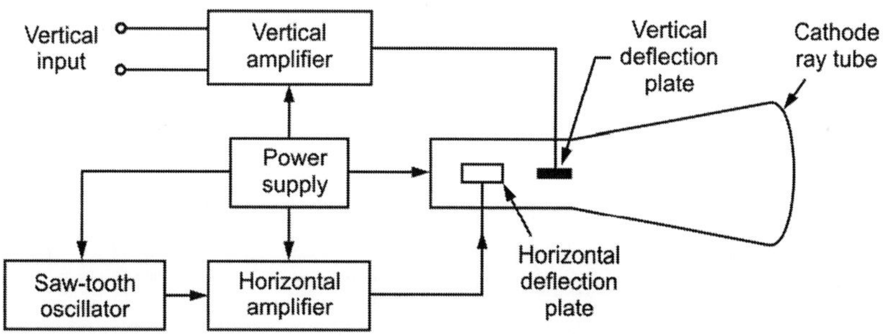

Fig. 3.13 Cathode-ray oscilloscope.

Cathode ray tube (CRT):

A cathode ray tube is the 'heart' of an oscilloscope and is very similar to the picture tube in a television set.

Figure 3.14 shows the cross-sectional view of a general-purpose electrostatic C.R.T.

It has the following four major *components:*

1. ***Electron gun*** .... it produces a stream of electrons.
2. ***Focusing and accelerating anodes*** .... they produce a narrow and sharply-focused beam of electrons.
3. ***Horizontal and vertical deflecting plates*** .... for the path of beam.
4. ***An evacuated glass envelope with a phosphorescent screen*** .... produceds a bright spot when struck by a high velocity electron beam.

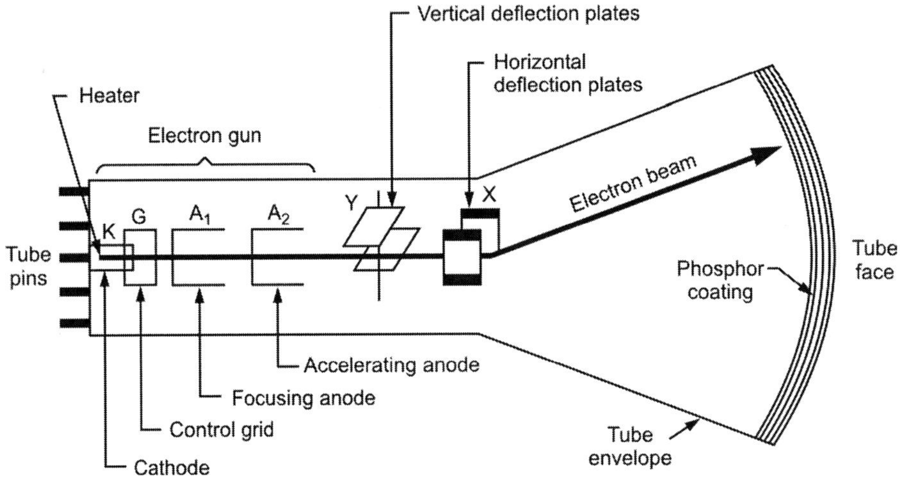

Fig. 3.14 Cathode ray tube.

Working of a C.R.O.:

When a signal is to be displayed or viewed on the screen it is applied across the Y-plates of a cathode ray tube. But to see its waveform or pattern, it is essential to spread it to horizontally from left to right. This is achieved by applying a sawtooth voltage wave to X-plate.

Under these conditions, the electron beam would move uniformly from left to right thereby graphing vertical variations of the input signal versus time. Due to repetitive tracing of the viewed waveform, we get a continuous display because of *persistence of vision.*

However, to get a stable stationary display on the screen, it is essential to *synchronize the horizontal sweeping of the beam* (sync) with the input signal across Y-plates. The signal will be properly synced only when its frequency *equals the sweap-generator frequency*. The usual method of synchronizing the input signal is to use a portion of the input signal to trigger the sweep generator so that the frequency of the sweep signal is locked or synchronized to the input signal. It is called internal sync because the synchronization is obtained by internal wiring connections as shown in Fig. 3.15.

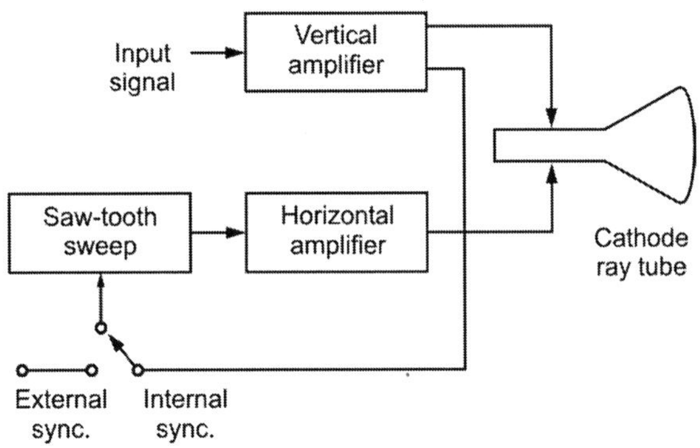

Fig. 3.15

Applications of C.R.O.:

1. Tracing of an actual waveform of current or voltage.
2. Determination of amplitude of a variable quantity.
3. Comparison of phase and frequency.
4. In televisions.
5. In radar.

6. For finding B.H. curves for hysteresis loop.
7. For engine pressure analysis.
8. For studying the heart beats, nervous reactions etc.
9. For tracing transistor curves.

HIGHLIGHTS

1. The *signal conditioning equipment* may be required to perform the following functions on the transduced signal:
   (*i*) Amplification
   (*ii*) Modification or modulation
   (*iii*) Impedance matching
   (*iv*) Data processing
   (*v*) Data transmission.
2. An *amplifier* is a device which is used to increase or augment the weak signal.
3. An *operational amplifier* (Op-Amp) is a linear integrated circuit (IC) that has very high voltage gain, a high input impedance and a low output impedance.
4. *Filtering* is process of attenuating unwanted components of a measurement while permitting the desired component to pass.
5. A *good display,* functionally, is one which permits the best combination of speed, accuracy and sensitivity when transferring the necessary information from the instrument to the operator.
6. The electrical indicating instruments may be classified as:
   (*i*) Analog instruments.
   (*ii*) Digital instruments.
7. Essential features of indicating instruments are:
   (*i*) Deflecting device.
   (*ii*) Controlling device.
   (*iii*) Damping device.
8. The *digital instruments* indicate the value of the measurand in the form of decimal numbers whereas the analog instruments display the quantity to be measured in terms of deflection of a pointer *i.e.,* an analog displacement or an angle corresponding to the electrical quantity.
9. The *digital meters* work on the principle of *"quantization".*
10. A *digital instrument* can be considered as a counter which counts the pulses in a predetermined time.
11. A numerical indicator tube (NIT) consists of a gas filled glass tube having ten cathodes in the form of numbers and an anode.
12. A *recorder* records electrical and non-electrical quantities as a function of time. Two types of recorders are:
    (*i*) Analog recorders.
    (*ii*) Digital recorders.
13. Instruments that record change of only one measured variable are called *single point recorders.* A *multipoint recorder* may have as many as 24 inputs, with traces displaced in 6 colours.
14. A *X-Y recorder* is an instrument which gives a graphic record of the relationship between two variables.
15. *Cathode ray oscilloscope* (*CRO*) is an instrument which presents signal waveforms visually. It is also useful for comparing two signals in phase, frequency or amplitude.

THEORETICAL QUESTIONS

1. What do you mean by the following terms as applied to instrumentation or measurement system ?
   (*i*) Detector-transducer stage.
   (*ii*) Signal conditioning stage.
2. State the limitations of mechanical amplification.
3. What are the advantages of electrical signal conditioning ?
4. Explain briefly the following functions of signal condioning equipment:
   (*i*) Amplification
   (*ii*) Modification or modulation
   (*iii*) Impedance matching
   (*iv*) Data processing
   (*v*) Data transmission.
5. Explain briefly the following :
   (*i*) D.C. signal conditioning system.
   (*ii*) A.C signal conditioning systems.
6. Describe briefly the term "Amplification".
7. Explain briefly any two of the following amplifiers:
   (*i*) Mechanical amplifiers
   (*ii*) Fluid amplifiers
   (*iii*) Electrical and electronic amplifiers.
8. State the generalities that can be listed for an ideal electronic amplifier.
9. What are A.C. and D.C. amplifiers? Explain briefly.
10. What do you mean by "Modulated and unmodulated signals"?
11. What is an Op-Amp? State its limitations as well.
12. Explain briefly the term "Common-mode rejection ratio (CMRR)".
13. State the applications of Op-Amp.
14. Enumerate some of the commonly used Op-Amp circuits.
15. Explain briefly the following:
    (*i*) Buffer amplifier.
    (*ii*) Differential amplifier.
16. State the advantages of differential amplifiers.
17. What is an attenuator? How are the attenuators classified ?
18. What do you mean by the terms "Filtering" and "Filter" ?
19. How are filters classified ?
20. What do you mean by "Signal transmission"?
21. Explain briefly any three of the following types of transmission ?
    (*i*) Mechanical transmission
    (*ii*) Hydraulic transmission
    (*iii*) Pneumatic transmission
    (*iv*) Magnetic transmission.
22. Explain briefly any two of the following:
    (*i*) Semiconductor light emitting diodes (LEO);
    (*ii*) Liquid crystal displays;
    (*iii*) Hot filament or bar tubes;
    (*iv*) Numerical indicator tubes (NIT).
23. What is a recorder?
24. Elaborate the difference between a display unit and a recorder.
25. What is meant by a direct reading instrument?

26. Explain the functioning of a basic type of strip chart recorder. Enumerate the different types of marking mechanisms used in it.
27. Distinguish between single point and multi-point recorders.
    What is a X-Y recorder? State its applications.
28. Explain the moving of an ultraviolet (U.V.) recorder. State its applications.
29. What are the basic components of a magnetic tape recorder for instrumentation applications. List its advantages and disadvantages.
30. Explain with heat diagram the construction and working of a cathode ray oscilloscope (CRO).
31. Describe the different parts of a cathode ray tube (CRT).
32. What are the applications of a CRO?

❑❑❑

Time Related Measurements

Chapter 4

4.1 MEASUREMENT OF DISPLACEMENT

4.1.1 Measurement of Linear Displacement

Linear displacement may be measured by the following *transducers* :

1. Resistive potentiometers.
2. Strain gauges.
3. Variable inductance transducers.
4. Linear variable differential transducers (LVDT).
5. Capacitive transducers.
6. Piezoelectric transducers.
7. Hall effect transducers.
8. Electronic transducers.
9. Ionization transducers.
10. Digital transducers.
11. Acoustic transducers.

- *Mechanical displacement* may be converted into an electrical variable by the simple expedient of adjusting resistance in an electrical circuit. A slide-wire resistor, having a movable contact attached to the part whose displacement is to be measured, may be connected through a 2-conductor circuit to a steady-voltage source in series with an ammeter (or milliammeter) calibrated in terms of the displacement. If the resistor is connected as a voltage divider, the need for a regulated supply is eliminated, and with a 3-conductor circuit the display instrument may be a *ratio-meter or potentiometer.* Such combinations are common and are available for both D.C. and A.C. operation.
- Where deflections are small, of the order of 0.25 mm, measurement may be made by use of a *differential transformer.*

Measurement of small displacements:

The linear or differential transformer is *the popular means for measuring small displacements of the order of 0.25 mm.* This device (see Fig. 4.1) is generally produced with a single primary winding and two secondaries, all disposed along a common axis and having in the common magnetic circuit a moveable iron core longitudinally despicable with the motion to be measured. The secondaries may be connected additively or differentially and may be included in the circuit of a null type instrument based either by shifting the core of a similar transformer excited from the same source or by the use of a slide wire potentiometer.

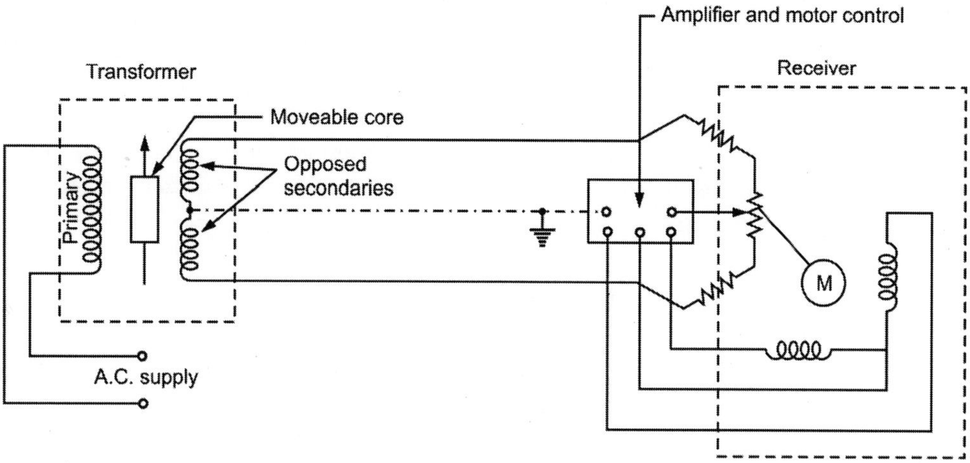

Fig. 4.1 Schematic diagram of differential transformer transducer with servo-activated receiver.

Linear transformers are regularly supplied for operation at all frequencies upto 30000 Hz. The sensitivity, of course, increases with the frequency.

Linear transformers may be interconnected in a great variety of arrangements to perform computations or to express desired mathematical functions of measured variables.

4.1.2 Measurement of Angular Displacement

The following transducers are used for angular displacement:

1. Resistive potentiometer.
2. Variable inductive transducers.
3. Synchros.
4. Capacitive transducers.
5. Rotary variable differential transformer (RVDT).
6. Variable reluctance transducer.

Variable reluctance transducer

Figure 4.2, shows a variable reluctance transducer for *measurement of angular displacement.*

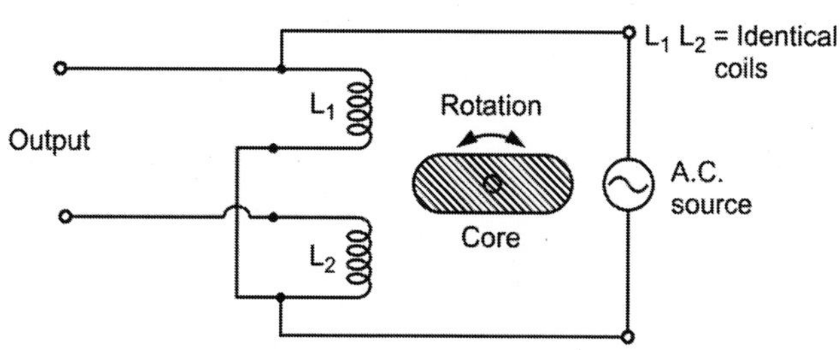

Fig. 4.2 Variable reluctance transducer for measurement of angular displacement.

L_1 and L_2 are the two identical coils connected in series opposition. These coils are so placed/ positioned with respect to a cam shaped iron core that, when the core is in the *null position, the impedances of both the coils are same.* Hence, voltage across the two coils are the same, though they are 180° out of phase. Any angular displacement from the null position results in a *differential voltage output proportional to the amount of displacement.* The clockwise and anticlockwise angular displacements produce different outputs which are *opposite in phase.* Thus the angular displacement may be measured in magnitude and direction.

In these transducers, angular changes from null position to ± 45° may be sensed.

4.2 MEASUREMENT OF VELOCITY/SPEED

4.2.1 Measurement of Linear Velocity

The following types of transducers are used (in the analogue methods) for measurement of *linear velocity* (first derivative of displacement).

1. Electro-magnetic transducers
   (*i*) Moving magnet type,
   (*ii*) Moving coil type.
2. Seismic type transducers.

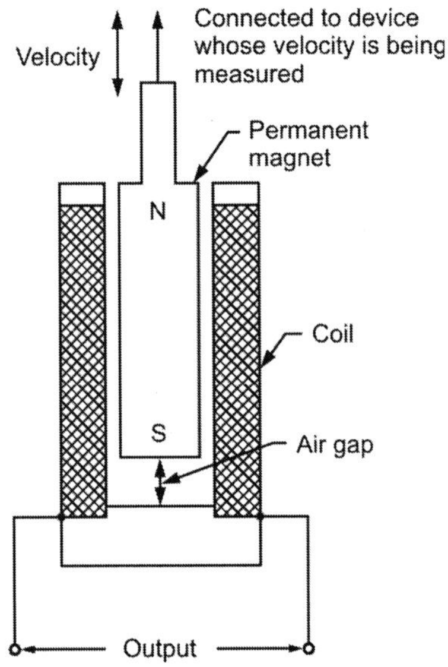

Fig. 4.3 Moving magnet type transducers.

Electro-magnetic transducers:

An electro-magnetic transducer utilizes the voltage produced in a coil on account of change in flux linkages resulting from change in reluctance. This is the most commonly used transducer for measurement of linear velocities.

(*i*) **Moving magnet type.** A moving type electro-magnetic transducer uses *permanent magnet which provides a constant polarizing field.* Figure 4.3 shows such a transducer. It consists of rod rigidly coupled to the device whose velocity is being measured. The rod is a permanent magnet which is surrounded by a coil.

Working. When the magnet moves a voltage is induced in the coil which is *directly proportional to the velocity.* The voltage induced in the coil (placed in a magnetic field) is given by:

$$e = BANV \quad ...(4.1)$$

i.e.

$$e \propto V \quad ...(4.2)$$

where, B = flux density, Wb/m^2,
A = Area of coil, m^2,
N = Number of turns of coil, and
V = Relative velocity of magnet with respect to coil.

The *direction of motion is determined by the polarity of the output voltage.*

Advantages:

1. The output voltage is linearly proportional to velocity.

2. Negligible maintenance (due to absence of mechanical surfaces/contacts).
3. Can be employed as event markers which are robust and inexpensive to manufacture.

Disadvantages:

1. Limited frequency response.
2. Unsuitable for measurement of vibrations.
3. Stray magnetic fields adversely affect their performance.

(*ii*) Moving coil type. This type of transducer operates essentially through the *action of a coil moving in a magnetic field; the voltage generated in the coil being proportional to the velocity of the coil.*

The *advantages* of this type of transducer are:

- Reduced stray magnetic fields effects due to the whole device being contained in an antimagnetic case.
- More satisfactory arrangement, as the system forms a closed magnetic circuit with a constant air gap.

4.2.2 Measurement of Angular Velocity

Angular velocity may be measured with the help of a *"tachometer"*.

A **tachometer** may be *defined* as follows:

"An instrument which either continuously indicates the value of rotary speed or continuously displays a reading of average speed over rapidly operated short-intervals of time."

Or

"An instrument used for measurement of angular velocity, as of shaft either by registering the total number of revolutions during the period of contact, or by indicating directly the number of revolutions per minute."

Classification of Tachometers:

The tachometers are *classified* as follows:

I. *Mechanical tachometers:*

(*i*) Hand speed indicator.
(*ii*) Revolution counter and timer.
(*iii*) Tachoscope.
(*iv*) Centrifugal tachometers.
(*v*) Vibrating reed tachometers.

II. *Electrical tachometers:*

(*i*) D.C. tachometer generators
(*ii*) A.C. tachometer generators
(*iii*) Photoelectric tachometers (or Speed-meter)
(*iv*) Eddy current or drag-cup tachometer.
(*v*) Capacitive tachometer,
(*vi*) Stroboscopic tachometers.

4.2.2.1 Mechanical tachometers

1. *Hand speed indicator:*

This tachometer has an integral stopwatch and counter with automatic disconnect.

- The spindle operates when brought in contact with the shaft; the counter, however, does not function until the start and wind button is pressed to start the watch and engage the automatic clutch. Depressing of the starting button also serves to wind the starting watch.
- The revolution counter automatically gets disengaged after a fixed interval, usually 3 to 6 seconds.

The *instrument indicates the average speed over the short interval;* the *dial is designed to indicate the rotational speed directly in r.p.m.*

These tachometers can be used to measure speeds within range of 20000 to 30000 r.p.m. These units have an accuracy of 1 per cent of the full scale.

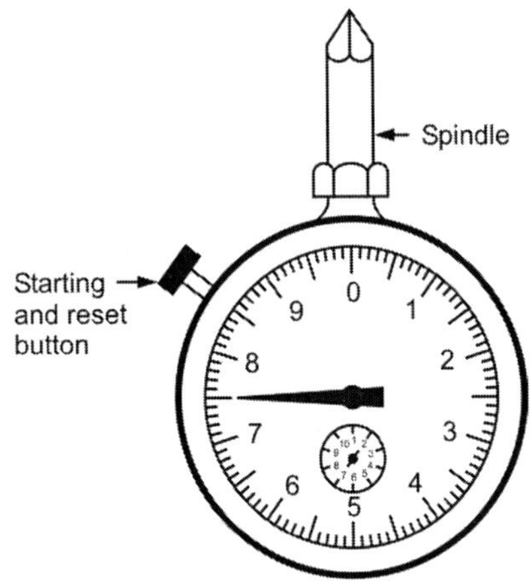

Fig. 4.4 Hand speed indicator.

2. *Revolution counter and timer:*

The revolution counter employs a timing device and counts the number of revolutions in a measured length of time. Thus this device measures an average rotational speed rather than an instantaneous rotational speed.

It consists of a worm gear which is also the shaft attachment and is driven by the speed source. The worm drives the spur gear which is turn actuate the pointer on a calibrated dial.

- Such speed counters are limited to low speed engines which permit reading the counter at definite time intervals. The speed upto 2000-3000 r.p.m. can be measured satisfactorily by a properly designed and manufactured revolution counter.

3. *Tachoscope:*

It consists of a revolution counter incorporating a built-in-timing device. The two components are integrally mounted and start simultaneously when the contact point is pressed against the rotating shaft. The rotational speed is computed from the readings of the counter and timer.

Tachoscopes can be used to measure speeds upto 5000 r.p.m.

4. *Centrifugal tachometer:*

A centrifugal tachometer operates on the principle that *centrifugal force is proportional to the speed of rotation.*

Figue 4.5 shows the schematic arrangement of such a tachometer. Two fly-balls (small weights) are arranged about a central spindle. Centrifugal force developed by these rotating balls works to compress the spring as a function of rotational speed and accordingly positions a pointer.

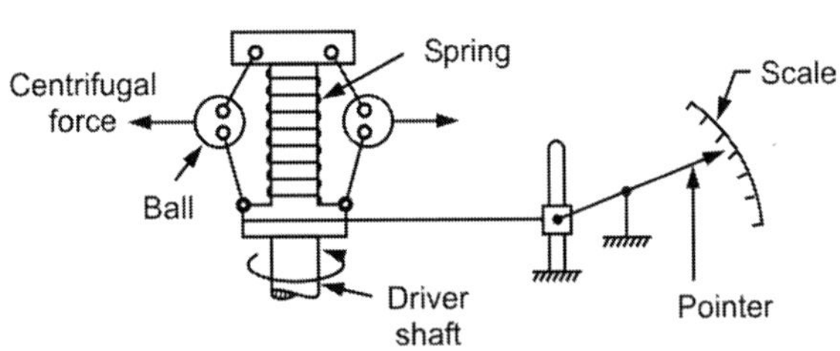

Fig. 4.5 Centrifugal tachometer.

- They are suitable to measure speeds upto 40000 r.p.m and are usually used to make and break circuit for speed control.
- The *accuracy of such tachometers is of the order of* ±1.

Other types of centrifugal type tachometers provide the constraining force by atmospheric pressure or gravity instead of a spring.

5. *Vibrating reed tachometers:*

In this type of tachometer, a series of consecutively timed steel rods are used to determine the speed on the basis of vibrations created by the machine. The rods are of different lengths. On one side the rods are fixed to a base which is kept in contact with any non-moving part of the machine and on the other side they are attached to a part at the bottom of which a scale is calibrated. The reed timed to resonance with the machine vibration frequency responds visibly and indicates speed on the scale.

- These tachometers can be used in the speed range of 600 to 10000 r.p.m.
- The accuracy of such tachometers is of the order of ±0.5%.

4.2.2.2 Electrical Tachometers

Electrical tachometers are preferred, as compared to mechanical tachometers, in view of advantages they offer as *electrical transducers.*

1. *D.C. tachometer generator:*

In a D.C. generator the e.m.f. generated depends upon the following two factors:

(*i*) Field excitation

(*ii*) Speed.

If for the field system permanent magnet pole pieces are used, then the generated voltage depends only on the speed. Hence the *speed can be computed by measuring the generated e.m.f.*

A D.C. tachometer generator is shown in Fig. 4.6.

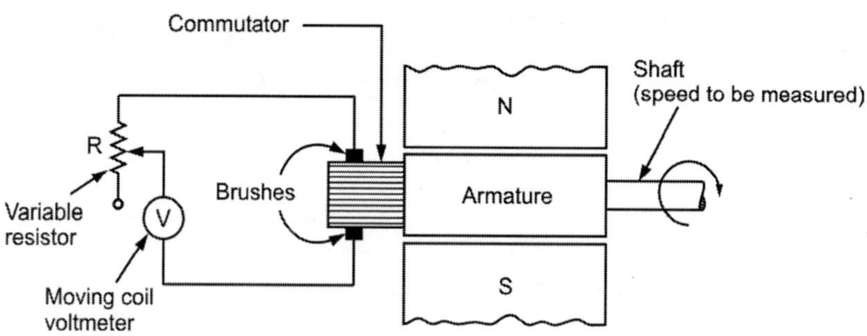

Fig. 4.6 D.C. tachometer generator.

- The shaft whose speed is to be measured is coupled to the armature.
- A moving coil voltmeter is connected across the brushes to measure the generated voltage. The variable resistance R is incorporated to limit the current through the voltmeter.

Since voltage is proportional to speed, the voltmeter may be calibrated in terms of speed (r.p.m.).

2. *A.C. Tachometer generator:*

The inherent demerits associated with D.C. tachometer generator, due to the provision of commuter and brushes, are eliminated in A.C. tachometer generator. An A.C. tachometer generator is shown in Fig. 4.7.

- It consists of, like an alternator, a stationary armature (stator) and a rotating field system (rotor). Owing to the generation of e.m.f. in a stationary coil on a stator, commutation problems no longer exist.
- The alternating e.m.f. induced in the stationary coil is *rectified* and the output D.C. voltage is measured with the help of a moving coil voltmeter (V).
- The *ripple content* of the rectified voltage is *smoothened* by the capacitor filter (C).

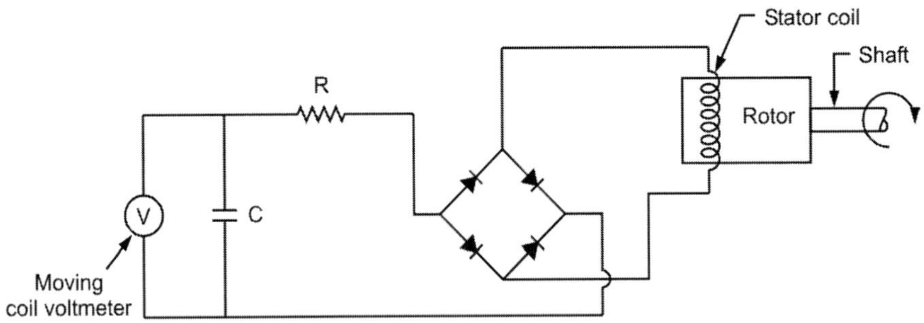

Fig. 4.7 A.C. tachometer generator.

- As the speed depends on both the amplitude of the voltage and frequency, any one of them can be used as a measure of the speed. In an A.C. tachometer, it is the *induced voltage* that is considered as the required parameter.

3. *Photoelectric tachometer:*

Figure 4.8 shows a photoelectric tachometer.

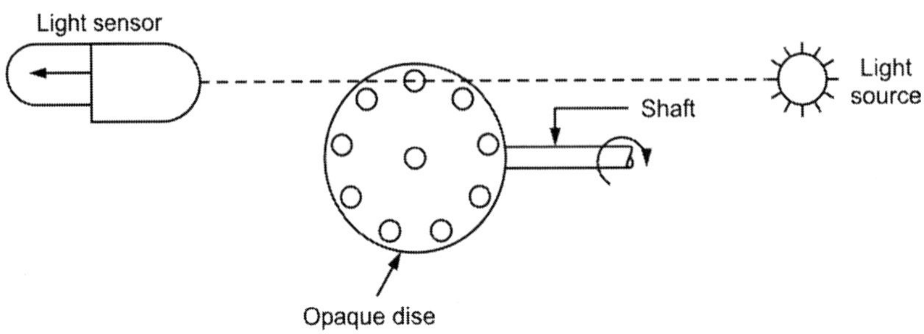

Fig. 4.8 Photoelectric tachometer.

- It consists of an *opaque disc* mounted on the shaft whose speed is to be measured. The disc has a number of equivalent holes around the periphery. On one side of the disc there is a source of light (L) while on the other side there is a light sensor (may be a photosensitive device or phototube) in line with it (light-source).
- On the rotation of the disc, holes and opaque portions of the disc come alternatory in between the light source and the light sensor. When a hole comes in between the *two,* light passes through the holes and falls on the light sensor, with the result that an *output pulse is generated.* But when the opaque portion

of the disc comes in between, the light from the source is blocked and hence there is no pulse output. Thus *whenever a hole comes in line with the light source and sensor, a pulse is generated.* These pulses are counted/measured through an *electric counter.*

The number of pulses generated depends upon the following *factors:*

(*i*) The number of holes in the disc;

(*ii*) The shaft speed.

Since the number of holes are fixed, therefore, the number of pulses generated depends on the speed of the shaft only. The electronic counter may therefore be calibrated in terms of speed (r.p.m.).

Advantage. It is a *digital instrument.*

Disadvantage. It is required to replace the light source periodically and if the grating period is small then errors might creep in the output.

4. *Eddy current or drag cup tachometer:*

In this type of tachometer (Fig. 4.9) the test-shaft rotates a permanent magnet and this induces eddy currents in a drag cup or disc held close to the magnet. The eddy currents produce a torque which rotates the cup against the torque of a spring. The disc turns in the direction of rotating magnetic field until the torque developed equals that of the spring. A pointer attached to the cup indicates the rotational speed on a calibrated scale.

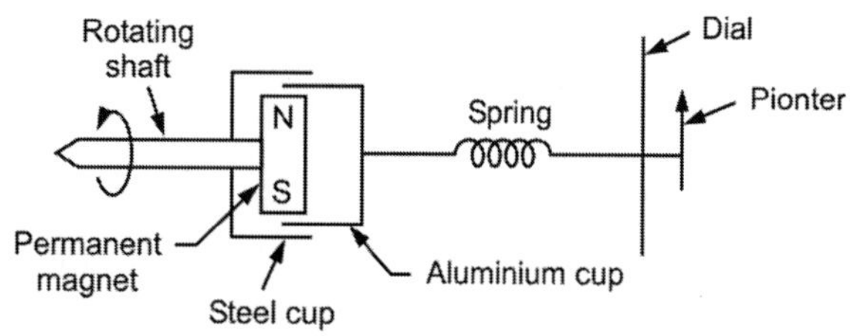

Fig. 4.9 Eddy current or drag type tachometer.

- The *automobile speedometers* operate on this principle.
- These tachometers are used for measuring rotational speeds upto 12000 r.p.m. with an accuracy of ± 3 per cent.

5. *Capacitive tachometer:*

Refer to Fig. 4.10. A capacitive pick-up tachometer consists of a vane attached to one end of the rotating machine shaft. When the shaft rotates between the field capacitive plates, there occurs a change in the capacitance. The capacitor forms a part of an oscillator tank so that *number of frequency changes per unit of time is a measure of the shaft speed.* The pulses thus produced are *amplified* and *squared,* and may then be *fed to frequency measuring unit* or to a digital counter so as to provide a digital analog of the shaft rotation.

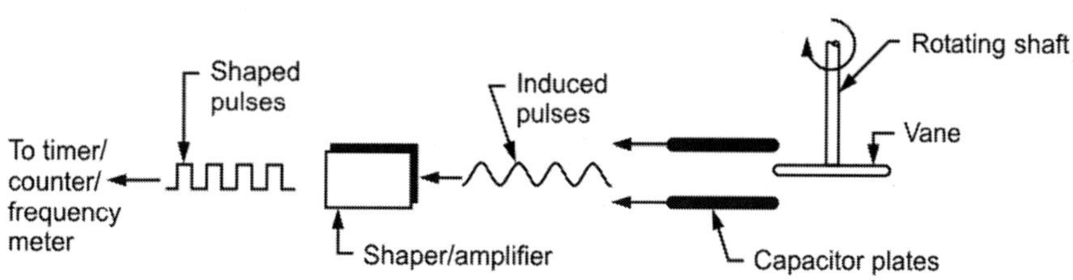

Fig. 4.10 Capacitive pick-up tachometer.

6. *Stroboscopic tachometers:*

Its working principle is to synchronise a flashing light with rotation of shaft, making it appear standstill.

Usually, a reference mark is put on the rotating body and the speed of oscillation of flash produced from stroboscope itself is a measure of r.p.m. of rotating body, *without requiring any contact.*

- These tachometers are most oftenly used in *experimental work.*
- Its disadvantage is that it cannot be used where ambient lighting is above a certain value; for it to be most effective, the surrounding light must be subdued.

HIGHLIGHTS

1. A *tachometer* is an instrument which either continuously indicates the value of rotary speed or continuously displays a reading of average speed over rapidly operated short intervals of time.
2. Electrical tachometers are preferred, as compared to mechanical tachometers, in view of advantages they offer as electrical transducers.

THEORETICAL QUESTIONS

1. How can linear displacement be measured? Explain briefly.
2. Explain briefly how small displacements can be measured by a linear or differential transformers.
3. Describe briefly with a neat sketch how angular displacement is measured by a variable reluctance transducer.
4. Name the transducers which are used for the measurement of linear velocity.
5. Explain with a neat diagram the method of measuring linear velocity with the use of moving magnet type transducer.
6. How is angular velocity measured ?
7. What is a tachometer ?
8. How are tachometers classified ?
9. Explain briefly any two of the following mechanical tachometers:
   (*i*) Hand speed indicator
   (*ii*) Vibrating reed tachometer
   (*iii*) Centrifugal tachometer.
10. Why are electrical tachometers preferred to mechanical tachometers?
11. Explain briefly any two of the following electrical tachometers:
    (*i*) D.C. tachometer generator
    (*ii*) A.C. tachometer generator
    (*iii*) Photoelectric tachometer.
12. Give the description of an eddy current tachometer, with a neat sketch.
13. Describe briefly a capacitive pick-up tachometer.

❑❑❑

Measurement of Pressure

Chapter 5

5.1 DEFINITION OF PRESSURE

Pressure *is defined as a force per unit area.*

Pressures are exerted by gases, vapours and liquids. The instruments that we generally use, however, record pressure as the difference between the two pressures. Thus, it is the *difference between the pressure exerted by a fluid of interest and the ambient atmospheric pressure.* Such devices indicate the pressure either above or below that of the atmosphere.

Atmospheric pressure:

The atmospheric air exerts a normal pressure upon all surfaces with which it is in contact, and it is known as **atmospheric pressure.** The atmospheric pressure is also known as *"Barometric pressure".*

The atmospheric pressure at sea level (above absolute zero) is called *standard atmospheric pressure.*

Note: The local atmospheric pressure may be a little lower than these values, if the place under question is higher than sea level, and higher values if the place is lower than sea level, due to the corresponding decrease of the column of air standing, respectively.

Gauge pressure:

It is the pressure, measured with the help of pressure measuring instrument, in which atmospheric pressure is taken as datum.

The *atmospheric pressure* on the scale is marked as *zero.*

Gauges record the pressure above or below the local atmospheric pressure, since they measure the difference in pressure of the liquid to which they are connected and that of surrounding air.

If the pressure of the liquid is below the local atmospheric pressure, then the gauge is designated as *'vacuum gauge'* and the recorded value indicates the amount by which the pressure of the liquid is *below* local atmospheric pressure, *i.e., negative pressure.*

(Vacuum pressure is defined as the pressure *below the atmospheric pressure).*

Absolute pressure:

It is necnessary to establish an absolute pressure scale which is independent of the changes in atmospheric pressure. A *pressure of absolute zero can exist only in complete vacuum.*

Any pressure measured above the absolute zero of pressure is termed as an **'absolute pressure'**

A schematic diagram showing the gauge pressure, vacuum pressure and the absolute pressure is given in Fig. 5.1.

Mathematically:

1. Absolute pressure = Atmospheric pressure + gauge pressure

   i.e., $p_{abs} = p_{atm} + p_{gauge}$

2. Vacuum pressure = Atmospheric pressure – absolute pressure

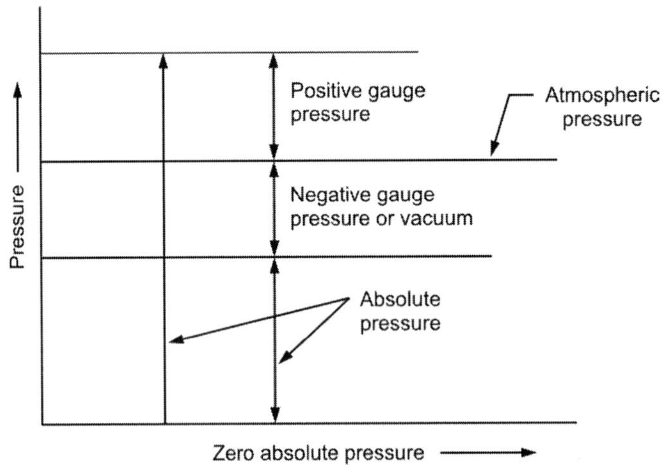

Fig. 5.1 Relationship between pressures.

Static pressure (p_s) and total pressure (p_t):

"*Static pressure*" *is defined as the force per unit area acting on the wall by a fluid at rest or flowing parallel to the wall in a pipeline.*

"*Total or stagnation pressure*" *is defined as the pressure that would be obtained if the fluid stream were brought to rest isentropically.* For an incompressible fluid or gas flowing at low velocities, the *dynamic pressure* equals $\frac{V^2}{2g}$ where, V is the velocity of the fluid flow.

Thus, Total pressure = Static pressure + dynamic (or velocity) pressure

$$\therefore \quad p_t = p_s + \frac{V^2}{2g} \qquad \text{...(5.1)}$$

5.2 UNITS FOR PRESSURE AND MEASURING INSTRUMENTS

The fundamental S.I. unit of pressure is pressure is newton per square meter (N/m^2). This is also known as *Pascal* (P_a).

"*Low pressures*" are often expressed in terms *of mm of water or mm of mercury*. This is an abbreviated way of saying that the pressure is such that will support a liquid column of stated height.

- *Standard atmospheric pressure has the following equivalent values:*
  101.3 kN/m^2 or 101.3 kPa; 10.3 m of water; 760 mm of mercury; 1013 mb (millibar); $\simeq$ 1 bar = 100 kPa = 10^5 N/m^2.
  (When the local atmospheric pressure is not given in a problem, it is taken as 100 kN/m^2 or 10 m of water for simplicity of calculations.)

Pressure Measurement Instruments

The pressure measurement instruments can be *categorized* as follows:

1. *Instruments for measuring low pressures (below 1 mm of Hg):*
   - Manometers
   - Low pressure gauges.

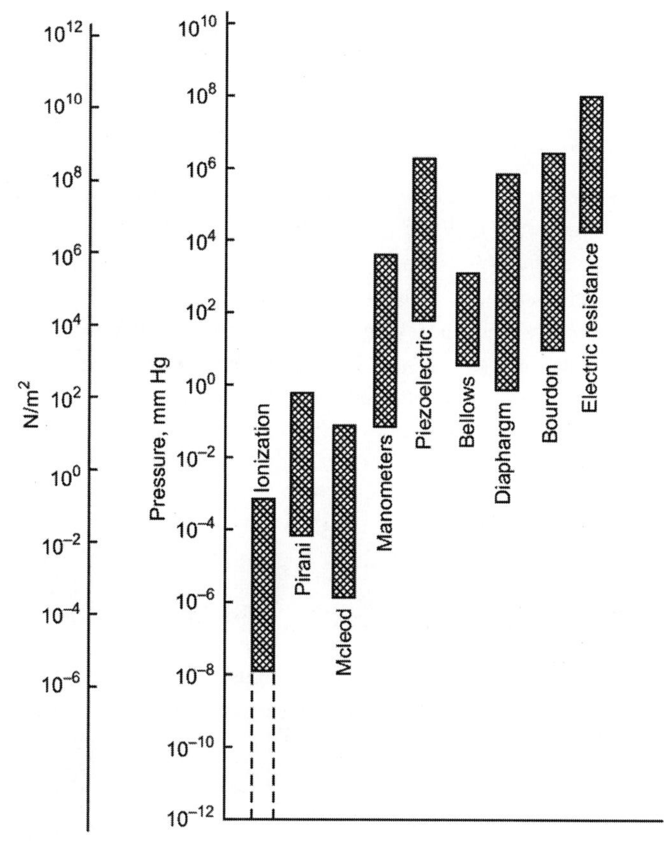

Fig. 5.2 Summary of applicable range of pressure gauges.

2. *Instruments for medium and high pressures (between 1 mm of Hg to 1000 atmospheres:*
   - Bourdon tube
   - Diaphragm gauges
   - Bellow pressure gauge
   - Dead-weight pressure gauge.

3. *Instrument for measuring low vacuum and ultra high vacuum (760 torr to 10^{-9} torr and beyond; 1 torr = 1 mm of Hg):*
   - Mcleod
   - Thermal conductivity
   - Ionization gauges.
4. *Instruments for measuring very high pressures (1000 atmospheres and above):*
   - Bourdon tube
   - Diaphragm gauges
   - Electrical resistance pressure gauges.
5. *Instruments for measuring varying pressure:*
   - Engine indicator
   - Cathode ray oscilloscope (CRO).

Figure 5.2. shows the summary of applicable range of pressure gauges.

5.3 MECHANICAL GAUGES

The manometers (discussed earlier) are suitable for *comparatively low pressure.* For high pressures they become unnecessarily larger even when they are filled with heavy liquids. Therefore for measuring medium and high pressures, we make use of elastic pressure gauges. They employ different forms of elastic systems such as tubes, diaphragms or bellows etc. to measure the pressure. The elastic deformation of these elements is used to show the effect of pressure. Since these elements are deformed within the elastic limit only, therefore these gauges are sometimes called *elastic gauges.* Sometimes they are also called *secondary instruments, which implies that they must be calibrated by comparison with primary instruments such as manometers etc.*

Some of the important types of these gauges are:

1. *Bourdon tube pressure gauge.*
2. *Diaphragm gauge.*
3. *Vacuum gauge.*

5.3.1 Elastic Pressure Elements

Elastic pressure elements or mechanical type of transducers are used for measurement of very high pressures up to about 700 MN/m^2. There are *three main types of pressure elements:*

1. Bourdon tube.
2. Diaphragms.
3. Bellows.

The action of these mechanical transducers depends upon the *displacement caused by the pressure.* The displacement produced may actuate a pointer whose deflection may be direct measure of the pressure applied or the displacement is measured with the help of a secondary transducer which is electrical in nature. The output of the secondary transducer which is electrical in format is a measure of the pressure applied.

1. Bourdon tubes/elements

Figure 5.3. shows Bourdon tube configurations.

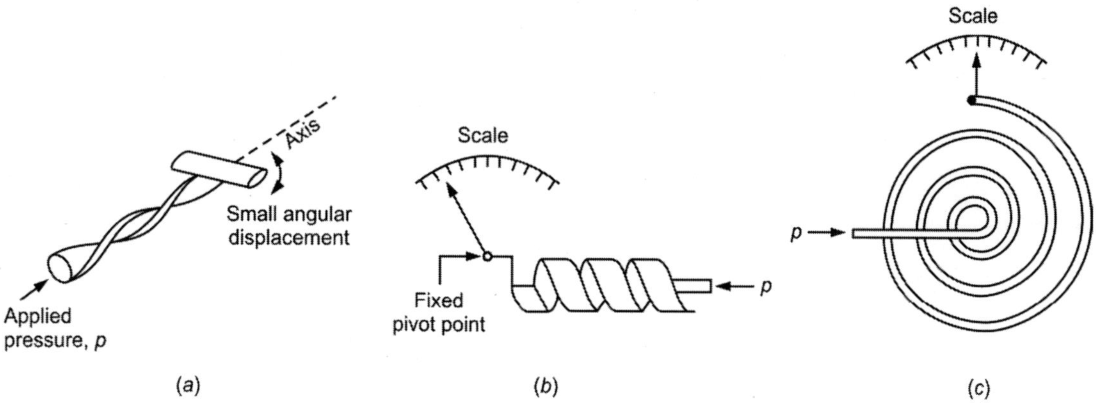

Fig. 5.3 (a) Twisted tube; (b) Helical (c) Spiral.

Advantages:

The Bourdon the tube element has the following *advantages:*

1. Simple in construction and cheap.
2. Available in several different ranges.
3. Capability to measure gauge, absolute and differential pressures.
4. The sensitivities of Bourdon tube may be changed by changes in their dimensions.
5. Excellent sensitivity.
6. Simple and straight forward calibration with dead weight tester.
7. Easily adapted to strain, capacitance, magnetic and other electrical transducers.

Disadvantages:

1. Susceptibility to shock and vibration.
2. Inherent hysteresis and slow response to pressure changes.
3. Unsuitable for low pressure applications.

2. Diaphragm Elements

Figure 5.4 shows basic diaphragm types.

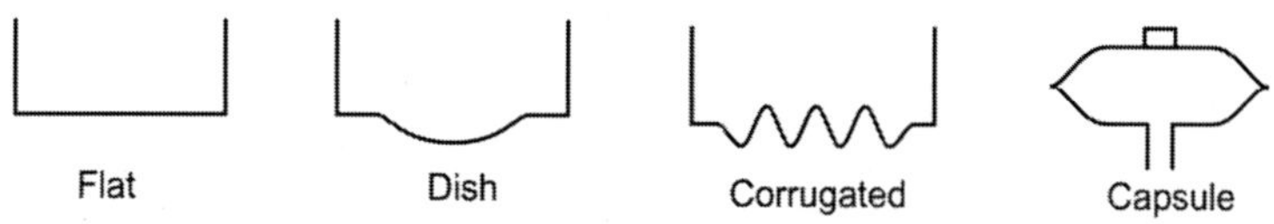

Fig. 5.4 Basic diaphragm types.

Advantages:

1. Small in size and moderately priced.
2. Wide linear range.
3. Can withstand high over pressures and under pressures.
4. Small hysteresis.

5. Can be used for measurement of absolute and differential pressures as also vacuum.

- The diaphragm elements find *expensive use in applications where measurement of low pressures including vacuum is involved.*

  Ranges: 0-50N/m^2 to 0-200 kN/m^2

  Accuracies range: ± 0.5% to ± 1.25% of full span.

Disadvantages:

1. Need protection against shocks and vibrations.
2. Cannot be used to measure high pressures.
3. Difficult to repair.

3. Bellow gauges/elements

Figure 5.5 show some bellow gauges.

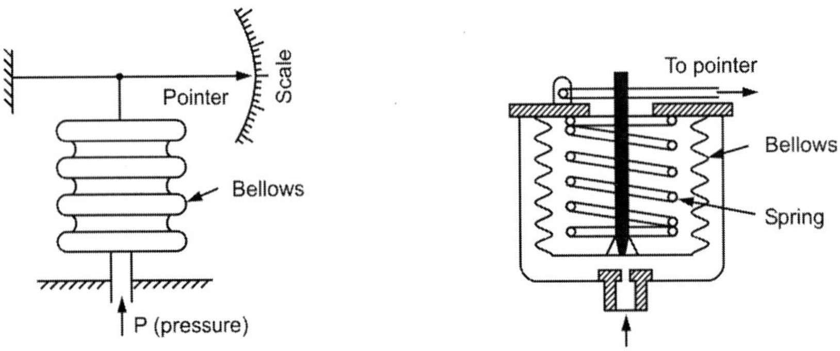

Fig. 5.5 Bellow gauges.

The flexibility of bellows is directly *proportional to:*

- The number of convolutions,
- The square of outside diameter of the bellows and (inversely proportional to the cube of wall thickness), and
- Young's modulus of elasticity of material.

Advantages:

1. Simple and rugged construction.
2. Useful for measurement of low and medium pressures.
3. Moderate cost.
4. Can be used for measurement of absolute, gauge and differential pressures.

Disadvantages:

1. Need spring for accurate characterisation.
2. Greater hysteresis and zero drift problems.
3. Unsuitable for transient measurement due to longer relative motion and mass.
4. Require compensation for ambient temperature changes.

5.3.2 Pressure Gauges

1. Bourdon tube pressure gauge:

Bourdon tube pressure gauge *is used for measuring high as well as low pressures.*

A simple form of this gauge is shown in Fig. 5.6. In this case, the pressure element consists of a metal tube of approximately elliptical cross-section. This tube is bent in the form of a segment of a circle and responds to pressure changes.

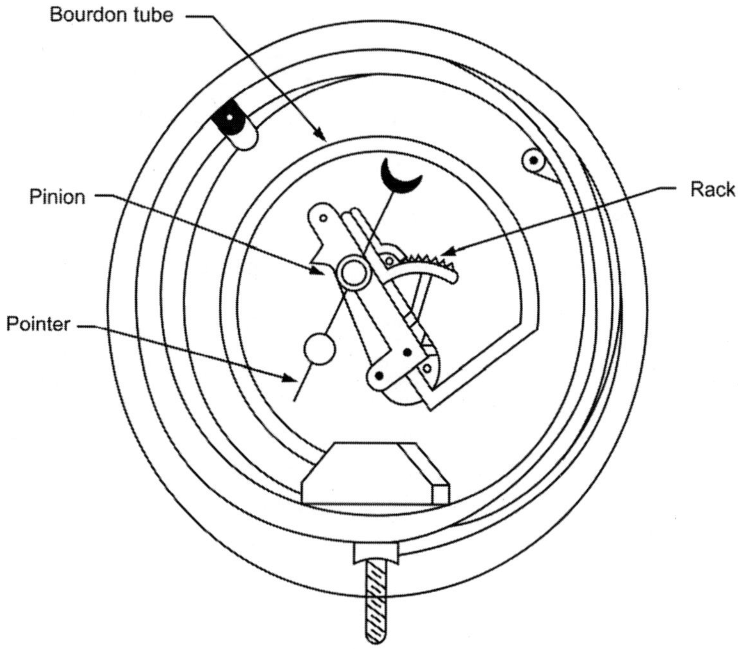

Fig. 5.6 Bourdon tube pressure gauge.

When one end of the tube which is attached to the gauge case, is connected to the source of pressure, the internal pressure causes the tube to expand, whereby circumferential stress *i.e.,* hoop tension is set up. The free end of the tube moves and is in turn connected by suitable levers to a rack, which engages with a small pinion mounted on the same spindle as the pointer. Thus the pressure applied to the tube causes the rack and pinion to move. The pressure is indicated by the pointer over a dial which can be graduated in a suitable scale.

The Bourdon tubes are generally made of *bronze or nickel steel.* The *former* is generally used for *low pressures and the latter for high pressures.*

Depending upon the purpose for which they are required Bourdon tube gauges are made in different forms, some of them are:

(*i*) *Compound Bourdon tube* — used for measuring pressures both *above and below atmospheric pressure.*

(*ii*) *Double Bourdon tube* — used where *vibrations are encountered.*

Advantages and disadvantages of Bourdon tube gauges:

Advantages:

(*i*) High pressure range.
(*ii*) Low cost.
(*iii*) Simple construction.
(*iv*) Good accuracy except at low pressures.
(*v*) They are *easily adopted for designs for obtaining electrical output.*

Disadvantages:

(*i*) They are susceptible to shocks and vibration and are subject to hysteresis.
(*ii*) Low spring gradient which limits their use for precision measurements upto a pressure of 3 MN/m^2.

Errors in Burdon gauges and their rectification

Following *three types of errors* are commonly found in Bourdon gauges:

(*i*) Zero error,
(*ii*) Multiplication error,
(*iii*) Angularity error.

1. ***Zero error.*** The zero (or constant) error remains constant over the entire pressure range.
   Cause. The pointers or hands having become loose on spindle.
   Remedy. Rectified by keeping the pointer in correct position.
2. ***Multiplication error.*** Owing to this error the gauge may tend to give progressively *a higher or low reading.*
   Cause. Wrong setting in the multiplying mechanism between the Bourdon tube and the spindle.
   Remedy. The multiplication screw is loosened and the connecting link is moved either a little inwards when the gauge reads low or little outwards when the gauge reads high.
3. ***Angularity error.*** Due to this error, the gauge may read correctly both at the maximum and minimum readings but may give an *inaccurate reading at the mid point.*
   Cause. Distortion of the tube.
   Remedy. Angularity screw is adjusted and the pointer is set at some other point so that the gauge reads correctly at the mid-point as well.

5.4 PRESSURE TRANSDUCERS

The following *pressure transducers* will be discussed:

1. Resistance type pressure transducers.
2. Pressure voltage type.
3. Inductive type.
4. Capacitive type.
5. Carbon pile type.
6. Piezoelectric type.
7. Photoelectric type.
8. Electromagnetic type.

5.4.1 Resistance Type Pressure Transducers

Resistance type devices used in pressure transducers include:

1. Strain gauges.
2. Moving contacts.

1. *Strain gauge:*

A strain gauge is simply a fine wire in the form a grid (Fig. 5.7). When the grid is distorted, the resistance (R) of the wire change according the formula:

$$R = k\frac{L}{A}$$

where, k = A constant for the particular kind of wire

L = Length of the wire, and

A = Cross-section area.

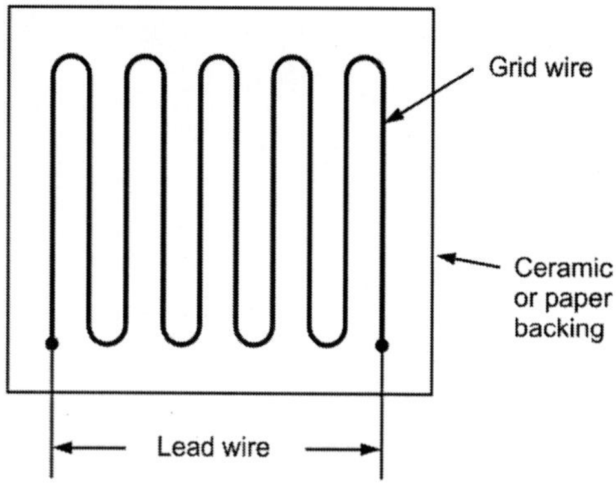

Fig. 5.7 Strain gauge.

As the strain gauge is distorted by the elastic deformation element, its *length is increased and its cross-section area is reduced.* Both of these changes *increase the resistance.*

- Since little distortion is required to change the resistance of a strain gauge through its total range, this type of transducer can be used to detect *very small movement* and, therefore, *very small pressure changes.*
- Since current flows in the strain gauge, some heat will be generated which must be compensated for. One method of compensation employs a second strain gauge placed close to the measuring unit. These two are used in a bridge circuit so that the *heating effect of the two resistive elements is cancelled out,* and the only resistance change in the bridge is due to the deformation.

Most often, strain gauge bridges are A.C. bridges.

2. *Moving contact type resistance transducer:*

This type of transducer is most often used with bellows because of the force required. Fig. 5.8 shows a typical circuit for such a transducer.

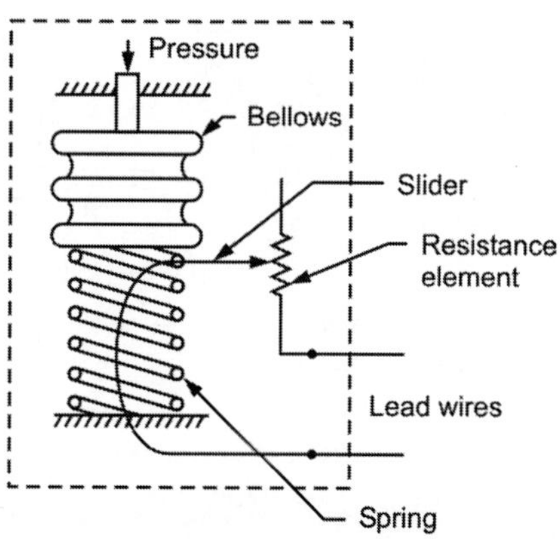

Fig. 5.8 Moving contact type resistance transducer.

5.4.2 Pressure-voltage Transducer (Potentiometer Devices)

One of the simplest of pressure-voltage transducer uses a potentiometer. *The potentiometer is used to convert a pressure signal into an electrical voltage as shown* in Fig. 5.9.

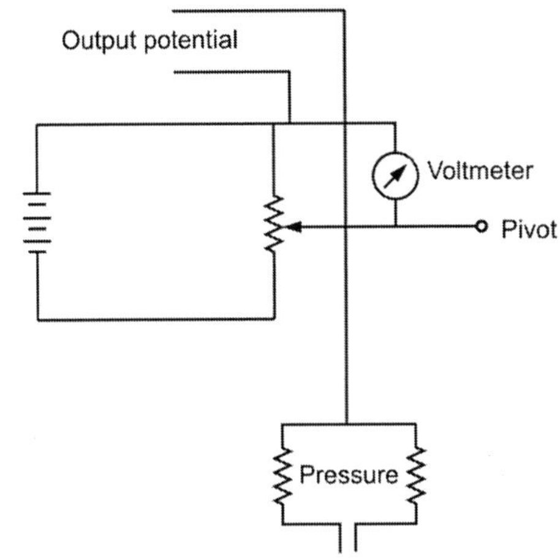

Fig. 5.9 Pressure-voltage transducer.

The potentiometer should have *little friction.* Levers may be used to match the required movement of the potentiometer with that of the bellows, the signal applied to potentiometer may be relatively larger of the order of several volts. By the use of potentiometer method the *signal can be sent relatively long distance.* In that case a remote reading systems may employ servomechanism system. In that case the galvanometer of the potentiometer is brought to the zero position.

5.4.3 Inductance-type Pressure Transducer

Inductance-type pressure transducers consist of the following three parts:

1. A coil
2. A movable magnetic core.
3. The elastic deformation element. The element is attached to the core.

When the pressure varies, the element causes the core to move within the coil. An alternating current is passed through the coil, and as the core moves the inductance of the coil changes. *The current passing through the coil increases as the inductance decreases.* This type of transducer is used in a *current-sensitive coil.*

To get increased sensitivity the coil may be divided into two, using a centre tap (see Fig. 5.10). This actually provides two coils. As the core moves inside the coils, the *inductance of one coil decreases as the inductance of the other increases.*

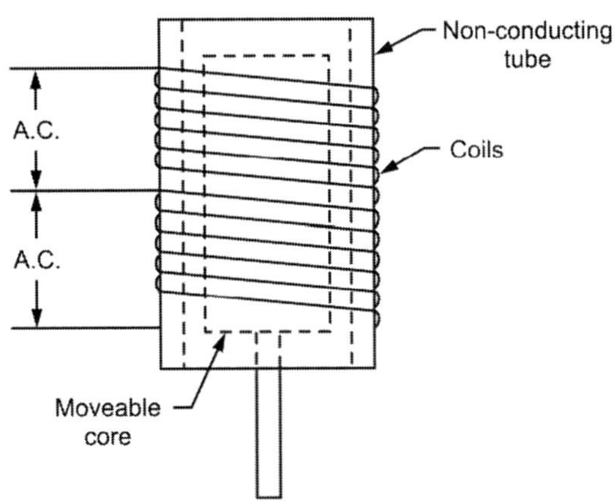

Fig. 5.10.

- *Another type of inductance unit used for pressure measurement employs a mutual inductance unit, or linear voltage differential transducer* (LVDT). In this type, three coils are wound on a single tube (see Fig. 5.11). The centre coil is connected to an A.C. source. Voltage is induced in the two outside coils. *When the movable core is centered the induced voltages in each of the outside coils are equal.* Moving the core unbalances these voltages, and the difference between them can be measured.

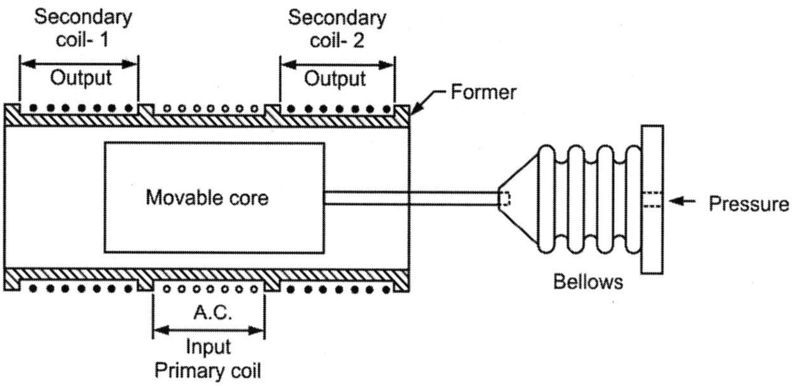

Fig. 5.11 Linear voltage differential transducer (LVDT).

Reluctance-type pressure transducer:

Similar to the inductance type pressure transducer is the reluctance-type. Figure 5.12 shows a reluctance-type pressure transducer. Here the electrical energy is provided by the field of a permanent magnet rather than A.C. source.

For pressure measurement, the elastic deformation element moves an armature closer to or farther from the permanent magnet. *The reluctance of a coil wrapped around the permanent magnet varies with the movement.*

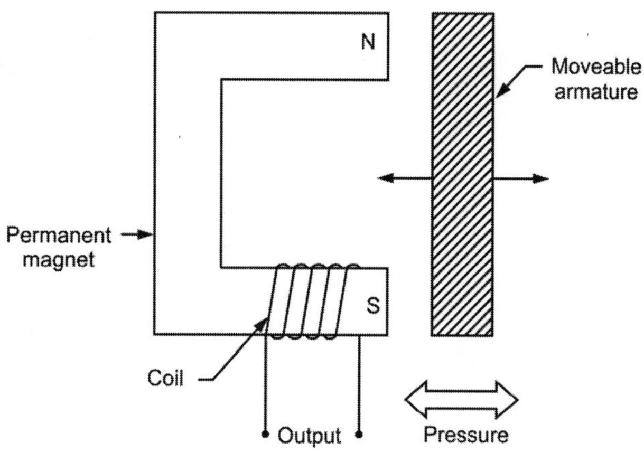

Fig. 5.12 Reluctance-type pressure transducer.

5.4.4 Capacitive-type Pressure Transducer

Capacitive pressure transducers consist of two conductive plates and a dielectric (See Fig. 5.13). As the pressure increases, the plates move farther apart, *changing the capacitance.* The fluid whose pressure is being measured serves at the dielectric.

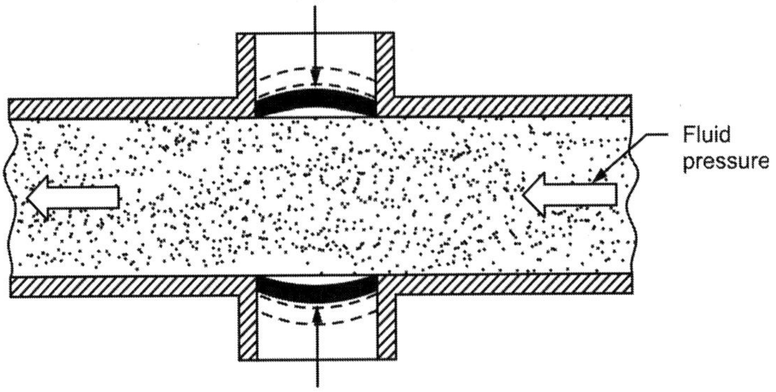

Fig. 5.13 Capacitive-type pressure transducer.

5.4.5 Carbon Pile Pressure Transducers

A carbon pile pressure transducer is shown in Fig. 5.14. It translates pressure change to resistance change. *As pressure change reduces the volume of the carbon pile the resistance decreases.*

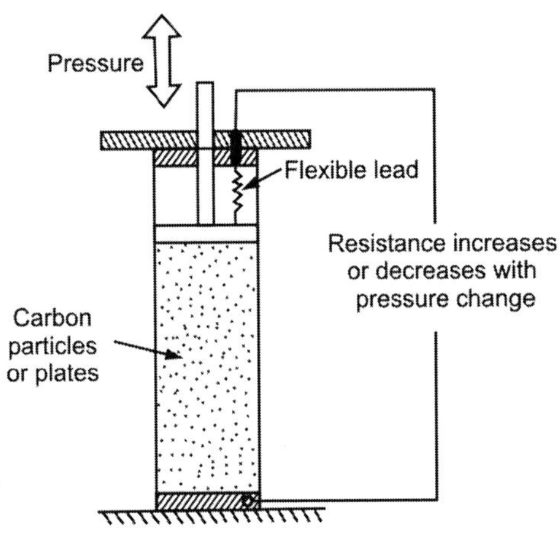

Fig. 5.14 *Carbon pile pressure transducer.*

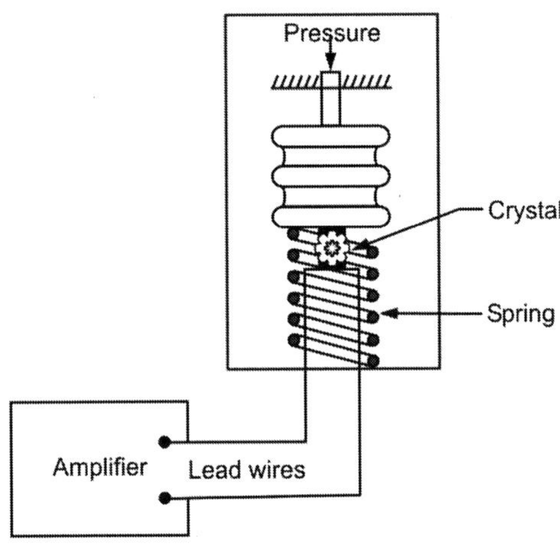

Fig. 5.15 *Piezoelectric pressure transducer.*

5.4.6 Piezoelectic Pressure Transducers

Figure 5.15 shows a piezoelectric pressure transducer. As the pressure is applied to crystals to certain materials a *difference in voltage across particular points of their structure occurs.*

5.4.7 Photoelectric Pressure Transducers

A photoelectric pressure transducer is shown in Fig. 5.16. It uses a photo-tube and a light source separated by a small window, whose aperture is controlled by the force summing member of the pressure transducers.

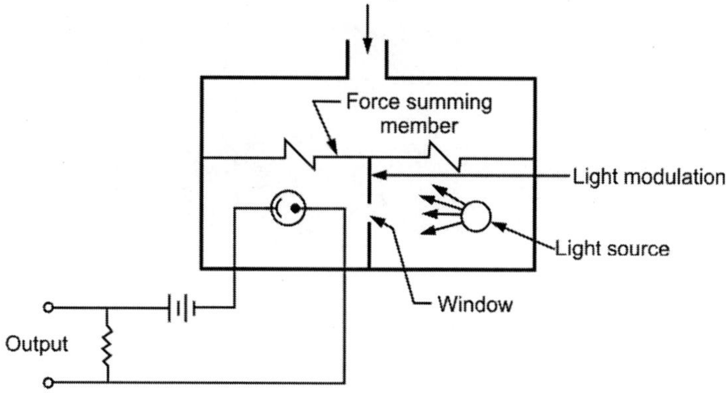

Fig. 5.16 *Photoelectric pressure transducer.*

5.4.8 Electromagnetic Pressure Transducers

For many years a technique has been used which consists of coil in plate or grid circuit of an oscillating tube (one that is generating alternating current) to deliver current to some other circuit. The energy may be transferred into some other winding having a transformer action. The coil has to deliver more power to the secondary winding, the additional power must come from the winding in the plate. The increase in plate current causes relays to operate at some predetermined values of current. It can also be used to develop voltages across known resistors. It can cause electromagnets to establish forces used for measuring the force applied which is obtained by pressure bellows.

5.5 MEASUREMENT OF HIGH PRESSURES

Conventional pressure-measuring devices, such as strain-gauge pressure cells and Bourdon-tube gauges may be used at pressure as high as 70000 MPa. *Bourdon tubes* for such pressures are nearly round in section and have a high ratio of wall thickness to diameter. They are, therefore, quite *stiff*, and *deflection per turn is small.* For this reason, high-pressure Bourdon tubes are often made of a number of turns.

For the measurement of very high pressures (above 70000 MPa), the *"electrical gauges'" based on the principle of change of resistance with change of pressure are used.*

It is known that resistance of fine wires changes with pressure according to the following linear relationship.

$$R = R_1 (1 + b\,\Delta p) \quad \text{...(5.1)}$$

where, R = Resistance at a pressure of 1 atm., ohms,

b = The pressure coefficient of resistance, and

Δp = The gauge pressure.

A pressure transducer based upon this principle is called a **Bridgman gauge.**

Figure 5.17 shows a Bridgman gauge. The sensing element does not actually contact the process medium but is separated therefrom by a kerosene-filled bellows. One end of the sensing coil is connected to a central terminal, as shown, while the other is grounded, thereby completing the necessary electrical circuit.

The fine wire used in a typical gauge is made of manganin or gold chrome. Manganin has a pressure coefficient of resistance $25 \times 10^{-12}\ Pa^{-1}$ and the total resistance of the wire is 100 ohms.

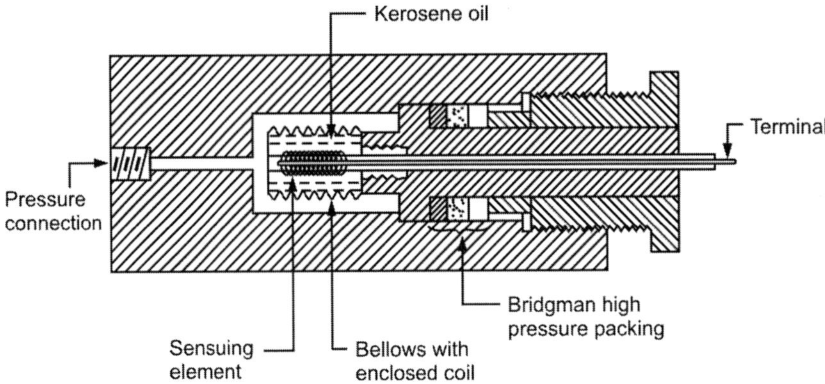

Fig. 5.17 Bridgman gauge.

Although pressure coefficient resistance of gold is 10×10^{-12} Pa^{-1} (it gives a change in resistance which is $\frac{1}{4}$ th of that of manganin for the same change in pressure), it is preferred to manganin in many applications as it has *resistance temperature coefficient* which is one half of that of manganin and, therefore, its use results in *lower errors on account of temperature changes.*

This gauge claims the following **advantages:**

1. Commercially available with scale upto to pressures of 15 MPa with accuracies of 0.1 to 0.5% when properly calibrated.
2. Good dynamic response (the resistance wire itself can respond to variations in MHz range but the overall frequency response of the pressure measurement system is limited to *much lower values because of the acoustic response of the transmitting fluid).*

5.6 MEASUREMENT OF LOW PRESSURE (VACUUM)

In general, pressures below atmospheric may be called **low pressures or vacuums.**

It is impossible to reach an absolute pressure of zero, since a *positive* magnitude of absolute pressure exists at all times, even in a vacuum.

The *micrometer, which is one-millionth of a meter (0.001 mm) of mercury column, is a common unit of "low pressure".*

Very low pressure**.......** any pressure *below* 1 mm of mercury (or torr)

Ultra low pressure**.......** *less than* a nanometer (10^{-3} μm).

The two basic methods of measurement of low pressure are:

1. **Direct methods.** Here, the displacement deflection caused by the pressure is measured and is correlated to the pressure.

   The following devices are included in this *category:*

   - Spiral Bourdon tubes.
   - Flat and corrugated diaphragms.
   - Capsules.
   - Manometers.

2. **Indirect or inferential methods.** In these methods, pressure is determined through the measurement of certain other pressure-controlled properties, including volume and thermal conductivity. Devices included in this category are:
   - McLeod gauge.
   - Thermal conductivity gauges.
   - Ionization gauges.
   - Radioactive vacuum meters.

5.6.1 McLeod Gauge

This gauge is considered as a *vacuum standard since the pressure can be computed from the gauge dimensions.* It cannot be used directly below about 10^{-4} torr, however, pressure dividing techniques allows its use as a calibration standard for considerably lower ranges.

The **principle** *of working* of all McLeod gauges (various versions) is the *compression of a sample of the low-pressure gas to a pressure sufficiently high to read with a simple manometer.*

The McLeod gauge consists of a system of glass tubings in which a known volume of gas at unknown pressure is trapped and then *isothermally compressed by a rising column.* This amplifies the unknown pressure and allows its measurement by conventional manometers.

Figure 5.18 shows the basic from of McLeod gauge. The process of measurement involves the following *steps:*

The plunger is withdrawn [See Fig. 5.18 (*a*)] lowering the mercury level below the opening thereby admitting the gas at the unknown pressure p into the system. Let V be the volume of the gas admitted into the measuring capillary, bulb and into the tube down to the opening. The bulb and the measuring capillary are then at the same pressure p (vacuum source).

2. The plunger is then pushed in and the mercury level goes up, sealing off a gas sample of known volume V in the bulb and measuring capillary.

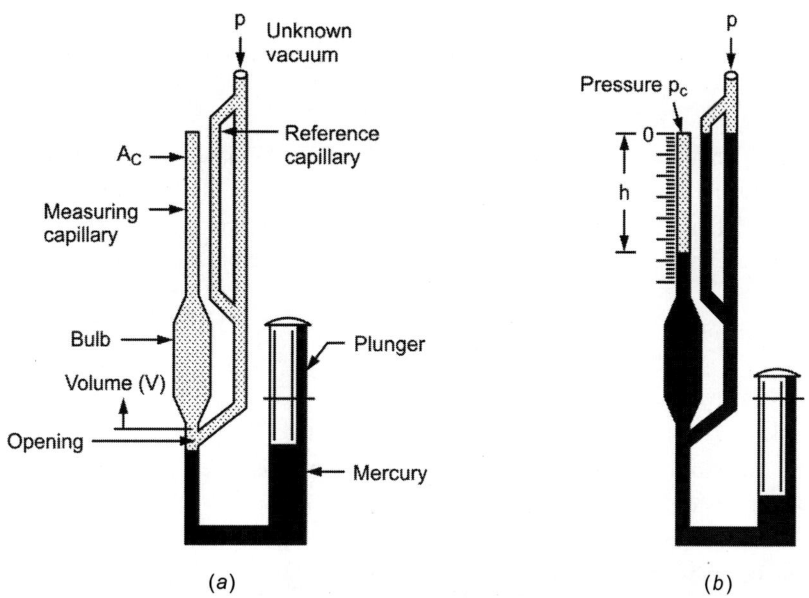

Fig. 5.18 McLeod gauge.

The unknown pressure (p) is calculated using Boyle's law (assuming isothermal compression) as follows:

$$pV = p_c A_c h \quad ...(5.2)$$

where, p_c = Pressure of gas in measuring capillary tube, P_a (pascal), and
A_c = Area of cross-section of measuring, capillary tube, m^2.

Now,
$$p_c = p + h\rho_m \quad ...(5.3)$$

where, h = Height of mercury column in reference capillary, m, and
ρ_m = Density of mercury, kg/m^3.

Substituting for p_c from equation (5.3) in equation (5.2), we get:

$$pV = (p + h\rho_m)A_c h = pA_c h + h^2 \rho_m A_c$$

or,
$$p(V - A_c h) = h^2 \rho_m A_c$$

∴
$$p = \frac{h^2 \rho_m A_c}{V - A_c h} Pa \quad ...(5.4)$$

If $A_c h << V$, as is usually the case, then:

$$p = \frac{h^2 \rho_m A_c}{V} \quad ...(5.5)$$

If the pressure p and p_c and can be considered in *mm of Hg* instead of in *Pa,* eqns. (5.4) and (5.5) can be modified as:

$$p = \frac{h^2 A_c}{V - A_c h} \quad ...(5.6)$$

and,
$$p = \frac{h^2 A_c}{V} \quad ...(5.7)$$

...*(Approximate formula)*

Hence, the measuring capillary tube can be directly calibrated in terms of pressure.

Advantages:

(*i*) The unknown pressure is calculated in terms of the physical dimensions of the gauge; this aspect makes the gauge an absolute instrument suitable for calibrating other low pressure gauges.

(*ii*) The gauge is not influenced by the composition of the gas (except for the effect that if the measured gas contains any vapours that are condensed by the compression process, then the pressure will be in error).

Limitations:

(*i*) Lack of a continuous output reading.

(*ii*) Not suitable on systems where mercury cannot be tolerated.

(*iii*) There are limitations to the lowest pressure that can be measured with this gauge unless special techniques are used.

(*iv*) Scale is of square law type; for a linear relation between p and h the system needs modification so as to keep the compression ratio constant for all pressures.

5.6.2 Thermal Conductivity Gauges

A *thermal conductivity gauge* works on the following *"principle"* :

The temperature of a given wire through which an electric current is flowing depends on two factors: (*i*) The magnitude of current, and (*ii*) The rate at which heat is dissipated. The current can be kept constant and the rate at which heat is dissipated will depend on the conductivity of the surrounding media. Further as the density of a given medium is reduced, *i.e., pressure is lowered,* its *conductivity* will also reduce and the wire will become hotter for a given current flow. Thus it is seen that the *temperature of the wire is directly dependent upon the pressure of the surrounding medium.*

The most commonly used conductivity gauges are:

1. Thermocouple gauge.
2. Pirani gauge.

5.6.2.1 Thermocouple Vacuum Gauge

A thermocouple vacuum gauge operates on the principle that at low pressure the thermal conductivity of a gas is a function of pressure.

Figure 5.19 shows basic elements of a *thermocouple vacuum gauge.* It consists of a heater element (heated to a temperature of 50° to 400°C by a known constant current) having a thermocouple in contact with its centre as shown in Fig. 5.19. The heater element and thermocouple are enclosed in a glass or metal envelope which is sealed into the vacuum system. The heater elements is supplied with a constant electric energy and its temperature (which is a function of the heat loss and hence thermal conductivity or pressure of the surrounding gas) is measured by a thermocouple. *The voltage measuring instrument can be directly calibrated to read the pressure of the gas.*

- The thermocouple gauges of one type or another are available to measure pressure in the range 10^{-4} to 1 torr.

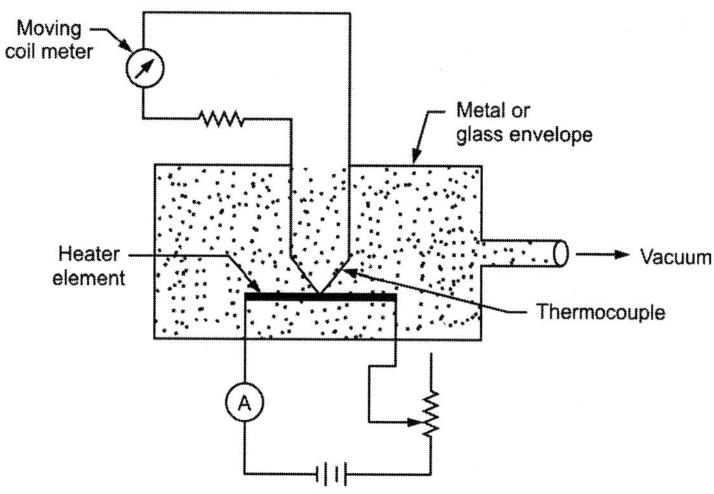

Fig. 5.19 Thermocouple vacuum gauge.

These gauges have the following *advantages* and *disadvantages:*

Advantages:

1. Inexpensive and rugged construction.
2. No departure from linearity in the range 0.02 to 1 mm of Hg.
3. Convenient and continuous reading.
4. Possibility of remote reading from the system
5. Possibility of process control with meter relay.

Disadvantages:

1. Required electric power.
2. Narrow reading range.
3. Need for individual and frequent calibration for different gases.

5.6.2.2 *Pirani vacuum gauge*

In this gauge, the temperature of the wire is determined by measuring the *change of resistance.*

The pirani gauge employs a single filament (in the form of four coiled wires of tungsten or platinum connected in parallel) enclosed in a glass tube/chamber, whose pressure is to be measured (see Fig. 5.20).

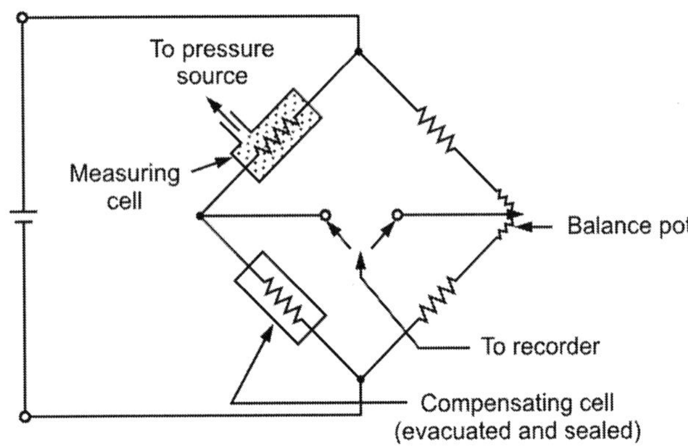

Fig. 5.20 Pirani vacuum gauge.

As the surrounding pressure changes, the filament temperature and hence its resistance also changes. A *compensating cell* is also employed to minimise variation caused by ambient temperature changes (its filament material is same as that of the measuring cell). The resistance change of the filament in the measuring cell is measured by the use of a resistance bridge which is *calibrated in terms of pressure.*

These gauges cover range from about 10^{-5} to 1 torr.

Advantages:

(*i*) These gauges are rugged, inexpensive and usually more accurate than thermocouple gauges.
(*ii*) The pressure reading range is wider.
(*iii*) Fast response to changes in pressure.
(*iv*) Possibility of process control and remote reading.

Disadvantages:

(*i*) Calibration is non-linear and varies from one gas to another.
(*ii*) Require calibration against some pressure standards.
(*iii*) Poor transient response.
(*iv*) Operation requires electrical power.

5.6.3 Ionization Gauges

These gauges are used for measurement of pressures as low as 0.000001 micron (one billionth of a mm of Hg); the maximum pressure which an ionization gauge can measure is about 1 micron.

The ionization gauge (Fig. 5.21) is very similar to the ordinary electronic tube. This gauge therefore, mainly consists of an envelope which is evacuated by the pressure to be measured and contains a heated filament, a positively biased grid and a negative biased plate. The grid draws electrons from the heated filament, and collision between them and gas molecules causes ionization (Ionization is the process of knocking off an electron from an atom and thus producing a free electron and a positively charged ion) of the molecules. As the plate is negatively charged, the positively charged ions are attracted to the plate of the tube, causing current (I_1) flow in the external circuit; the electrons are collected by the grid and a current I_2 is produced in the grid circuit.

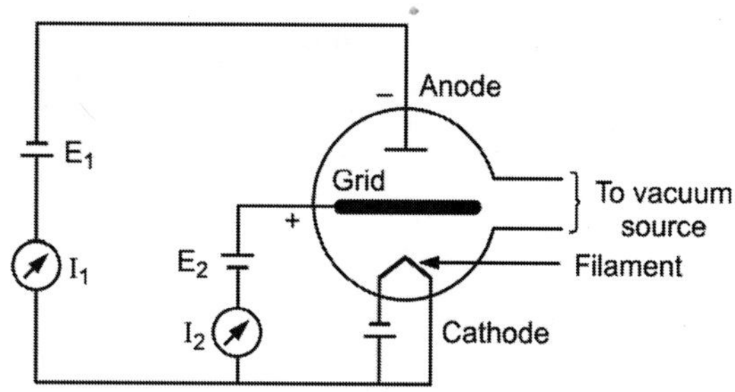

Fig. 5.21 Hot-filament ionization gauge.

The rate of ion production is proportional to the number of electrons available to ionize the gas and the amount of gas present. Thus, the ratio of + *ve* ions, *i.e.*, anode current I_1 to electrons *i.e.*, grid current I_2 is a measure of gas pressure. The following relation holds good:

$$p = \frac{1}{S}\frac{I_1}{I_2} \qquad ...(5.8)$$

where S (the proportionality constant) is called the *sensitivity* of the gauge and its value is determined by the calibration of the particular gauge. Sensitivity is a function of *nature of gas, tube geometry* and the *operating voltages*.

Advantages:

(*i*) Fast response to changes in pressure.
(*ii*) Possibility of process control and remote indication.
(*iii*) Constant sensitivity for a given gas over a wide range of pressures.
(*iv*) Wide pressure range.

Disadvantages:

(*i*) Some gases get decomposed by hot filament.
(*ii*) Contamination of the measured gas by gases forced out of the hot filament.
(*iii*) The filament can burn out quickly if exposed to air when hot or before the pressure is at low value (This disadvantage is eliminated by using a radioactive source of alpha-particles as the ionizing agent).
(*iv*) High cost.
(*v*) Complex electric circuitry.
(*vi*) These gauges must be provided with automatic cut out in order to protect them in case of system leak or break.
(*vii*) These gauges must be carefully installed-in particular absorbed gases must be eliminated.
(*viii*) Calibration varies with gases.

5.6.4 Radioactive Vacuum Meters

A radioactive vacuum meter works on the principle that *the current flowing through an ionization chamber depends upon the pressure, or more precisely upon the density of the gas in the chamber.*

- This type of meter permits the measurement and recording of the absolute pressure from 10×10^{-8} to 10 mm of mercury.

5.7 CALIBRATION OF PRESSURE MEASURING EQUIPMENT

The following are the *laboratory standards for calibrating pressure measuring instruments:*

1. Manometers.
2. Dead weight testers.
   - *Manometers and dead weight testers are suitable for static calibration only:*

5.7.1 Manometers

The calibration of low pressure gauges such as *manometers* and *straight bellows types* is usually done by comparison method using *precision manometer.* For calibration, simple U-tube manometers with antiparallax devices and moving reservoir type manometers with vernier scales are employed.

- Manometers are used upto 200 kN/m^2; *above this pressure dead weight testers are used.*

5.7.2 Dead Weight Testers

This tester is *a primary standard for pressure measurement.* It provides good calibration facility over a wide pressure range (0.00007 to 70 N/mm^2 gauge in steps as small as 0.01 % of range with a calibration uncertainty of 0.01 to 0.05 % of the reading).

A dead weight tester is shown in Fig. 5.22. It consists of an accurately machined and finished piston which is inserted into a close fitting cylinder; both of known cross-sectional areas. At the top of the piston is provided a platform to hold standard weights of known accuracy.

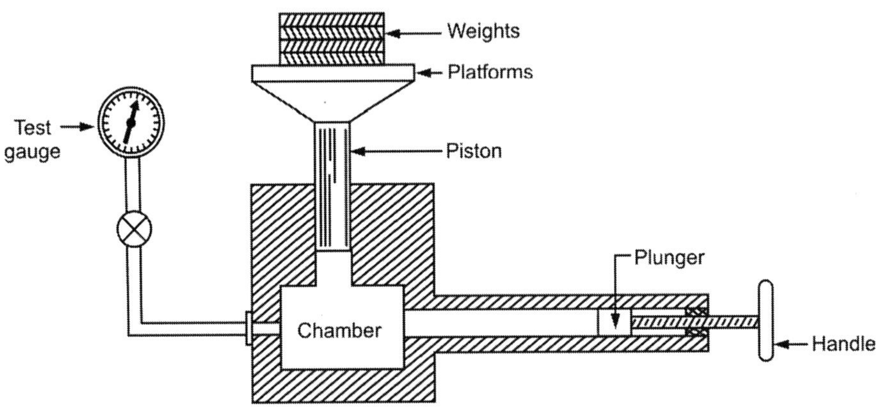

Fig. 5.22 Dead weight tester.

Working:

- The chamber and the cylinder of the tester are filled with clean oil.
- The gauge to be tested is fixed at its appropriate place and piston is inserted in the cylinder.

- A weight is placed on the piston. Consequently the pressure is exerted on the oil and this pressure is transmitted to the gauge deflecting its pointer. In order to get the desired setting of the gauge pointer, the weights are adjusted.
- Then pressure is varied by forcing the plunger in, by moving the handle. The movement of the handle is continued till enough pressure is built up to lift the piston-weight combination. The piston-weight combination floats freely *under these conditions*. Thus the equilibrium is established *with fluid force balanced against the gravitational force of the weights plus the friction drag.*

i.e.,
$$p \cdot A = Mg + F_d$$

Hence, pressure,
$$p = \frac{Mg + F_d}{A} \quad ...(5.9)$$

where, M = Mass, kg

g = Acceleration due to gravity, m/s^2

F_d = Friction drag, N, and

A = Effect area of piston-cylinder combination, m^2.

The effect of equivalent area (A) depends on such factors *as piston cylinder clearance, pressure level, temperature,* and is normally taken as the *mean of the cylinder and piston areas:*

Many kinds of pressure-measuring devices may be calibrated such as:

- *Engine indicators;*
- *Industrial pressure gauges;*
- *Piezoelectrical pressure transducers etc.*

The following *factors affect the accuracy of the dead weight tester:*

- Friction between the piston and the cylinder.
- Uncertainty of value of effective area A.
- Uncertainty of gravitational constant g.

HIGHLIGHTS

1. *Pressure* is defined as a force per unit area.
2. *Gauges* record the pressure above or below the local atmospheric pressure, since they measure the difference in pressure of the liquid to which they are connected and that of surrounding air.
3. *Manometers* are defined as the devices used for measuring the pressure at a point in a fluid by balancing the column of fluid by the same or another column or liquid. They are treated as *standards* for pressure and differential pressure measurements because they are virtually free from any friction or hysteresis effects.
4. Some important *mechanical gauges* are:

   (*i*) Bourdon tube pressure gauge. (*ii*) Diaphragm gauge.

   (*iii*) Vacuum gauge.
5. For the measurement of *very high pressures* (above 70000 MPa), the "*electrical gauges*" based on the principle of change of resistance with change of pressure are used.
6. The two basic methods of measurement of low pressure are:

   (*i*) Direct methods. The following devices belong to this category:

   – Spiral bourdon gauges;
   – Flat and corrugated; diaphragms;
   – Capsules;
   – Manometers.

(*ii*) **Indirect methods.** Devices included in this category are:

– McLeod gauge;

– Thermal conductivity gauges;

– Ionization gauges;

– Radioactive vacuum meters.

7. The following are the laboratory standards for calibrating pressure measuring instruments:

(*i*) Manometers. (*ii*) Dead weight testers.

THEORETICAL QUESTIONS

1. Define the following terms:

   (*i*) Pressure
   (*ii*) Atmospheric pressure
   (*iii*) Gauge pressure
   (*iv*) Absolute pressure
   (*v*) Static pressure
   (*vi*) Total or stagnation pressure.

2. What are the units of pressure ?
3. Explain briefly any two of the following mechanical gauges:

   (*i*) Bourdon tube pressure gauge
   (*ii*) Diaphragm gauge
   (*iii*) Vacuum gauge.

4. Explain briefly the following types of pressure elements:

   (*i*) Bourdon tube
   (*ii*) Diaphragm
   (*iii*) Bellows.

5. State the advantages and disadvantages of "Bourdon tube element."
6. What are the advantages and disadvantages of "Diaphragm elements" ?
7. List the advantages and disadvantages of bellow gauges/elements.
8. Explain briefly any three of the following pressure transducers:

   (*i*) Resistance type pressure transducers;
   (*ii*) Pressure voltage type;
   (*iii*) Piezoelectric type;
   (*iv*) Capacitance type;
   (*v*) Electromagnetic type.

9. How are very high pressures measured ? Explain briefly with a neat sketch the construction and working of a Baridgman gauge used for measuring high pressures.
10. What do you mean by low pressures or vacuums?
11. Draw diagrams to show how LVDTs can be used with bellows, elements and Bourdon tubes for measurement of pressure. Give their advantages and disadvantages.
12. Explain how capacitive type of transducers can be used for pressure measurement.
13. Describe the applications of piezoelectric transducers for measurement of pressure. List their advantages and disadvantages.
14. Describe the construction, working and theory of McLeod gauge for measurement of vacuum. List its advantages and disadvantages.
15. Explain briefly the construction and working of a thermocouple vacuum gauge. State its advantages and disadvantages.
16. Describe briefly the construction and working of a Pirani gauge. Give its advantages and disadvantages.
17. Give the construction and working of a hot-filament ionization gauge. List-its advantages and disadvantages.
18. (*a*) Explain how calibration of pressure measuring equipment is carried out.

    (*b*) Describe the construction and working of a "Dead Weight Tester". State the factors which affect the accuracy of dead weight testers.

❑❑❑

Strain Measurement

Chapter 6

6.1 STRAIN GAUGES

6.1.1 Introduction

Strain is *defined as the change in length of a line segment between two points divided by the original length of the line segment. For* the strain measurement it is the usual practice to make measurements over shortest possible gauge lengths. This is owing to the fact that the measurement of a change in a finite length does not give the strain at any fixed point, but rather the average value over the length. The strain at various points might be different depending upon the strain gradient along the gauge length. When the strain gradient is constant over a small length, the average strain will be point strain at the middle point of the gauge length.

***A* strain gauge** *is a device which is used to measure dimensional change on the surface of a structural member under test.*

Strain gauges give indication of strain at only one point

The strain measurement is of significant importance in a variety of applications due to the following *reasons:*

(*i*) To avoid the use of large factors of safety in the design of aircraft and automatic control equipment due to consideration of mass/inertia.

(*ii*) Strain measurement utility as a means to determine maximum stress values or in specialised transducers to measure pressure, acceleration, force, torque etc.

(*iii*) Necessity to verify experimentally the strain in complex physical systems (*e.g.* human skulls and legs) where strain only be estimated approximately even by the use of most rigorous analytical methods.

6.1.2 Type of Strain Gauges

Four types of strain gauges are:

1. Wire-wound strain gauges.
2. Foil-type strain gauges.
3. Semiconductor strain gauges.
4. Capacitive strain gauges.

6.1.2.1 *Wire-wound strain gauges*

There are two main classes of wire-wound strain gauges:

1. Bonded strain gauge.
2. Unbounded strain gauge.

Bonded strain gauge

It is composed of fine wire, wound and cemented on a resilient insulating support, usually a wafer unit. Such units may be mounted upon or incorporated in mechanical elements or structures whose deformations under stress are to be determined. While there are no limits to the basic values which may be selected for strain-gauge resistances, a typical example may be taken as of the order or 100 to 500 Ω.

Figure 6.1 shows the commonly used form of resistance wire strain gauges.

Unbonded strain gauge

Figure 6.2 shows *an unbounded strain gauge. M* and *N* are attached by rods *m* and *n*, respectively, to points between which *displacement is to be measured.* Pick-up and measuring networks are energized from similar but isolated sources. Unbalance originating in pick-up is detected and balanced by servo-actuated measuring network, *providing a reading of strain on graduated scale.*

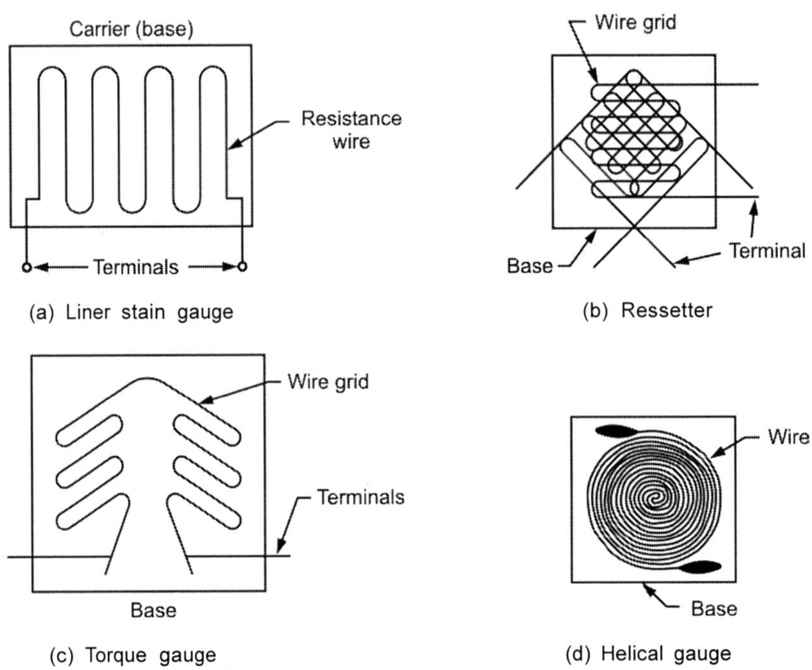

Fig. 6.1 Resistance wire strain gauges.

In the unbonded strain gauge the resistance structure comprises of fine wire winding stretched between insulating supports mounted alternately on the two members between which displacement is to be measured (see Fig. 6.2). These wires comprise the four arms of a Wheatstone-bridge network of *which two opposite arms are tightened and the other two slackened by the displacement.*

— *While a bonded gauge tends to respond to me average strain in the surface to which it is cemented, the unbonded form measures displacement between the two points to which the respective supports are attached.*

— *Unbonded wire strain gauges* are usually operated on input potentials ranging upto 35 V direct or alternating current. Under conditions of extreme balance, corresponding to full operating range, the open-circuit e.m.f. may be of the order of 8 to 10 mV and closed circuit current upto 100 μA.

Strain gauges for use on A.C. circuits are supplied in both capacitive and inductive forms, wherein *I* the corresponding characteristics of A.C. circuit components are varied by the displacements to be measured.

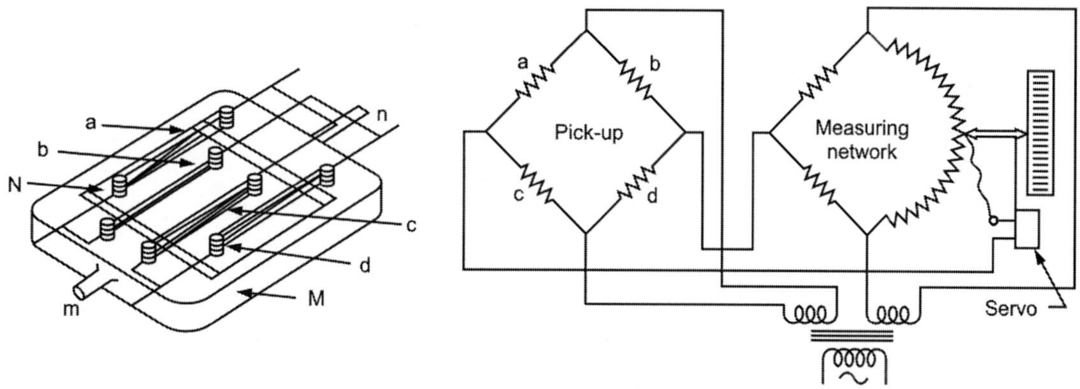

Fig. 6.2 Unbonded strain gauge

Requirements/Characteristics of resistance wire strain gauges

The resistance wire strain gauges should have the following characteristics to have excellent and reproducible results.

1. The strain gauge should have a *high value of gauge factor.* A high value of gauge factor indicates a large change in resistance for a particular strain resulting in high sensitivity.
2. The *resistance of strain gauge should be as high as possible* since this minimizes the effects of undesirable variations of resistance in the measurement circuit.
3. The strain gauges should have a *low resistance temperature co-efficient.* This is essential to minimize errors on account of temperature variations which affect the accuracy of measurements.
4. The strain gauge *should not have any hysteresis effect in its response.*
5. In order to maintain constancy of calibration over the entire range of strain gauge, it should have *linear characteristics, i.e.,* the variations in resistance should be a linear function of the strain.
6. The strain gauges are frequently used for dynamic measurements and hence their frequency response should be good. The linearity should be maintained within accuracy limits over the entire frequency range.

6.1.2.2 Foil strain gauges

In these gauges *the strain is sensed with the help of metal foil.* Foil gauges have a *much greater dissipation capacity as compared with wire wound gauges on account of their greater surface area for the same volume.* Due to this reason they can be employed for *higher operating temperature range.*

In these gauges, the bonding is better due to large surface area of the foil. The *bonded foil* gauges *find* a wider field of action.

Figure 6.3 shows a typical foil gauge.

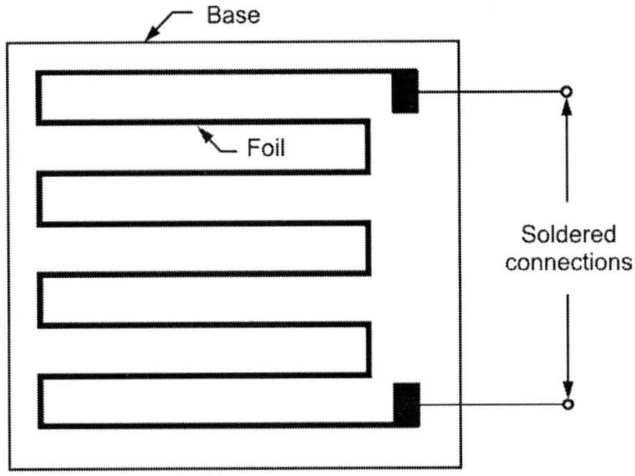

Fig. 6.3 Foil gauge.

- The characteristics of foil types strain gauges are similar to those of wire wound strain gauges and their gauge factors are typically the same as that of wire wound strain gauges.
- The resistance value of foil gauges which are commercially available is between 50 and 1000 Ω.

The *advantage* of foil type strain gauges is that they can *be fabricated economically on a mass scale.*

6.1.2.3 Semiconductor strain gauges

- *Semiconductor strain gauges depend for their action upon piezo-resistive effect, i.e., the change in value of the resistance due to change in resistivity.*
- These gauges are used where a *very high gauge factor and small envelope are required.*
- For semiconductor strain gauges semiconducting materials such as *silicon* and *germanium* are used.
- A typical strain gauge consists of a *strain sensitive crystal material and leads that are sandwiched in a protective matrix*. The production of these gauges employs conventional semiconductor technology using semiconducting wafers or filaments which have a thickness of 0.05 mm and bonding them on suitable insulating substances, such as teflon. Gold, lead are generally applied for making the contacts.

Figure 6.4 shows some typical semiconductor strain gauges.

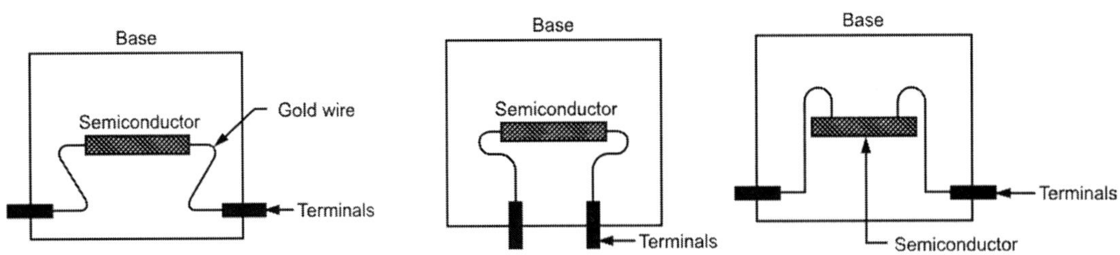

Fig. 6.4 Semiconductor strain gauges.

Advantages:

1. These gauges have high gauge factor.
2. Excellent hysteresis characteristics.
3. Fatigue life is in excess of 10×10^5 operations and the frequency response is upto 10^{12} Hz.
4. These gauges can be very small ranging in length from 0.7 to 7 mm. *They are very useful for measurement of local strains.*

Disadvantages:

1. The *major and serious disadvantage is that these gauges are very sensitive to change in temperature.*
2. Linearity of these gauges is *poor.*

6.1.2.4 Capacitive strain gauges

Figure 6.5 shows a capacitive strain gauge. It uses *the principle of variation of capacitance with variation of distance between electrodes.* The electrodes are flexible metal strips of about 0.1 mm thickness. The strain to be measured is applied to the top plate. This changes the distance between the curved electrodes resulting in change of capacitance.

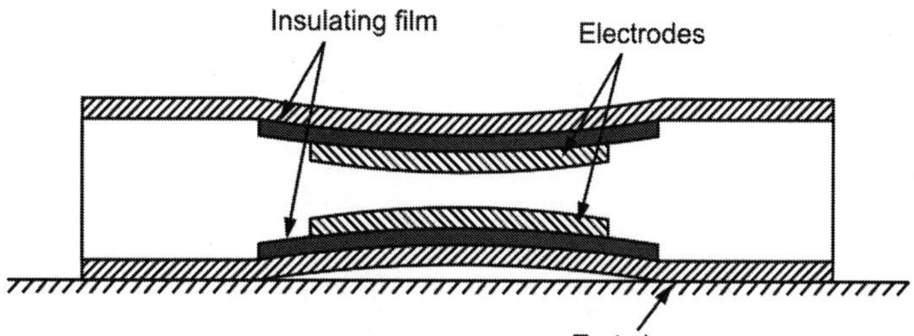

Fig. 6.5 Capacitive strain gauge.

The strain-capacitance relationship, in general, is *not linear* but variations in dimensions and shape allow gauge characteristics to be chosen so as to match the range of capacitance to be measured with a good degree of accuracy.

- A capacitance strain gauge has a capacitance of about 0.5 pF.
- Its overall size is 5 mm × 17 mm × 1 mm.
- It uses a polyamide film of insulating material.
- It can be used upto a temperature of 300°C.

6.2 THEORY OF STRAIN GAUGES

When a strain gauge is subjected to tension (*i.e.,* positive strain) its length *increases* while its cross- sectional area *decreases.* Since the resistance of a conductor is proportional to its length and inversely proportional to its area of cross-section, the resistance of the gauge increases with positive strain. The change in the value of resistance of strained conductor is more than what can be accounted for an increased in resistance due to dimensional changes. The extra change is the value of resistance is attributed to a *change in the value of resistivity of a conductor when strained; this property is known as* **piezo-resistive effect.**

Strain gauges are most commonly used in wheatstone bridge circuits to measure the change of resistance of grid of wire for calibration proposes; the **"gauge factor"** *is defined as the ratio of per unit change in resistance to per unit change in length.*

i.e.,
$$\textit{Gauge factor } (G_f) = \frac{\Delta R/R}{\Delta L/L} \qquad ...(6.1)$$

where, ΔR = Corresponding change in resistance R, and

ΔL = Change in length per unit length L.

The resistance of the wire of strain gauge R is given by:

$$\rho = \frac{\rho L}{A}$$

where, ρ = Resistivity of the material of wire (of strain gauge),

L = Length of the wire, and

A = Cross-sectional area of the wire,

$= KD^2$, K and D being a constant and diameter of the wire respectively.

As earlier stated, when the wire is strained its length increases and lateral dimension is reduced as a function of Poisson's ratio (μ); consequently there is an increase in resistance.

Now,
$$R = \frac{\rho L}{KD^2}$$

Differentiating it, get we:

$$dR = \frac{KD^2(\rho \cdot dL + L \cdot d\rho) - \rho L(2KD \cdot dD)}{(KD^2)^2}$$

$$= \frac{1}{KD^2}\left[(\rho \cdot dL + L \cdot d\rho) - 2\rho L \cdot \frac{dD}{D}\right]$$

$$\frac{dR}{R} = \frac{\frac{1}{KD^2}\left[\rho \cdot dL + L \cdot d\rho - 2\rho L \cdot \frac{dD}{D}\right]}{\frac{\rho L}{KD^2}}$$

$$= \frac{dL}{L} + \frac{d\rho}{\rho} - 2\frac{dD}{D}$$

Now, Poisson's ratio,
$$\mu = \frac{\text{Lateral strain}}{\text{Longitudinal strain}} = \frac{-dD/D}{dL/L}$$

or,
$$\frac{dD}{D} = -\mu \times \frac{dL}{L}$$

For small variations, the above relationship can be written as:

$$\frac{\Delta R}{R} = \frac{\Delta L}{L} + 2\mu\frac{\Delta L}{L} + \frac{\Delta\rho}{\rho} \quad \text{...(6.2)}$$

Gauge factor,
$$G_f = \frac{\Delta R/R}{\Delta L/L}$$

or,
$$\frac{\Delta R}{R} = G_f \cdot \frac{\Delta L}{L} = G_f \times e \quad \text{...(6.3)}$$

where, $e = \text{strain} = \dfrac{\Delta L}{L}$

The *gauge factor* can be written as:

$$G_f = 1 + 2\mu + \frac{\Delta\rho/\rho}{e} \quad ...(6.4)$$

$$= \underset{\substack{\text{Resistance}\\\text{change due to}\\\text{change of length}}}{1} + \underset{\substack{\text{Resistance}\\\text{change due}\\\text{to change}\\\text{in area}}}{2\mu} + \underset{\substack{\text{Resistance}\\\text{change due to}\\\text{piezo-resistive}\\\text{effect.}}}{\frac{\Delta\rho/\rho}{e}}$$

or,
$$G_f = 1 + 2\mu + \frac{\Delta\rho/\rho}{\Delta L/L} \quad ...(6.5)$$

The strain is usually expressed in terms of *microstrain;* 1 micro strain = 1 μm/m

If the change in the value of resistivity of a material when strained is *neglected,* the gauge factor can be rewritten as:

$$G_f = 1 + 2\mu \quad ...(6.6)$$

Equation (6.6) is valid only when *piezo-resistive effect* (*i.e.*, change in resistivity due to strain) is *almost negligible.*

- The Poisson's ratio for all metals lies between 0 and 0.5. This give G_f as 2 approximately. In case of wire wound strain gauges where the common value for Poisson's ratio is 0.3, the value of G_f amounts to 1.6.
- The value of the gauge factor varies from material-to-material but it is generally assumed that it remains constant in the working range of strain required. *Its value is determined experimentally.*
- *Knowing the gauge factor (G_f), the strain in the member can be directly found out by the change of resistance.*

Example 6.1: *The gauge factor of a resistance wire strain gauge using a soft iron wire of small diameter is 4.2. Neglecting the piezo-resistive effect, calculate the Poisson's ratio.*

Solution: *Given:* $G_f = 4.2$

When the *piezo-resistive effect* is neglected, the gauge factor is given by:

$$G_f = 1 + 2\mu \quad ...(\text{Eqn. } 6.6)$$

or,
$$4.2 = 1 + 2\mu$$

∴
$$\mu = \frac{4.2 - 1}{2} = \mathbf{1.6} \quad \textbf{(Ans.)}$$

Example 6.2: *A simple electrical strain gauge of resistance 120Ω and having a gauge factor of 2 is bonded to steel having an elastic limit stress of 400 MN/m² and modulus of elasticity 200 GN/m². Calculate the change in resistance due to:*

(i) Change in stress equal to $\frac{1}{10}$ of the elastic range.

(ii) Change of temperature of 20°C if the material is advance alloy; the resistance temperature coefficient of advance alloy is 20 × 10^{-6} /°C.

Solution: *Given*: R = 120Ω; G_f = 2; Elastic limit stress = 400 MN/m²; Modulus of elasticity = 200 GN/ m²; Resistance temperature coefficient, $\alpha_0 = 20 \times 10^{-6}$/°C.

Change in resistance:

(*i*) Change in stress $= \frac{1}{10} \times 400 \text{ MN/m}^2 = 40 \times 10^6 \text{ N/m}^2$

Modulus of elasticity $= 200 \text{ GN/m}^2 = 200 \times 10^{12} \text{ N/m}^2$

$$\text{Strain, } e = \frac{\text{Stress}}{\text{Modulus of elasticity}} = \frac{40 \times 10^6}{200 \times 10^{12}} = \frac{1}{5} \times 10^{-6}$$

$$\text{Gauge factor, } G_f = \frac{\text{Per unit change in resistance}}{\text{Per unit change in length}}$$

or, $$G_f = \frac{\Delta R/R}{e}, \quad \text{or,} \quad \Delta R = RG_f e$$

or, $$\Delta R = 120 \times 2 \times \frac{1}{5} \times 10^{-6} = 48 \times 10^{-6}\ \Omega = \mathbf{48\ \mu\Omega}$$ **(Ans.)**

(*ii*) $$R_{t2} = R_{t1}\,[1 + \alpha_0(t_2 - t_1)]$$

∴ Change is resistance $R_{t2} - R_{t1} = R_{t1}\,\alpha_0(t_2 - t_1)$

or, $$\Delta R = R_{t2} - R_{t1} = 120 \times 20 \times 10^{-6}\,(20)$$
$$= 48 \times 10^{-3}\ \Omega = 48 \text{ m}\Omega$$ **(Ans.)**

6.3 STRAIN GAUGE CIRCUITS

The following strain gauge circuits will be discussed:

1. Ballast circuit.
2. Wheatstone bridge circuit,
   (*i*) Balanced (null) condition
   (*ii*) Unbalanced (Deflection) condition
   — Quarter bridge
   — Half bridge
   — Full bridge.

6.3.1 Ballast-circuit (voltage-sensitive potentiometric circuit)

Figure 6.6 shows a ballast circuit-voltage-sensitive potentiometric circuit. Here,

v_i = Input supply voltage,
v_0 = Output voltage,
R_b = Ballast resistance, and
R_g = Resistance of the unstrained resistance gauge.

The output voltage, when *no stress* is applied to the strain gauge, is given by:

$$v_0 = \left(\frac{R_g}{R_g + R_g}\right) v_i \quad ...(6.7)$$

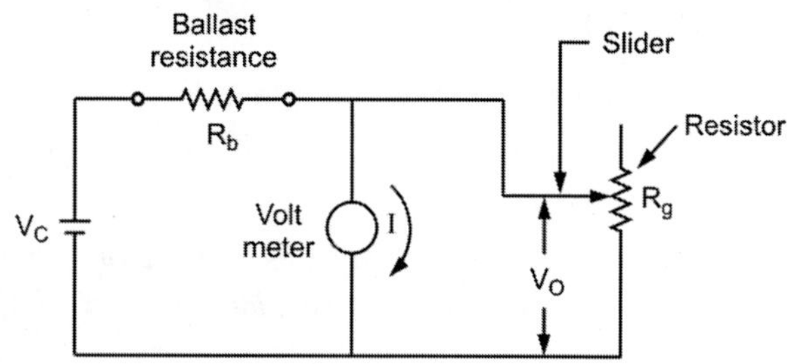

Fig. 6.6 Ballast circuit-voltage-sensitive potentiometric circuit.

When the gauge is strained, the gauge resistance changes to $(R_g + dR_g)$ and the output voltage becomes,

$$v_0 + dv_0 = \left[\frac{(R_g + dR_g)}{(R_g + dR_g) + R_b}\right] v_i \quad \text{...(6.8)}$$

$\therefore$ The *change in the output voltage*,

$$dv_0 = \left[\frac{(R_g + dR_g)}{(R_g + dR_g) + R_b} - \frac{R_g}{R_g + R_b}\right] v_i$$

$$= \left[\frac{(R_g + dR_g)(R_g + R_b) - R_g\{(R_g + dR_g) + R_b\}}{\{(R_g + dR_g) + R_b\}(R_g + R_b)}\right] v_i$$

$$= \left[\frac{(R_g^2 + R_g R_b + dR_g R_g + dR_g R_b - R_g^2 - dR_g R_g - R_g R_b)}{(R_g + R_b)^2}\right] v_i$$

...($\because dR_g < R_g$)

$$= \left[\frac{dR_g \cdot R_b}{(R_g + R_b)^2}\right] v_i = \frac{R_g \cdot R_b}{(R_g + R_b)^2} \cdot \frac{dR_g}{R_g} v_i \quad \text{...(6.9)}$$

... Multiplying numerator and denominator by R_g.

Also, *condition of maximum sensitivity* is given by: $R_g = R_b$

Hence, $$dv_0 = \frac{v_i}{4} \cdot \frac{dR_g}{R_g}$$

Also, $$\frac{dR_g}{R_g} = G_f \cdot \in$$

$\therefore$ $$dv_0 = \left(\frac{G_f}{4}\right) \in v_i$$ ...where, G_f denotes the gauge factor ...(6.10)

From eqn. (6.9) it is evident that *change in output-voltage when gauge is strained is directly proportional to strain.*

- The ballast circuit is used for *dynamic strain measurements where static strain components are ignored.*

Limitations of potentiometeric circuits:

(*i*) No possibility of compensation for temperature variations.

(*ii*) High sensitivity to A.C. interference giving hum due to ground loops, induction from high current lines and poor connections.

Example 6.3: *A strain gauge of 120 Ω resistance and a gauge factor 2 is used in a strain gauge potentiometeric circuit. The current flow through the system, when working under maximum sensitivity conditions, is limited to 18 mA. Calculate the following when a strain of 400 micro-strain is impressed upon the gauge:*

(i) The input voltage.

(ii) The change in output voltage.

Solution: *Given*: $R_g = 120\ \Omega$; $G_f = 2$; $I = 18$ mA; $\in = 400$ micro-strain.

Refer to Fig. 6.6.

(*i*) The input voltage, v_i:

The current flowing through the system,

$$I = \frac{v_i}{R_g + R_b}$$

For maximum sensitivity: $R_g = R_b$,

$$\therefore \quad I = \frac{v_i}{2R_g}$$

or, $$v_i = 2I \cdot R_g = 2 \times (18 \times 10^{-3}) \times 120 = \mathbf{4.32\ V\ (Ans.)}$$

(*ii*) The change in supply voltage, dv_0:

$$dv_0 = \left(\frac{G_f}{4}\right) \in \cdot\, v_i \qquad ...(6.10)$$

or, $$dv_0 = \frac{2}{4} \times 400 \times 10^{-6} \times 4.32$$

$$= 8.64 \times 10^{-4}\text{ V or } \mathbf{0.864\ mV\ (Ans.)}$$

Example 6.4: *On a steel cantilever beam is mounted a single strain gauge. The strain gauge has a resistance of 120 Ω and a gauge factor 2 and is connected in series with a 120 Ω ballast resistor and a 12 V battery. If the bending stress at the gauge fluctuates from 0 to 1800 MN/m^2, calculate the corresponding variations in output voltage.*

Take modulus of elasticity of steel = 210 GN/m^2.

Solution: *Given:* $R_g = 120\ \Omega$; $G_f = 2$; $R_b = 120\ \Omega$; $v_i = 12$ V;

Bending stress (σ_b) varies from 0 to 1800 MN/m^2;

Modulus of elasticity (E) = 2106 N/m^2.

Variation in output voltage:

Corresponding to the given bending strain, the strain is given by:

$$\in = \frac{\sigma_b}{E} = \frac{1800 \times 10^6}{210 \times 10^9} = 0.008571$$

Output voltage due to change in resistance because of applied strain,

$$v_0 = \frac{v_i}{2}\left(1 + \frac{1}{2} G_f \cdot \in\right)$$

or, $$v_0 = \frac{12}{2}\left(1 + \frac{1}{2} \times 2 \times 0.008571\right) = \mathbf{6.051\ V\ (Ans.)}$$

6.3.2 Wheatstone Bridge Circuit

The wheatstone bridge technique can be used in the following two ways:

(*i*) Null mode;
(*ii*) Deflection mode.

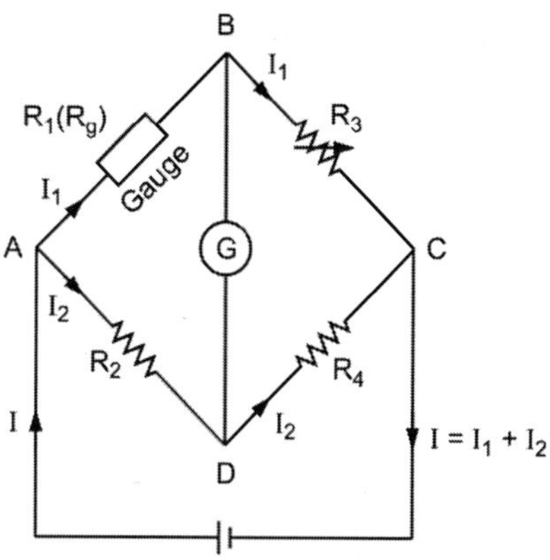

Fig. 6.7 Wheatstone bridge circuit.

1. Null mode: Refer to Fig. 6.7. The resistance, with *no straining*, are so arranged that $v_B = v_D$ and the galvanometer gives zero deflection.

Then, $$\frac{R_1}{R_3} = \frac{R_2}{R_4} \quad \text{...(6.11)}$$

where, $R_1 = R_g$ = Unstrained resistance of the gauge.

In measurement of strains, generally R_1 is the strain gauge, R_2 and R_4 are the fixed resistances and R_3 is a variable resistor.

When the gauge is strained, its resistance R_1 changes by an amount dR_1. This change unbalances the bridge resulting into the deflection of the galvanometer. The balance is then regained by Adjusting R_3 by an amount dR_3. The rebalanced condition gives:

$$\frac{R_1 + dR_1}{R_3 + dR_3} = \frac{R_2}{R_4}$$

or, $$R_1 + dR_1 = (R_3 + dR_3)\frac{R_2}{R_4}$$

or, $$R_1 + dR_1 = R_3 \times \frac{R_2}{R_4} + dR_3 \times \frac{R_2}{R_4}$$

or, $$R_1 + dR_1 = R_1 + dR_3 \times \frac{R_2}{R_4} \qquad \left[\because R_1 = \frac{R_3}{R_4} \times R_2 \text{ from eqn. (6.11)}\right]$$

or, $$dR_1 = dR_3 \times \left(\frac{R_2}{R_4}\right) \quad \text{...(6.12)}$$

If the resistances of all the limbs of the wheatstone bridge are equal, then:

$$R_1 = R_2 = R_3 = R_4 = R_g$$

and, $$dR_1 = dR_3 \quad \text{...(6.13)}$$

The change in resistance dR_1, in terms of strain, is given by:

$$dR_1 = G_f \in R_g, \quad \text{where } G_f \text{ is the gauge factor and } \in \text{ is the strain}$$

$$\therefore \quad dR_3 = G_f \in R_g \qquad ...(6.14)$$

Equation (6.14) indicates that the *change in the value of resistance R_3 is direct measurement of strain.*

Example 6.5: *The following data relate to the strain gauge Wheatstone bridge circuit shown in Fig. 6.7. $R_1 = (R_g) = 79.5\Omega$; $R_2 = 80\Omega$; R_3 = variable; $R_4 = 800\ \Omega$.*

(i) Calculate the value of variable resistance R_3 at initial balance with zero strain.

(ii) If the above calculated resistance is increased to 7970 Ω after the strain was applied, calculate the change in resistance of the gauge and the strain applied if the gauge factor is 2.

Solution: *Given:* $R_1 (= R_g) = 79.5\ \Omega$; $R_2 = 80\ \Omega$; R_3 = variable; $R_4 = 8000\ \Omega$.

When the bridge is *balanced* (zero strain), we have:

(i) Variable resistance, R_3:

$$\frac{R_1}{R_3} = \frac{R_2}{R_4} \qquad \text{(from eqn. 6.11)}$$

$$\therefore \quad R_3 = \frac{R_1 \times R_4}{R_2} = \frac{79.5 \times 8000}{80}$$

$$= \mathbf{7950\ \Omega\ (Ans.)}$$

(*ii*) Change in gauge resistance (dR_1) and strain ($\in$):

When strain is applied, let dR_3 be the change in variable resistance R_3 to rebalance the bridge. Then,

$$dR_3 = 7970 - 7950 = 20\ \Omega$$

Change in resistance $dR_l (= dR_g)$ of the strain gain $(R_1 = R_g)$ is then given by:

$$\text{dR}_1 = dR_3 \left(\frac{R_2}{R_4}\right) = 20 \times \frac{80}{8000}$$

$$= \mathbf{0.2\ \Omega\ (Ans.)}$$

Also,
$$\frac{dR_1}{R_1} = G_f \cdot \in$$

$$\therefore \quad \in = \frac{1}{G_f} \cdot \frac{dR_1}{R_1}$$

or,
$$\in = \frac{1}{2} \times \frac{0.2}{79.5}$$

$$= \mathbf{1.2579 \times 10^{-3}\ (Ans.)}$$

The strain ($\in$) would be a *tensile strain* since R_3 and hence $R_1 (= R_g)$ have increased.

2. Deflection mode: Initially the bridge resistances are so adjusted that the bridge is balanced. The equilibrium gets disturbed when the gauges are strained. Then, the voltage v_0 is measured under this unbalanced condition.

***(i) Quarter-bridge*:**

Figure 6.8 shows single gauge used for strain measurement (quarter-bridge). In this arrangement only one strain gauge is used and the other three elements of the bridge are fixed resistors.

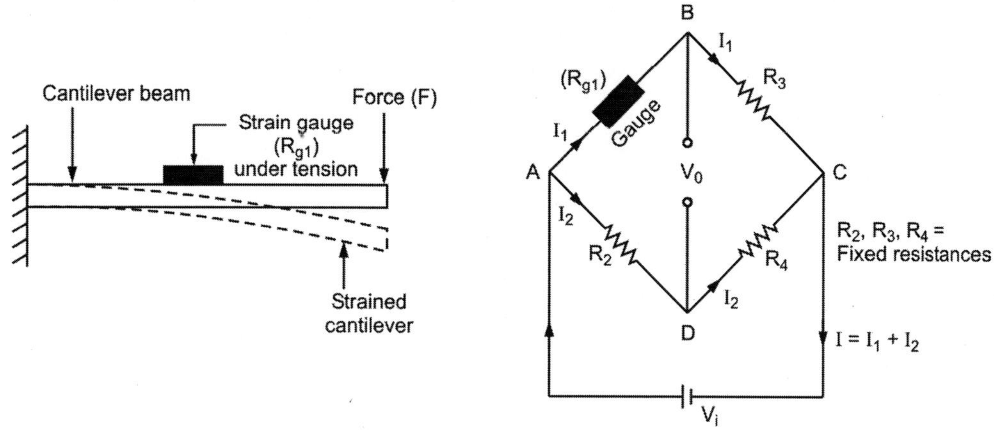

Fig. 6.8 Single gauge used for strain measurement (Quarter-bridge).

Let us assume that the galvanometer (measuring instrument) has infinite impedance and therefore no current flows through it. Then,

Current flowing through the limbs AB and BC, $I_1 = \dfrac{v_i}{R_{g1} + R_3}$ ...(6.15)

Voltage drop in limb AB (or voltage at terminal B), $v_{AB} = I_1 R_{g1} = \dfrac{R_{g1}}{R_{g1} + R_3} \cdot v_i$ ...(6.16)

Similarly, $$I_2 = \frac{v_i}{R_2 + R_4} \quad \text{...(6.17)}$$

and, $$v_{AD} = \frac{R_2}{R_2 + R_4} \cdot v_i \quad \text{...(6.18)}$$

Initially, $$R_{g1} = R_2 = R_3 = R_4 = R$$

$\therefore$ $$v_{AB} = v_{AD} = \frac{v_i}{2} \quad \text{...(6.19)}$$

and, $$v_0 = \text{Voltage across the terminals } B \text{ and } D$$
$$= v_{AB} - v_{AD} = 0$$

Obviously, the bridge is balanced under *unstrained conditions.*

When the gauge is strained (see Fig. 6.8), the resistance R_{g1} changes by an amount dR_{g1}. Then,

$$v_{AB} = \left[\frac{R_{g1} + dR_{g1}}{(R_{g1} + dR_{g1}) + R_3}\right] v_i = \left(\frac{R + dR}{2R + dR}\right) v_i$$

[$\because R_{g1} = R_3 = R$ and $dR_{g1} = dR$]

$$v_{AD} = \left(\frac{R_2}{R_2 + R_4}\right) v_i = \frac{v_i}{2}$$

($\because R_2 = R_4 = R$)

The changed output voltage,

$$v_0 + dv_0 = \left(\frac{R + dR}{2R + dR} - \frac{1}{2}\right) v_i$$

$$= \left(\frac{2R + 2dR - 2R - dR}{2(2R + dR)}\right) v_i = \left(\frac{dR}{4R + 2dR}\right) v_i$$

Since $dR << R$ and $v_i = 0$ (under unstrained conditions), therefore:

$$dv_0 = \frac{v_i}{4} \cdot \frac{dR}{R} \quad ...(6.20)$$

or,

$$dv_0 = \left(\frac{G_f}{4}\right) \in v_i \quad ...(6.21)$$

... in terms of gauge factor G_f and applied strain $\in$

From eqn. (6.21) it is obvious that the *output voltage is directly proportional to the applied strain.*

Example 6.6: *A strain gauge ($G_f = 2.0$) has been applied a tensile strain of 0.00095. The value of the gauge resistance and each of the other three resistances comprising the limbs of a wheatstone bridge circuit is 100 Ω initial. Calculate the following:*

(i) Change in the value of variable resistance R_3 to rebalance the null-type bridge.

(ii) Output voltage of a voltage sensitive bridge if the voltmeter resistance is very high and the input voltage is 6 V.

Solution: *Given:* Gauge factor, $G_f = 2.0$, $\in = 0.00095$,

$$R_1 (= R_g) = R_2 = R_3 = R_4 = 100\ \Omega;\ v_i = 6\ \text{V}.$$

(*i*) Change in the value of variable resistance R_3; dR_3:

We know that,

$$\frac{dR_1}{R_1} = G_f \cdot \in$$

$\therefore$

$$dR_1 = G_f \cdot \in \cdot R_1 = 2.0 \times 0.00095 \times 100 = 0.19\ \Omega$$

Also,

$$\frac{dR_1}{dR_3} = \frac{R_2}{R_4}$$

$\therefore$

$$dR_3 = dR_1 \times \frac{R_4}{R_2}$$

or,

$$dR_3 = 0.19 \times \frac{100}{100} = \mathbf{0.19\ \Omega\ (Ans.)}$$

(*ii*) Output voltage, v_0:

The output voltage of a voltage sensitive bridge is given by the relation:

$$v_0 = \frac{v_i}{4}\left(\frac{dR_1}{R_1}\right) \quad \text{(from Eqn. 6.20)}$$

$$= \frac{6}{4} \times \frac{0.19}{100} = 2.85 \times 10^{-3}\ \text{V or } \mathbf{2.85\ mV\ (Ans.)}$$

(*iii*) Half-bridge:

Figure 6.9 shows two gauges used for strain measurement (Half-bridge). In this arrangement two of the bridge elements are strain gauges and the other two are fixed resistors. The strain gauge-1 is bonded to the upper surface of the cantilever beam and a second strain gauge-3 is bonded to the lower surface and located precisely underneath the gauge-1. These gauges are connected electrically to form adjacent limbs of the Wheatstone bridge circuit.

The temperature effects are cancelled out by having $R_2 = R_4$ and using *two identical gauges* in the opposite arms of the bridge.

Suppose, $R_{g1} = R_{g3} = R_2 = R_4 = R$

Under *no strain condition:*

$$v_{AB} = v_{AD} = \frac{v_i}{2};\ v_B = v_D \text{ and } v_0 = 0.$$

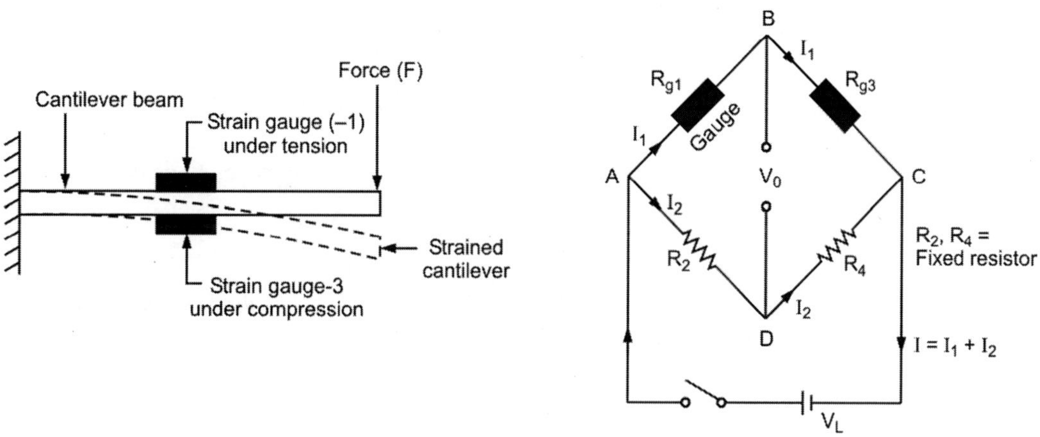

Fig. 6.9 Two gauges used for strain measurement (Half-bridge).

On the application of load to the cantilever beam, the resistance of the gauge R_{g1} *increases* due to *tensile load* whilst R_{g3} *decreases* due to equal *compressive strain* so that,

Resistance of gauge $R_{g1} = R_{g1} + dR_{g1}$

and, Resistance of gauge $R_{g3} = R_{g3} - dR_{g3}$

Now, $$v_{AB} = \frac{R_{g1}}{R_{g1} + R_{g3}} \cdot v_i$$

$$= \frac{R + dR}{(R + dR) + (R - dR)} v_i = \frac{R + dR}{2R} \cdot v_i \quad \text{...(6.22)}$$

and, $$v_{AD} = \frac{R_2}{R_2 + R_4} v_i = \frac{v_i}{2} \quad \text{...(6.23)} \quad (\because R_2 = R_4)$$

The changed output voltage,

$$v_0 + dv_0 = \frac{R + dR}{2R} \cdot v_i - \frac{v_i}{2}$$

$$= v_i\left(\frac{R + dR}{2R} - \frac{1}{2}\right) = v_i\left(\frac{2R + 2dR - 2R}{4R}\right)$$

or, $$v_0 + dv_0 = \frac{v_i}{2} \cdot \frac{dR}{R} \qquad ...(6.23)$$

Since under unstrained conditions $v_0 = 0$, therefore, change in output voltage due to applied strain becomes:

$$dv_0 = \frac{v_i}{2} \cdot \frac{dR}{R}$$

or $$dv_0 = \left(\frac{G_f}{2}\right) \in \cdot v_i \qquad ...(6.24)$$

which is *twice* the output of Wheatstone bridge using one gauge only.

The eqn. (6.24) can be rewritten as:

$$dv_0 = \frac{v_i}{4}\left[\frac{dR}{R} - \left(\frac{-dR}{R}\right)\right]$$

or, $$dv_0 = \frac{v_i}{4}\left[\begin{Bmatrix}\text{Fractional change in} \\ \text{resistance of gauge in limb } AB\end{Bmatrix} - \begin{Bmatrix}\text{Fractional change in} \\ \text{resistance of gauge in limb } BC\end{Bmatrix}\right] \qquad ...(6.25)$$

The –ve sign with fractional change in resistance of the gauge in limb BC is due to the fact that compressive and tensile strain are of *opposite signs.*

In general, for the two gauges connected in the adjacent limbs of a bridge circuit, we have:

$$dv_0 = \frac{v_i}{4}\left(\frac{dR_{g1}}{R} - \frac{dR_{g3}}{R}\right) \qquad ...(6.26)$$

Thus, when both the gauges are mounted on the top of the cantilever beam, the two effects cancel each other and the output voltage is zero.

(iv) Full-bridge:

Figure 6.10 shows four gauges used for strain measurement (Full-bridge). In this arrangement all the four elements of the bridge are strain gauges.

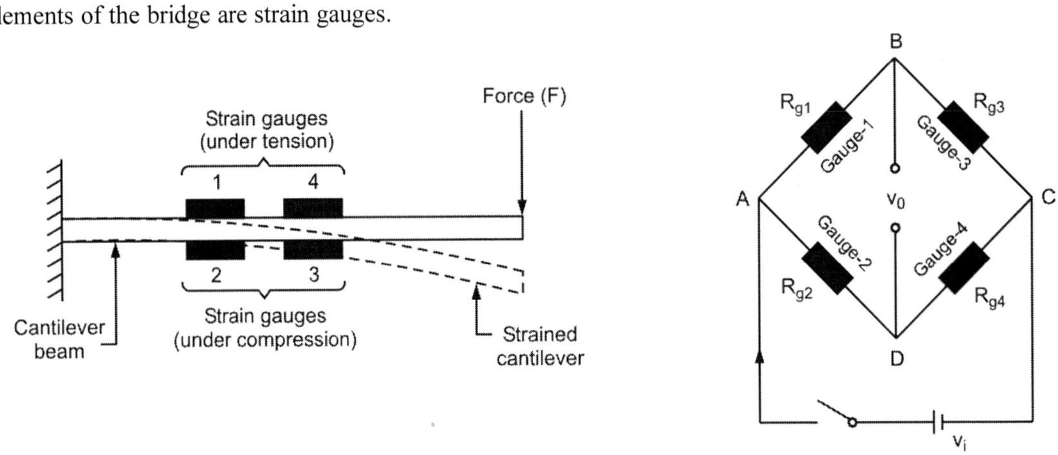

Fig. 6.10 Four gauges used for strain measurement (Full-bridge).

All the four gauges are *similar* and have *equal resistances when unstrained, i.e.,*

$$R_{g1} = R_{g2} = R_{g3} = R_{g4} = R.$$

Under no-strain condition:

$$v_{AB} = v_{AD} = \frac{v_i}{2};\ v_B = v_D \text{ and } v_0 = 0.$$

When the load is applied to the cantilever beam, the resistances R_{g1} and R_{g4} *increase* due to *tensile load* whilst-resistances R_{g2} and R_{g3} *decrease* due to equal *compressive strain*. When strained the resistances of the various gauges are:

$$R_{g1} = R_{g4} = R + dR \text{ (tension)}$$

$$R_{g2} = R_{g3} - R - dR \text{ (compression)}$$

and,
$$v_{AB} = \frac{R_{g1}}{R_{g1} + R_{g3}} \cdot v_i$$

or,
$$v_{AB} = \frac{R + dR}{(R + dR) + (R - dR)} v_i = \frac{R + dR}{2R} \cdot v_i \qquad ...(6.27)$$

and,
$$v_{AD} = \frac{R_{g2}}{R_{g2} + R_{g4}} v_i$$

or,
$$v_{AD} = \frac{R - dR}{(R - dR) + (R + dR)} \cdot v_i = \frac{R - dR}{2R} \cdot v_i \qquad ...(6.28)$$

The changed output voltage,

$$v_0 + dv_0 = \frac{R + dR}{2R} \cdot v_i - \frac{R - dR}{2R} \cdot v_i$$

$$= v_i \left(\frac{R + dR}{2R} - \frac{R - dR}{2R} \right) = v_i \frac{dR}{R}$$

Since the output voltage under unstrained conditions, $v_0 = 0$, therefore, change in output voltage due to applied strain becomes:

$$dv_0 = v_i \frac{dR}{R}$$

i.e.,
$$dv_0 = G_f \in \cdot v_i \qquad ...(6.29)$$

which is *the four times* the output of Wheatstone bridge using *one gauge only*.

- It may be noted that all the relations derived above are subject to the following conditions: (*i*) the values of the resistances of all the four limbs of the bridge are initially equal, and (*ii*) the galvanometer has infinite impedance and no current flows through it.

6.4 TEMPERATURE COMPENSATION

In order to measure static strains accurately, it is absolutely necessary that adequate temperature compensation may be provided in the Wheatstone bridge. The errors due to change of temperature arise due to the following two *reasons*:

(*i*) Resistance change of the wire in the strain gauge with change in temperature.

(*ii*) Differential expansion existing between the gauges and the metal to which they are bonded, if the temperature coefficients of the two are not same.

Temperature effects may be dealt with as follows :

(*i*) *By compensation or cancellation.* This method is extensively used for both metallic as well as semiconductor gauge.

(*ii*) *By evaluation as a part of the reduction problem.* This method is employed only for semiconductor gauges.

Temperature compensation or cancellation techniques:

Temperature compensation may be provided by:

- Adjacent arm balancing or compensating gauge.
- Self-temperature compensation.
- Use of special external control circuitry.

1. Adjacent arm compensating gauge:

(i) Use of dummy gauge. The arrangement is shown in Fig. 6.11. The active strain gauge is installed on the test specimen (cantilever beam in this case) while the dummy or compensating gauge is installed on a like piece of material and is *not subjected to strain.*

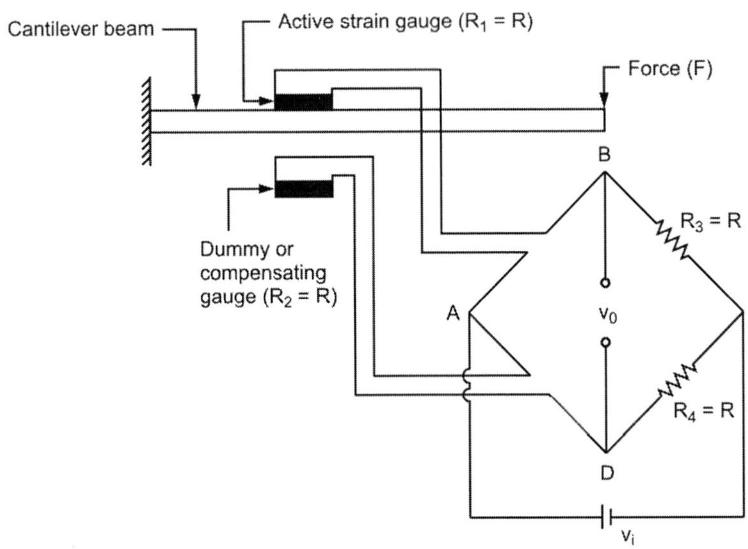

Fig. 6.11 Use of dummy gauge for temperature compensation.

The bridge is initially balanced and therefore $\frac{R_1}{R_2} = \frac{R_3}{R_4}$. Suppose a change in temperature occurs, and the resistances of active and dummy gauges change by dR_1 and dR_2 respectively. In order that there should be no output the bridge should be balanced and this requires:

$$\frac{R_1 + dR_1}{R_3} = \frac{R_2 + dR_2}{R_4}$$

or,

$$\frac{dR_1}{R_3} = \frac{dR_2}{R_4}$$

Let $R_3 = R_4 = R$ and, therefore, for bridge balance dR_1 should be equal dR_2. Now if active and dummy gauges are identical $R_1 = R_2 = R$ and if dummy gauge is placed in the same environments as the active strain gauge $dR_1 = dR_2$, therefore there will be no output due to changes in ambient temperature.

(ii) ***Use of two active gauges in adjacent arms.*** In some applications, where equal and opposite strains are known to exist, it is possible to attach similar gauges in such a way that one gauge experiences a positive strain and the other a negative strain (see Fig. 6.9).

- The arrangement shown in Fig. 6.9, *not only gives increased sensitivity but also eliminates the effects caused by temperature variations.*

(iii) ***Use of four active gauges.*** This arrangement is illustrated in Fig. 6.10.

- Four active-arm bridges are extensively used *when strain gauges are used as secondary transducers to give maximum sensitivity combined with full temperature compensation.* The effect of increasing the number of active gauges is the same as if a low impedance detector is used.

(iv) ***Poisson's method.*** In this method of temperature compensation two active gauges R_{g1} and R_{g3} bonded at right angles are used as shown in Fig. 6.12. Here gauge R_{g1}, is in tension but the gauge R_{g3} is reduced in length depending upon Poisson's ratio of the material.

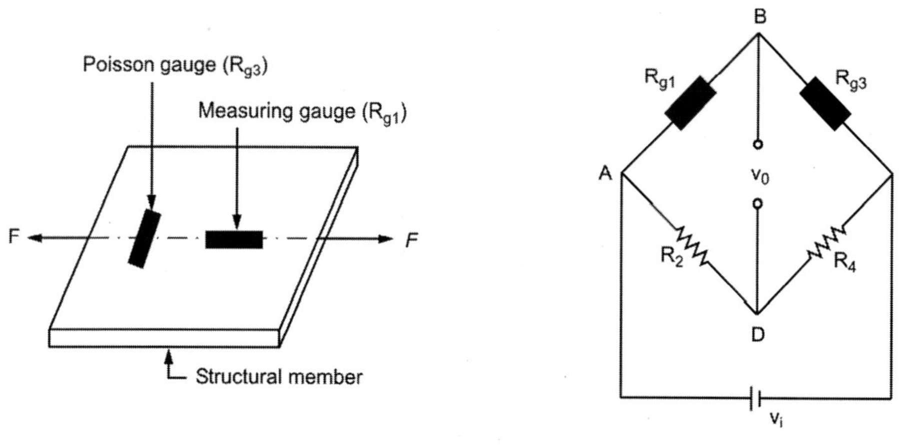

Fig. 6.12 Poisson's method for temperature compensation.

The *advantage* of this method is that *in addition to temperature compensation, the output of the bridge is increased by a factor* $(1 + \mu)$ *as compared to a bridge using a single active gauge.* The bridge sensitivity is increased by a factor $(1 + \mu)$ accordingly.

- This arrangement finds application *where it is not convenient or possible to mount the gauge in corresponding position on the underside of the test-member.*

2. Self temperature compensation:

In several applications it may be difficult or impossible to obtain temperature compensation by using adjacent arm compensating gauge acting either as a dummy gauge or a compensating gauge. Under these circumstances, it is desirable to use *self temperature compensation.*

Following two types of self-compensated gauges are available :

(i) Selected melt gauge:

- This method of self-compensation is based upon the fact that through proper manipulation of alloy and processing, usually through cold working, some control over the temperature sensitivity of the grid material may be obtained.
- Both wire and foil grid materials may be prepared which show very low apparent strain variation with temperature over a certain temperature range.

(ii) Dual element-gauge:

- In this case, two wire elements are connected in *series to make one strain gauge assembly*.
- *The two elements have different temperature characteristics and are so chosen that the net strain produced on account of temperature changes is minimised when the gauge is mounted on the test surface.*

6.5 STRAIN GAUGE ROSETTES

Rosette gauges are used for measuring strain in complex parts. These gauges have *three* or *four* separate grids with various angular orientations and they can be cemented to the part with no particular attention being paid to the overall gauge orientation. The resultant strain on each of the grids can then be recorded and the true magnitudes and directions of the significant surface strains in the part can be calculated from these data.

Depending on the arrangement of the grids, we have the following types of rosettes:

- Rectangular rosettes.
- Delta rosettes.
- T-delta rosettes.

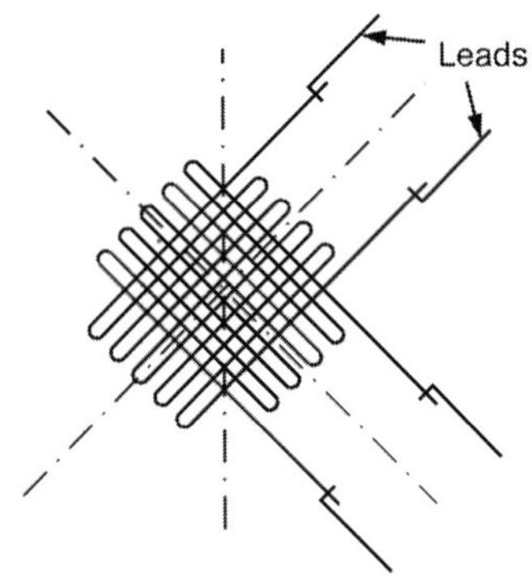

Grids do not make electrical contract.

Fig. 6.13 Two element rosette. Two grid elements are mounted in between two thin papers.

The formulae for calculation of principal strains and stresses for various arrangements are given below:

1. Two element rosette. Refer to Fig. 6.13.

- *Principal strains* : $\in_1$; $\in_2$
- *Principal stresses* :

$$\sigma_1 = \frac{E}{1-\mu^2}(\in_1 + \mu \in_2) \quad ...(6.30)$$

$$\sigma_2 = \frac{E}{1-\mu^2}(\in_2 + \mu \in_1) \quad ...(6.31)$$

- *Maximum shear stress:*

$$\tau_{max} = \frac{E}{2(1-\mu)}(\in_1 - \in_2) \quad ...(6.32)$$

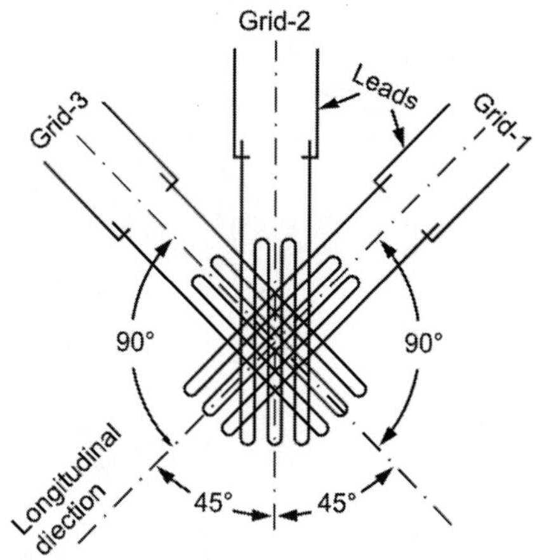

Fig. 6.14 Rectangular rosette. Grids do not make electrical contact. Grids are kept in-between two thin papers.

2. Three-element rectangular rosette. Refer to Fig. 6.14.

- *Principal strains:*

$$\epsilon_{max}, \epsilon_{min} = \frac{\epsilon_1 + \epsilon_3}{2} \pm \frac{1}{\sqrt{2}}[(\epsilon_1 - \epsilon_2)^2 + (\epsilon_2 - \epsilon_3)^2]^{1/2} \quad ...(6.33)$$

- *Principal stresses:*

$$\sigma_{max}, \sigma_{min} = \frac{E(\epsilon_1 + \epsilon_3)}{2(1-\mu)} \pm \frac{E}{\sqrt{2}(1+\mu)}[(\epsilon_1 - \epsilon_2)^2 + (\epsilon_2 - \epsilon_3)^2]^{1/2} \quad ...(6.34)$$

- *Maximum shear stress:*

$$\tau_{max} = \frac{E}{\sqrt{2}\,(1+\mu)}[(\epsilon_1 - \epsilon_2)^2 + (\epsilon_2 - \epsilon_3)^2]^{1/2} \quad ...(6.35)$$

The orientation of principal stress is given by:

$$\tan 2\theta = \frac{2\,\epsilon_2 - \epsilon_1 - \epsilon_3}{\epsilon_1 - \epsilon_3} \quad ...(6.36)$$

This is the axis at which τ_{max} occurs.

3. Three-element delta rosette. Refer to Fig. 6.15.

- *Principal strains:*

$$\epsilon_{max}, \epsilon_{min} = \frac{\epsilon_1 + \epsilon_2 + \epsilon_3}{3} \pm \frac{\sqrt{2}}{3}[(\epsilon_1 - \epsilon_2)^2 + (\epsilon_2 - \epsilon_3)^2 + (\epsilon_3 - \epsilon_1)^2]^{1/2} \quad ...(6.37)$$

- *Principal stresses:*

$$\sigma_{max}, \sigma_{min} = \frac{E(\epsilon_1 + \epsilon_2 + \epsilon_3)}{3(1-\mu)} \pm \frac{\sqrt{2}E}{3(1+\mu)}[(\epsilon_1 - \epsilon_2)^2 + (\epsilon_2 - \epsilon_3)^2 + (\epsilon_3 - \epsilon_1)^2]^{1/2} \quad ...(6.38)$$

- *Maximum shear stress:*

$$\tau_{max} = \frac{\sqrt{2}E}{3(1+\mu)}[(\epsilon_1 - \epsilon_2)^2 + (\epsilon_2 - \epsilon_3)^2 + (\epsilon_3 - \epsilon_1)^2]^{1/2} \quad ...(6.39)$$

The orientation of principal stress is given by

$$\tan 2\theta = \frac{\sqrt{3}(\epsilon_3 - \epsilon_2)}{2\epsilon_1 - \epsilon_2 - \epsilon_3} \quad ...(6.40)$$

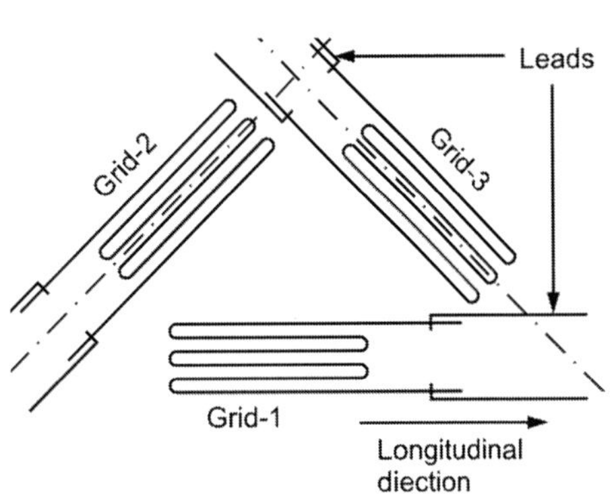

Fig. 6.15 Delta rosette.

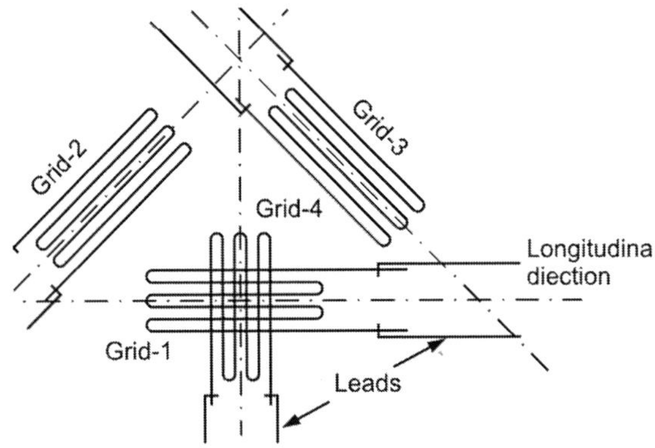

Fig. 6.16 T-delta rosette. Grids do not make electrical contacts.

4. **T-delta rosette.** Refer to Fig. 6.16.
   - *Principal strains:*

$$\epsilon_{max}, \epsilon_{min} = \frac{1}{2}\left[\epsilon_1 + \epsilon_4 \pm (\epsilon_1 - \epsilon_4)^2 + \frac{4}{3}(\epsilon_2 - \epsilon_3)^2\right]^{1/2} \quad ...(6.41)$$

   - *Principal stresses:*

$$\sigma_{max}, \sigma_{min} = \frac{E}{2}\left[\frac{\epsilon_1 + \epsilon_4}{1-\mu} \pm \frac{1}{1+\mu}\sqrt{(\epsilon_1 - \epsilon_4)^2 + \frac{4}{3}(\epsilon_2 - \epsilon_3)^2}\right] \quad ...(6.42)$$

   - *Maximum shear stress:*

$$\tau_{max} = \frac{E}{2(1+\mu)}\sqrt{(\epsilon_1 - \epsilon_4)^2 + \frac{4}{3}(\epsilon_2 - \epsilon_3)^2} \quad ...(6.43)$$

The principal axis is located θ according to,

$$\tan 2\theta = \frac{2}{\sqrt{3}}\left(\frac{\epsilon_3 - \epsilon_2}{\epsilon_1 - \epsilon_4}\right) \quad ...(6.44)$$

The *T*-delta rosette has the following *advantages* over others due to the fourth gauge element (three strain measurements are essential to completely define any stress situation):

(*i*) The fourth gauge gives a *check on the results* obtained from other three elements.

(*ii*) Due to four readings, averaging effect results in *better accuracy*.

Example 6.7: *In a rectangular rosette, the maximum and minimum strains are given by:*

$$\epsilon_{max}, \epsilon_{min} = \frac{\epsilon_1 + \epsilon_3}{2} \pm \frac{1}{\sqrt{2}}[(\epsilon_1 - \epsilon_2)^2 + (\epsilon_2 - \epsilon_3)^2]^{1/2}$$

and then are located at an angle, $\tan 2\theta = \dfrac{2\epsilon_2 - \epsilon_1 - \epsilon_3}{\epsilon_1 - \epsilon_3}$

Show that: $\epsilon_{max} = \epsilon_1$, *and,* $\epsilon_{min} = \epsilon_3$ *when* $\theta = 0$.

Solution: When $\theta = 0$, $\tan 2\theta = 0$ and hence:

$$\frac{2\epsilon_2 - \epsilon_1 - \epsilon_3}{\epsilon_1 - \epsilon_3} = 0, \quad \text{or,} \quad \epsilon_2 = \frac{\epsilon_1 + \epsilon_3}{2}$$

$$\therefore \qquad \epsilon_{max} = \frac{\epsilon_1 + \epsilon_3}{2} + \frac{1}{\sqrt{2}}[(\epsilon_1 - \epsilon_2)^2 + (\epsilon_2 - \epsilon_3)^2]^{1/2}$$

Now,

$$\epsilon_2 = \frac{\epsilon_1 + \epsilon_3}{2}$$

or,

$$2\epsilon_2 = \epsilon_1 + \epsilon_3, \quad \text{or,} \quad \epsilon_1 - \epsilon_2 = \epsilon_2 - \epsilon_3$$

$$\therefore \qquad \epsilon_{max} = \frac{\epsilon_1 + \epsilon_3}{2} + \frac{1}{\sqrt{2}}[(\epsilon_1 - \epsilon_2)^2 + (\epsilon_1 - \epsilon_2)^2]^{1/2}$$

or,

$$\epsilon_{max} = \epsilon_2 + \frac{\sqrt{2}}{\sqrt{2}}(\epsilon_1 - \epsilon_2) = \epsilon_1 \quad \textbf{Proved.}$$

Similarly, we can get, $\epsilon_{min} = \epsilon_3$.

6.6 STRAIN-GAUGE CALIBRATION

In most applications, a theoretical objection to the use of strain gauges is that it is impossible to check the accuracy of the readings obtained from them. Once the strain gauge is bonded in its measuring position, it cannot be removed, transferred or subjected to a known strain or calibration.

The calibration, in most of the cases, is done by introducing an accurately known sample of the variable and then observing response of the system. But this practice cannot be extended to bonded resistance strain-gauge because of the nature of transducer. In the case of strain gauges, the manufacturer has to maintain strict quality control and ensure that all the strain gauges produced by him are identical. Generally the gauge factors are usually quoted with a tolerance limit of about ± 0.2 percent or less. Thus if gauge factor and gauge resistance are known, any resistance strain-gauge system can be easily calibrated.

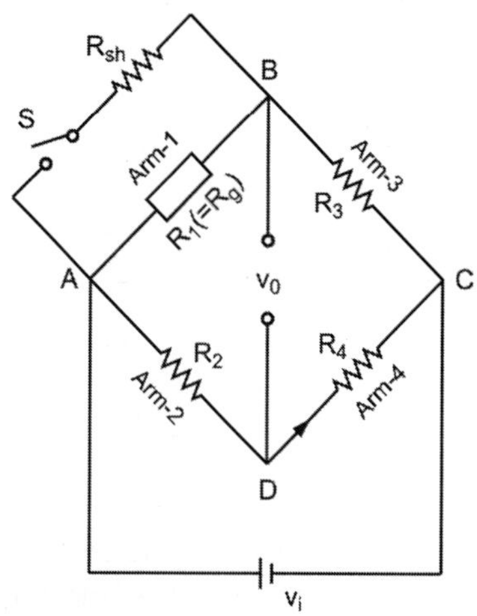

Fig. 6.17 Calibration of strain gauges.

The method of strain gauge calibration consists of determining the response of the system to the introduction of a known small resistance change at the gauge and of calculating an equivalent strain therefrom. The small resistance change is introduced by shunting a relatively high value precision resistance R_{sh} across the gauge (gauge resistance R_g) and to introduce this resistance R_{sh} in parallel to R_g, a switch *S* is used, as shown in Fig. 6.17.

Resistance of arm-1 of the bridge *before* switch S is closed = R_g

Resistance of arm-1 of the bridge *after* switch is closed

= Resistance of parallel combination of R_g and R_{sh}

$$= \frac{R_g \cdot R_{sh}}{R_g + R_{sh}}$$

∴ The change in resistance of arm-1 resulting from introduction of R_{sh},

$$dR = R_g - \frac{R_g \cdot R_{sh}}{R_g + R_{sh}} = \frac{R_g^2}{R_g + R_{sh}} \qquad ...(6.45)$$

Also, strain $\in = \frac{1}{G_f} \cdot \frac{dR}{R}$, where G_f is the gauge factor.

Substituting dR_g for dR in eqn. 6.46, we get the equivalent strain as,

$$\in_e = \frac{1}{G_f} \cdot \frac{dR_g}{R_g} = \frac{1}{G_f} \cdot \frac{1}{R_g} \cdot \frac{R_g^2}{(R_g + R_{sh})} = \frac{1}{G_f}\left(\frac{R_g}{(R_g + R_{sh})}\right) \qquad ...(6.46)$$

∴ Shunt resistance, $$R_{sh} = \frac{R_g(1 - G_f \cdot \in_e)}{G_f \cdot \in_e} \qquad ...(6.47)$$

In practice $G_f \cdot \in_e << 1$ and therefore the term $G_f \cdot \in_e$ in the numerator of eqn. 6.47 can be *neglected without-introducing any significant error.*

Then, $$R_{sh} = \frac{R_g}{G_f \in_e} \qquad ...(6.48)$$

The effective strain with *n-active strain gauges* in the bridge is given as:

$$\in_e = \frac{R_g}{nG_f(R_g + R_{sh})} \qquad ...(6.49)$$

The value of shunt resistance is so chosen that the read out system gives a full scale deflection.

- Following are *some other methods of electrical calibration:*
  - (*i*) The strain gain bridge is replaced with a substitute load, initially adjusted to equal the bridge load. A series resistance is then used for calibration.
  - (*ii*) An accurately known voltage is injected into the bridge circuit.

HIGHLIGHTS

1. *Strain* is defined as the change in length of a line segment between two points divided by the original length of the line segment.
2. A *strain gauge* is a device which is used to measure dimensional change on the surface of a structural member under test.
3. Temperature compensation may be provided by
   - (*i*) Adjacent arm balancing or compensating gauge.
   - (*ii*) Self-temperature compensation.
   - (*iii*) Use of special external control circuitry.

THEORETICAL QUESTIONS

1. Define the terms "Strain" and "Strain gauge".
2. What are the techniques of strain measurements ? Explain briefly.
3. What are the requirements of a strain gauge ?
4. Explain briefly the mechanical strain gauge.
5. What are optical strain gauges ? State their advantages and uses.
6. How are electrical strain gauges classified ?
7. State the characteristics which an ideal strain gauge wire should possess.
8. State the factors on which gauge factor depends.
9. Give the classification of resistance strain gauges.
10. Explain briefly "unbonded strain gauges".
11. Describe briefly the following bonded strain gauges:
    (*i*) Wire type strain gauges (*ii*) Metal foil gauges.
12. Explan briefly the semiconductor gauges, stating clearly their characteristics and uses.
13. Give the advantages and disadvantages of unbonded strain gauges.
14. What is temperature compensation and how is it achieved when using bonded strain gauge for the measurements of axial thrust, bending loads and torque?
15. Explain the function of a dummy gauge in a strain gauge load cell.
16. Explain briefly various temperature compensation or cancellation techniques.
17. How are strain gauges calibrated?
18. What are "Rosette gauges"? What are the various types of "Rosettes"?

❑❑❑

Measurement of Force, Torque, Acceleration and Vibrations

Chapter 7

7.1 MEASUREMENT OF FORCE

Force *is defined as any cause that produces, stops or changes the motion of a body or tends to produce these effects.*

An unknown force may be measured by the methods incorporating the following principles:

1. *Balancing the force against a known gravitational force on a standard mass.* **...Scales and balances**
2. *Translating the force to a fluid pressure and then measuring the resulting pressure.* **...Hydraulic and pneumatic load cells**
3. *Applying the force to some elastic member and then measuring the resulting deflection.* **...Proving ring**
4. *Applying the force to a known mass and then measuring the resulting acceleration.*
5. *Balancing the force against a magnetic force developed by interaction of a magnet and a current carrying coil.*

7.1.1 Scales and Balances

- A weighing machine, commercially known as *scales* is used for the measurement of force and torque by comparison of weights. It is *based on the principle of equilibrium, produced by two torques as a result of forces acting at equal or different distances from the fulcrum.* The *"even-arm balance"* and *"even-arm scale"* are *seldom used for force measurement because of the time consumed in the measuring operation.*
- The ***beam scale*** *is used for spring-testing devices because its accuracy permits testing a wide range of springs with high accuracy.*
- The use *of **pendulum scales** requires that the force should be steady for 2 to 3 seconds for the scale to come to rest.*

7.1.2 Springs

- The springs are frequently used for the measurement of force.
- The spring scales are available in capacities from a few grams to 20000 kg or more.

- Spring alloys are used for spring materials and *refinements in construction are effected in order to attain high accuracy* over the ambient temperature range from approximately 0°C to 50°C.
- The ordinary springs are capable for giving results accurately within the accuracy of 0.5 to 1.0%.

7.1.3 Proving Rings

The *proving rings* are **steel** *rings which are used as force standards.* They are particularly useful for *calibration of material testing machines in situations where dead weight standards are impracticable to use on account of their physical bulk.*

- The proving rings are most commonly used in *industries* for measurement of *very heavy forces.*
- The proving (stress) ring is a ring of *known physical dimensions and mechanical properties.* On the application of an external compressive or tensile load to the lugs or external bosses,the ring changes in diameter; the *change being proportional to the applied force.* The *deflection* of the ring is measured by means of a micrometer screw and a vibrating reed which are attached to the internal bosses. During use the tip of the micrometer is advanced and its *contact* with the reed is indicated by *considerable damping of the reed vibration.* The difference in readings of the micrometer before and after the application of load gives the amount of compression or the elongation of the ring.
- The deflection of the ring can also be measured by using LVDT which senses the movement of a core which is attached to the ring and moves due to deflection of the ring.
  - The maximum deflection is typically of the order of 1 percent of the outside diameter of the ring.
  - Proving rings are normally used for force within the range of 2 kN to 2MN.
  - The rings are very stable and accurate to ± 0.1 percent but *because deflection is not linear the actual force is determined by reference to a calibration curve.*

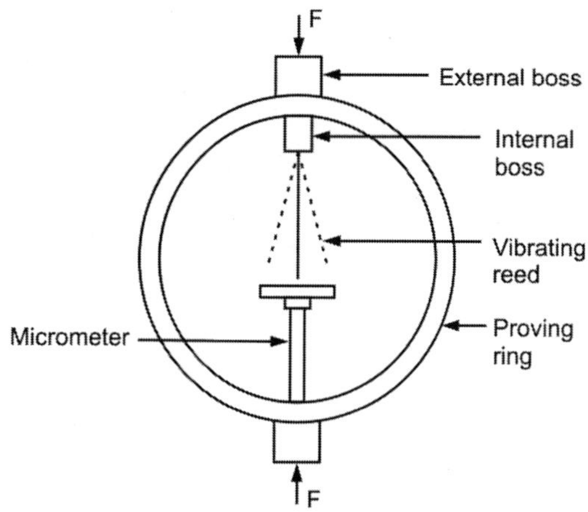

Fig. 7.1 Proving ring.

7.1.4 Load Cells

Load cells *are elastic devices that can be used for measurement of force through indirect methods i.e., through use of secondary transducers.*

Load cells utilize an elastic member as the primary transducer and strain gauges as secondary transducer. When the combination of the *strain gauge-elastic member* is used for weighing, it is called a *"load cell".*

While designing load cells using strain gauges the following *factors* should be considered:

(*i*) Stiffness of the elastic element.
(*ii*) Optimum positioning of gauges on the element.
(*iii*) Provision for compensation of the temperature.

When large loads are to be measured, the direct tensile-compressive member may be used, whereas, in case of small loads, strain amplification provided by bending may be used with advantage.

7.1.4.1 Hydraulic Load Cell

Figure 7.2 shows a *hydraulic load cell.*

Here the force variable is impressed upon a diaphragm which deflects and thereby transmits the force to a liquid. The liquid medium contained in a confined space, has a preload pressure of 2 bar. On the application of the force the liquid pressure increases and *equals the force magnitude divided by the effective area of the diaphragm*. The pressure is transmitted to and read on an accurate pressure gauge calibrated directly on force units.

- These cells have been used to measure loads upto about 25 MN (with an *accuracy of 0.1% of full scale); resolution is about 0.02 per cent.*

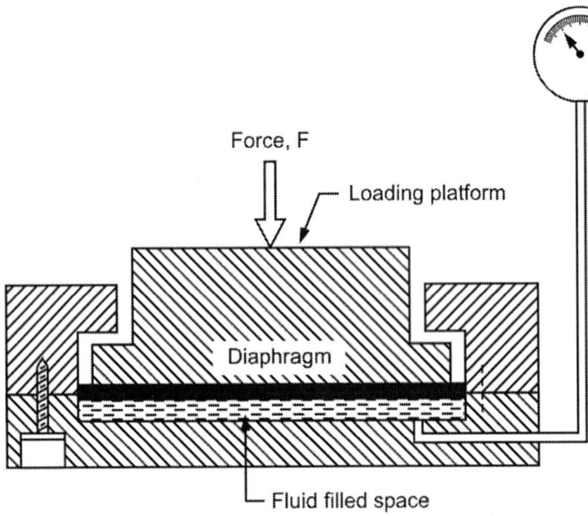

Fig. 7.2 Hydraulic load cell.

7.1.4.2 Pneumatic Load Cell

This cell operates on *the force-balance principle.* It employs a nozzle-flapper transducer similar to the conventional relay system. For any constant applied force, the system attains equilibrium at a specific nozzle opening and *corresponding pressure is indicated by the height of mercury column in a manometer.*

- The commercially available load cells (operating on this principle) can measure loads upto 25 kN with an accuracy of 0.5% of full scale.

7.1.4.3 Strain Gauge Load Cells

These cells convert weight or force into electrical outputs which are provided by the strain gauges; these outputs can be connected to various measuring instruments for indicating, recording and controlling the weight or force.

Usually the *strain gauges are directly applied to the force-developing device and the device is calibrated against strain-gauge output.*

- These are excellent force-measuring devices, particularly for *transient and non-steady forces.*
- These are used in conjunction with CRO (for display purposes) for measurement of *rapidly changing loads.*

7.1.4.4 Piezo-electric Load Cells

The principle of working of a piezo-electric load cell is that when *pressure is applied* across certain crystals (quartz etc.) then an *electrical charge proportional to the pressure is developed across certain well-defined faces of the crystal.*

- These are used for measuring forces, particularly of a *transient or rapidly changing variety.*
- Suitable for measuring forces from 5 N to 20 kN.

7.2 MEASUREMENT OF TORQUE

7.2.1 Introduction

To transmit energy by rotation it necessary to apply a turning force. In case of a shaft if the force is applied tangentially and in the plane of transverse cross-section the *torque or twisting moment* may be calculated by multiplying the force with the radius of the shaft. If the shaft is subjected to two opposite turning moments it is said to be in *pure torsion* and it will exhibit the tendency of shearing off at every cross-section which is perpendicular to the longitudinal axis.

The measurement of torque is carried out because of the *reasons:*

(*i*) It is of considerable interest for its own sake;
(*ii*) It is required to obtain load information necessary for stress and deflection analysis;
(*iii*) It is often associated with determination of mechanical power.

7.2.2 Torque Measurement Methods

Torque may be measured by the following *methods:*

1. Gravity balance method.
2. Mechanical torsion meter.
3. Optical torsion meter.
4. Electrical torsion meter.
5. Strain-gauge torsion meter.

1. Gravity balance method

This method is illustrates in Fig. 7.3. A mass m is moved along an arm until the value of the torque exerted by the mass balances the unknown torque.

$$\therefore \quad \text{Unknown torque, } T = F \times r \qquad ...(7.1)$$

where, $F = mg$ = force exerted by the mass

This method utilizes the movement of a constant mass, m, over a variable distance. Alternatively, magnitude of the mass may be varied, keeping the radius (r) constant.

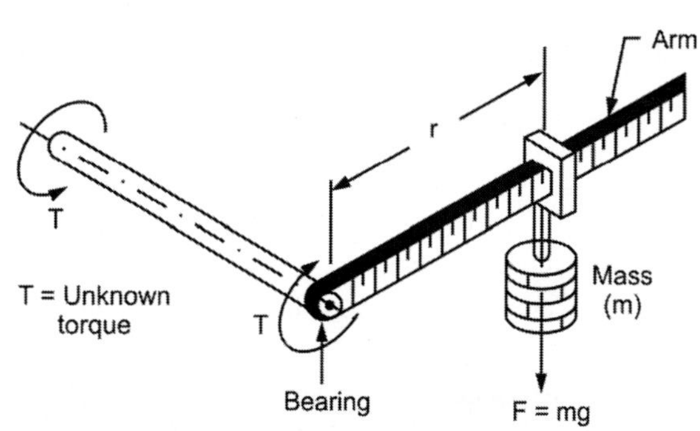

Fig. 7.3 Gravity balance method.

In both of the above cases, the *arm must be kept horizontal* so that arm distance is perpendicular to the line of action of force. Since the shaft is supported at the bearing, there may be a friction torque (due to the force acting on the bearing) leading to error in the measurement of torque. This *error may be eliminated by arranging to apply equal and opposite force.*

2. Mechanical torsion meter:

The following relationship holds good (within elastic limits) for the angular deflection of a shaft subject to torque:

$$\frac{T}{I_p} = \frac{\tau}{r} = \frac{C\theta}{l} \qquad ...(7.2)$$

where, T = Torque transmitted by the shaft,
I_p = Polar moment of inertia of the shaft section,
τ = Maximum shear stress induced at the outside surface,
r = Radius at which maximum shear stress occurs,
C = Modulus of rigidity of the shaft material,
θ = Angular twist, and
l = Length of the shaft over which the twist is measured.

The above relation gives:

$$T = \left(\frac{I_p}{r}\right) \times \tau, \text{ or, } T = \text{constant} \times \tau \qquad ...(7.3)$$

and

$$T = \left(\frac{I_p C}{l}\right) \times \theta, \text{ or, } T = \text{constant} \times \theta \qquad ...(7.4)$$

Thus, torque (T) can be calculated by measuring either the maximum shear stress (τ) or the angle of twist (θ).

In a *mechanical torsion meter* (Fig. 7.4) the angular deflection of a parallel length of shaft is used to measure torque. The angular twist over a fixed length of the bar is observed on a calibrated disc (attached to the rotating shaft) by using the stroboscopic effect of intermittent viewing and persistence of vision.

As the torque changes, the system gives a varying angle of twist between the driving engine and the driven load.

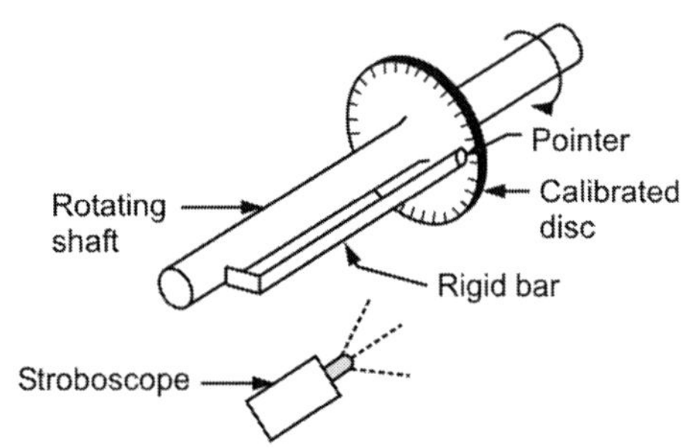

Fig. 7.4 Mechanical torsion meter.

3. Optical torsion meter:

In this meter the angular twist of a rotating shaft is detected by an *optical method.*

4. Electrical torsion meter:

Figure 7.5 shows such a system using two photoelectric or magnetic transducers, involving two sets of measurements:

(*i*) A count of the impulse from either slotted wheel; this count gives the *frequency or shaft speed.*

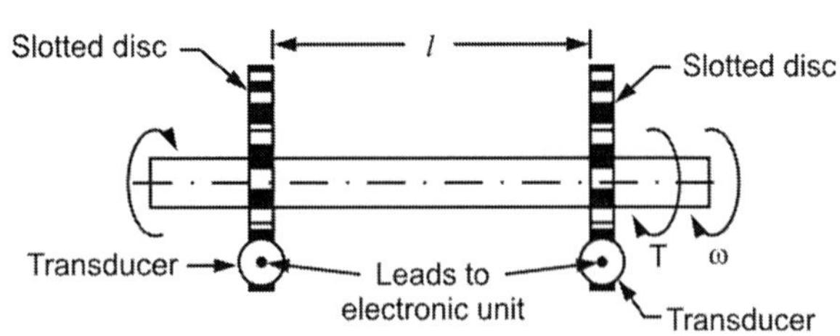

Fig. 7.5 Electrical torsion meter.

(*ii*) A measure of the time between pulses from the two wheels, this signal is proportional to the twist θ and hence *T*, in the shaft.

By combining the above two signals, *T* and ω, the power being transmitted by the shaft can be calculated.

5. *Strain-gauge torsion meter:*

Figure 7.6 shows a general configuration of a strain gauge bridge circuit widely employed for *torque measurement from a rotating shaft.*

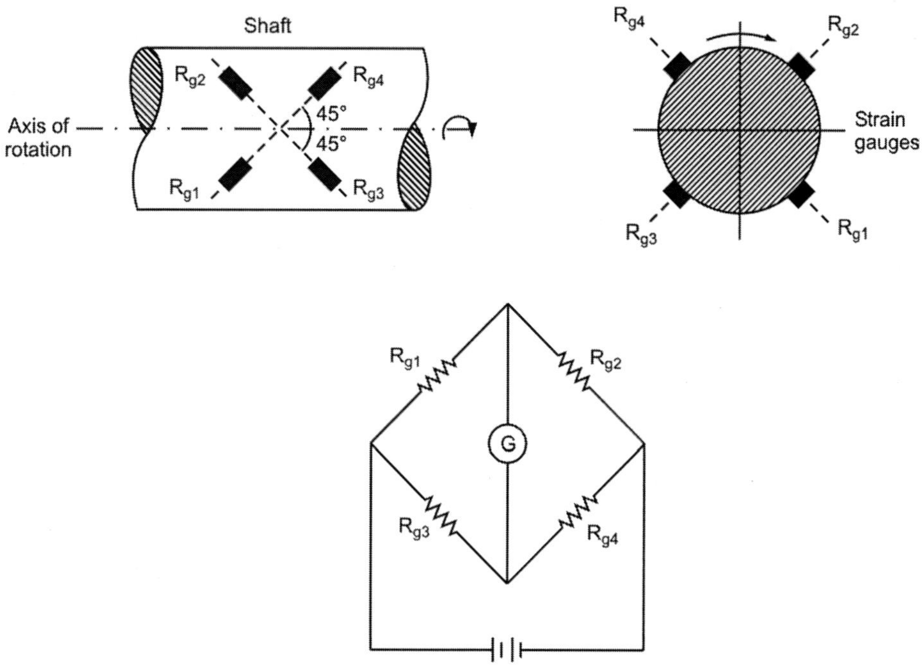

Fig. 7.6 Strain gauge torsion-meter.

Construction. Four bonded-wire strain gauges are mounted on a 45° helix with the axis of rotation; and are placed in *pairs diametrically opposite.* When the gauges are accurately placed and have matched characteristics, the system is temperature compensated and insensitive to bending and thrust or pull effects. Any change in the gauge circuit then results only from torsional deflection.

Working. When the shaft is under torsion, gauges 1 and 4 will *elongate* as a result of the tensile component of a pure shear stress on one diagonal axis, while 2 and 3 will *contract* due to compressive component on the other diagonal axis. These tensile and compressive principal strains can be measured and hence shaft torque calculated.

Advantages:

This arrangement has the following *advantages:*

1. Fully temperature compensated.
2. Gives the maximum sensitivity for a given torque.
3. Provides automatic compensation for bending and axial loads.

Limitation:

The main limitation/difficulty associated with the use of this arrangement is the connection of the bridge to its power source and display arrangement. *Slip rings* are used for this purpose. These are conducting rings attached to the shaft, but insulated from it, with one of the slip rings connected to each of the bridge terminals. Slip rings are mercury filled and transmit the signal to a stationary member where it is amplified and displayed or recorded.

- *Good results from the strain gauge method are available only when the shaft experiences sufficient strain to produce a measurable output signal.* In case the torque is small the gauge bridge will not give adequate output signal for measurement. In such cases a *flexible coupling* is introduced in the shaft. The coupling incorporates one or more elastic members (elastic member may be a different shaft or a commercial torque meter) to which the gauge may be attached. The elastic members are so designed that they *produce sufficiently large deflection even under light load conditions.* This large deflection produces large strains resulting in large output of the strain gauge bridge.

7.3 MEASUREMENT OF ACCELERATION

The acceleration of a moving body is generally measured by means of sensors called accelerometers. For measurement of acceleration, generally following types of accelerometers are used:

1. Piezoelectric type
2. Seismic type.

7.3.1 Piezo-electric Accelerometer

Refer to Fig. 7.7. A piezo-electric accelerator is probably the *simplest* and *most commonly used transducer for measuring acceleration.*

Construction. It consists of a *piezoelectric crystal* sandwitched between two *electrodes* and has *mass* placed on it. The unit is fastened to the base whose acceleration characteristics are to be obtained. The *can* threaded to the base acts as a spring and squeezes the mass against the crystal. *Mass exerts a force on the crystal and a certain voltage output is generated.*

Working. When the base is accelerated *downward* inertial reaction force on the base acts upward against the top of the can. This relieves stress on the crystal. According to Newton's second law of motion, *force = mass × acceleration,* since the mass is a fixed quantity, the *decrease in force is proportional to the acceleration.* Similarly, an acceleration in the upward direction would increase the force on the crystal in proportion to the acceleration. The resulting change in the output voltage is recorded and correlated to the acceleration imposed on the base.

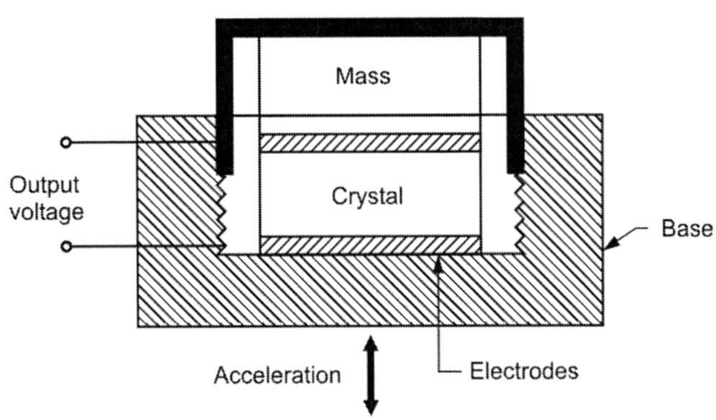

Fig. 7.7 Piezo-electric accelerometer.

Advantages:

1. Small size and a small weight.
2. High output impedance.
3. Can measure acceleration from a fraction of g to thousands of g.
4. High sensitivity.
5. High frequency response (10 Hz to 50 kHz).

Disadvantages:

1. Unsuitable for applications where the input frequency is lower than 10 Hz.
2. Subject to hysteresis errors.
3. Sensitive to temperature changes.

7.3.2 Seismic Accelerometer

In a *seismic* (displacement sensing) *accelerometer,* the displacement of a mass resulting from the applied force is measured and correlated to the acceleration.

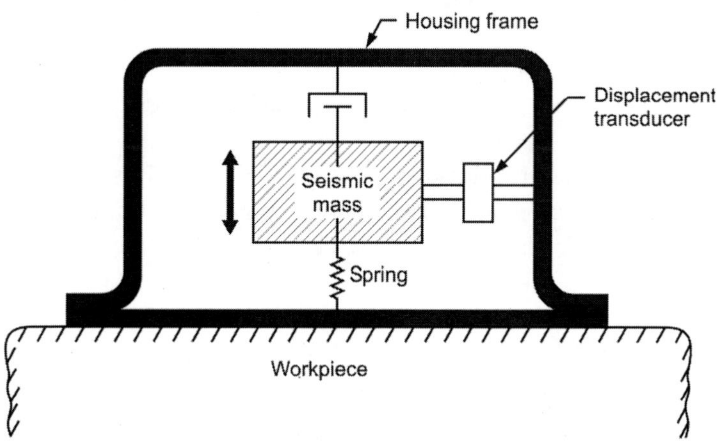

Fig. 7.8 Schematic diagram of a seismic transducer.

A schematic diagram of this accelerator is shown in Fig. 7.8. The mass is connected through, the parallel spring and damper arrangement to the housing frame. The housing frame is connected to the source of vibrations whose characteristics are to be measured. *The mass has tendency to remain fixed to its spatial position so that the vibrational motion is registered as a relative displacement between mass and housing frame. This displacement is sensed and indicated by an appropriate transducer.*

- This accelerometer may also be used as a *vibration sensor* (vibration refers to the repeated cyclic oscillations of a system).

7.4 MEASUREMENT OF VIBRATIONS

7.4.1 Introduction

If the displacement time variation is generally in continuous form with some degree of repetitive nature, it is called a vibration. When the action is of single event form, a transient, with the motion generally decaying or damping out before further action takes place, then this may be referred to a *shock.*

Measurements of vibrations may be made with a *transducer sensitive to amplitude* (*displacement*). *velocity or acceleration.* The output of this transducer after necessary signal conditioning is fed to the display which indicates the original measured variable or another variable derived from it either through *integration* or *differentiation.* Since the *output of the transducer is pulsating the frequency may be measured by a digital frequency meter at a suitable stage of signal conditioning process.*

7.4.2 Causes of Vibrations in Machines, Their Harmful Effects and Remedies

Causes of vibrations:

Some of the important *causes of vibrations* in machines are listed below:

1. Unbalanced reaprocating machine parts.
2. Unbalanced rotating machine parts.
3. Incorrect alignment of the transmission elements such as coupling etc.
4. Use of simple spur gears for power transmission.
5. Worn out teeth of the gears for power transmission.
6. Impact taking parts of the machines or explosion or impact of working fluids in prime movers.
7. Loose transmission belts and chains.
8. Loose fastenings of the moving parts.
9. Vibration waves from other sources and machines installed nearby, due to improper isolation of vibrations from them.
10. Due to more material contact such as bases plates on the foundations for the pedestal bearings.
11. Non-rigid machine foundations due to lack of compact soil below, causing settlement of foundation which may, in turn cause misalignment of machine components.

Harmful effects of vibrations:

Some of the *harmful effects of vibrations* in the machines are as follows:

1. Excessive wear of machine components resulting in reduction in their useful life.
2. Machine parts are subjected to excessive stresses and strains.
3. Machine components may fail due to fatigue failure.
4. Reduction in useful life of buildings and structures around the machines.
5. Workers and other human beings around the machines are subjected to discomfort in the form of physical and mental strains.
6. In case of machine tools and other production machinery poor finish on the work is produced.

Remedies:

Although it is impossible to eliminate the vibrations, yet these can be *reduced* by adopting various remedies, some of the *remedies* are listed below:

1. Partial balancing of reciprocating masses.
2. Balancing of unbalanced rotating masses.
3. Using helical gears instead of spur gases.
4. Proper tightening and locking of fastenings and periodically ensuring it again.
5. Correcting the mis-alignment of rotating components and checking it from time to time.

6. Timely replacement of worn out moving parts, slides and bearings with excessive clearance.
7. Isolating vibrations from other machines and sources by providing vibration insulation pads in the machine foundations.
8. Making machine foundations on compact ground and making them sufficiently strong, so that they do not yield or settle under the load of the machine.

7.4.3 Methods of Vibration Measurements

The methods of vibration measurement may be *classified* as follows:

1. *Methods based on sensory perception*—Vibrating wedge.
2. Mechanical methods—Dial type of micrometer.
3. Electrical methods.
4. Optical Methods.

The following are the ***instruments*** which are used for the *measurement of vibrations*:

(*i*) Absolute vibration sensor.
(*ii*) Electro-mechanical relative vibration pick-up for rectilinear vibrations.
(*iii*) Relative displacement pick-up with differential transformer.
(*iv*) Electromagnetic relative vibration pick-up.
(*v*) Strain gauges:

- Frequency meter
- Phase indicator
- Vibrometers.

HIGHLIGHTS

1. An *unknown force* may be measured by the following:
   (*i*) Scales and balances; (*ii*) Springs;
   (*iii*) Proving rings; (*iv*) Load cells.
2. *Load cells* are elastic devices that can be used for measurement of force through indirect methods, *i.e.*, through use of secondary transducers.
3. *Torque* may be measured by the following methods:
   (*i*) Gravity balance method (*ii*) Mechanical torsion meter
   (*iii*) Optical torsion meter (*iv*) Electrical torsion meter,
   (*v*) Strain-gauge torsion meter.
4. The following methods may be used to measure *vibrations*:
   (*i*) Vibrating wedge; (*ii*) Dial type micrometer;
   (*iii*) Electrical method; (*iv*) Optical methods.

THEORETICAL QUESTIONS

1. How is an unknown force measured ?
2. What is a proving ring ? How is it used to measure force ?

3. What is a load cell ?
4. Explain briefly the following load cells :
   (*i*) Hydraulic load cell
   (*ii*) Pneumatic load cell.
5. Describe briefly the construction and working of strain gauge load cell. State its fields of application as well.
6. What is the principle of working of a piezo-electric load cell ?
7. What is 'torque' ? How can it be measured ?
8. Explain briefly any two of the following torque measurement methods:
   (*i*) Gravity balance method
   (*ii*) Mechanical torsion meter
   (*iii*) Electrical torsion meter
9. Describe briefly how a strain gauge torsion meter is used to measure the torque.
10. What are accelerometers?
11. Explain briefly the following:
    (*i*) Piezo-electric accelerometer
    (*ii*) Seismic accelerometer.
12. Define the term "Vibration".
13. What are the causes of vibrations in machines, their harmful effects and remedies?
14. What are the methods of vibration measurements?

❑❑❑

Chapter 8

Temperature Measurements

8.1 INTRODUCTION

The **temperature** *is a thermal state of a body which distinguishes a hot body from a cold body*. The temperature of a body is proportional to the stored molecular energy, *i.e.*, the average molecular kinetic energy of the molecules in a system (A particular molecule does not have a temperature, it has energy; the gas as a system has temperature, The temperature may also be *defined* as:

- *The measure of the mean K.E. of the molecules of a substance.*
- *The degree of hotness and coldness of a body or an environment measured on a definite scale.*
- *The driving force or potential causing flow of energy as heat.*

Instruments for measuring ordinary temperatures are known as ***thermometers*** *and those for measuring high temperatures are known as* ***pyrometers***.

It has been found that a gas will not occupy any volume at a particular temperature. This temperature is known as *absolute zero temperature*. The temperatures measured with absolute zero as basis are called *absolute temperatures*. The point of absolute zero temperature is found to occur at 273.15°C below the *freezing* temperature of water.

Absolute temperature = Thermometer reading in °C + 273.15

Absolute temperature in degree centigrade is known as degree kelvin, denoted by K (SI unit).

8.2 TEMPERATURE MEASURING INSTRUMENTS

A. The temperature measuring instruments are *based on changes in a broad range of physical properties*, among which are the following:

1. Changes in physical dimensions:

(*i*) Liquid-in-glass thermometers,
(*ii*) Bimetallic elements.

- Liquid-in-glass thermometer continues to appear in both laboratory and household situations, primarily because of its ease of use and low cost.
- *"Bimetallic elements"* are used in various low-frequency, low cost applications.

2. ***Changes in gas pressure or vapour pressure:***
   (*i*) *Constant-volume gas thermometers.*
   (*ii*) *Pressure thermometers (gas, vapour and liquid filled).*

3. ***Change in electrical properties:***
   (*i*) Resistance thermometers (RTD, PRT).
   (*ii*) Thermistors.
   (*iii*) Thermocouples,
   (*iv*) Semiconductor-junction sensors
      (*a*) Diodes
      (*b*) Integrated circuits.
   - *"Electrical sensors"* are widely used particularly when automatic or remote recording is desired.

4. ***Changes in emitted thermal radiation:***
   (*i*) Thermal and photon sensors.
   (*ii*) Total-radiation pyrometers.
   (*iii*) Optical and two-colour pyrometers.
   (*iv*) Infrared pyrometers.
   - "*Radiant sensors*" are used for non-contact temperature sensing, either in high-temperature applications like combustors or for infra red sensing at low temperatures.

5. ***Changes in chemical phase:***
   (*i*) Fusible indicators.
   (*ii*) Liquid crystals.
   (*iii*) Temperature-reference (fixed-point) cells.

B. The temperature measuring instruments may also be *classified* into *two broad categories* as follows :

1. ***Non-electrical methods:***
   (*i*) By using change in volume of a liquid when its temperature is changed.
   (*ii*) By using change in pressure of a gas when its temperature is changed.
   (*iii*) By using changes in the vapour pressure when the temperature is changed.

2. ***Electrical method:***
   (*i*) By thermocouples.
   (*ii*) By change in resistance of material with change in temperature.
   (*iii*) By comparing the colours of filament and the object whose temperature is to be found out.
   (*iv*) By ascertaining the energy received by radiation.

C. The thermometers may also be *classified* as follows:
   1. Expansion thermometers
      (*i*) Liquid-in-glass thermometers
      (*ii*) Bimetallic thermometers.
   2. Pressure thermometers
      (*i*) Vapour pressure thermometers
      (*ii*) Liquid-filled thermometers
      (*iii*) Gas-filled thermometers.
   3. Resistance thermometers
   4. Thermocouple thermometers
   5. Radiation pyrometers
   6. Optical pyrometers.

8.3 EXPANSION THERMOMETERS

The expansion thermometers make use of the *differential expansion of two different substances.*

- In *"liquid-in-glass thermometers"*, it is the difference in expansion of liquid and the containing glass.
- In *"bimetallic thermometers"*, the indication is due to the difference in expansion of two solids.

8.3.1 Liquid-in-glass Thermometer

This is a very familiar type of thermometer. The mercury or other liquid fills the glass bulb and extends into the bore of the glass stem. Mercury is the most suitable liquid and is used from – 38.9°C (melting point) to about 600°C. The thermometer employed in the laboratory has the scale engraved directly on the glass stem.

A usual type of mercury-in-glass thermometer is shown in Fig. 8.1. An expansion bulb is usually provided at the top of the stem to allow room for expansion of mercury, in case the thermometer is subjected to temperature above its range. The upper limit for mercury-in-glass thermometer is about 600°C. *As the upper limit is far above the boiling point of mercury, some inert gas, i.e., nitrogen is introduced above the mercury to prevent boiling.*

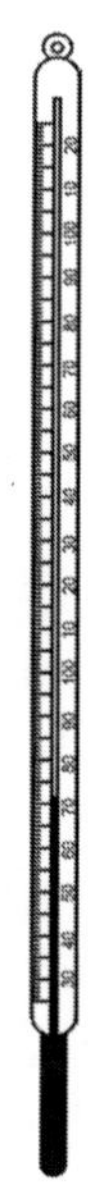

Fig. 8.1 Mercury-in-glass thermometer.

- ***Mercury,*** within its capabilities, is undoubtedly the best liquid for liquid-in-glass thermometers and is generally used in high grade instruments.
  - Alcohol is usually satisfactory.
  - Other liquids are also used primarily for the purpose of extending the useful ranges to lower temperatures.

Advantages and limitations:

Following are the *advantages and limitations of liquid-in-glass thermometers:*

Advantages:

(*i*) Low cost (relatively).
(*ii*) Simple to use.
(*iii*) Absence of need for auxiliary power,
(*iv*) Additional indicating instruments, not required.

Limitations:

(*i*) Portable but remote indication not possible.
(*ii*) Fragile construction.
(*iii*) Range limited to about 600°C.

8.3.2 Bimetallic Thermometers

In a bimetallic thermometer *differential expansion of bimetallic strips is used to indicate the temperature.*

In this type of thermometer two flat strips (Fig. 8.2) of different metals are placed side-by-side and are welded together. Many different metals can be used for this purpose. Generally one is *low expanding metal* and the other is *high expanding metal.* The bimetal strip is coiled in the form of a spiral or helix. *Due to rise in temperature, the curvature of the strip changes. The differential expansion of a strip causes the pointer to move on the dial of the thermometer.*

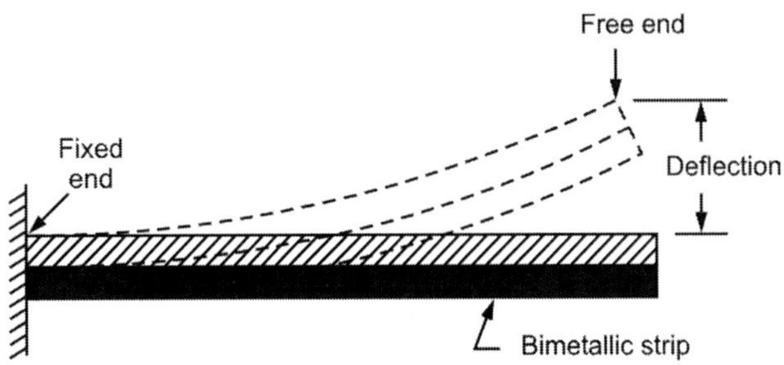

Fig. 8.2 Bimetallic strip fixed at one end.

The range over which a linear relationship exists between deflection and temperature depends upon the combination of metals used for the bimetallic strip.

Figure 8.3 shows the industrial type bimetallic thermometers. The *helical strip* is used for most process applications because of its ability to be managed into a small diameter sheath while maintaining ruggedness. The *spiral* strip often used in ambient temperature measurement devices and air-control thermostats.

- The bimetallic thermometers are made with metals which have widely different thermal expansion coefficients.
  - *Invar* (an alloy of nickel and iron) is the most commonly used *low expansion material.* Its thermal expansion coefficient remains stable over a wide temperature range.
  - *Nickel-iron alloys with chromium and manganese added* are often used for *thermal expansion materials.*
- In order to protect the bimetallic thermometers against wear and corrosion, they are usually mounted in wells.

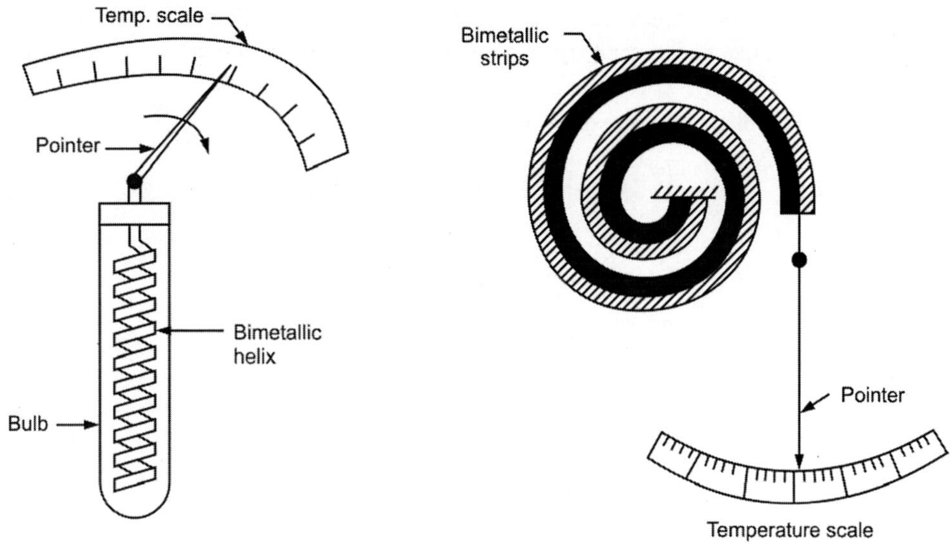

(a) Helical type (b) Spiral type

Fig. 8.3 Industrial type bimetallic thermometers.

Advantages:

(*i*) Simple and robust.
(*ii*) Relatively less costly.
(*iii*) Can withstand, in general, about 50% over range.
(*iv*) Their accuracy ranges from ± 0.5% for laboratory type to ± 2% for process type instruments.

Disadvantages:

(*i*) Temperatures indicated are not correct.
(*ii*) Not suitable for use at temperatures above 400°C for continuous duty or above 550°C for intermittent duty.

Applications:

(*i*) The bimetallic elements find wide applications in *simple thermometers* in which deflection of the element is made to open or close electrical contacts in the electrical heat supply to control a gas flow.
(*ii*) As *switching devices* used in domestic ovens, electric irons, car winkler lamps and the refrigerators.
(*iii*) Used as compensator for ambient temperature change in the filled system *thermometers*, aneroid barometers and in some watches as balance wheel compensators.

8.4 PRESSURE THERMOMETERS

In pressure thermometers liquids, gases and vapours can all be used.

The *principle* on which they work is quite simple. The fluid is confined in a closed system. In this case *pressure is a function of the temperature,* so that when the fluid is heated, the pressure will rise, and the temperature can be indicated by Bourdon type pressure gauge.

In general, the thermometer consists of a bulb which contains bulk of the fluid. The bulb is placed in the region whose temperature is required. A capillary tube connects the bulb to a Bourdon tube, which is graduated with a temperature scale.

The following pressure thermometers will be discussed:

1. *Vapour pressure thermometer.*
2. *Liquid-filled thermometer.*
3. *Gas-filled thermometer.*
    - These devices are commonly known as **Filled systems.**

Following are the *advantages* and *limitations of filled systems:*

Advantages:

(*i*) Most economical.
(*ii*) Versatile.
(*iii*) Rugged in construction,
(*iv*) Stable in operation,
(*v*) Accurate with ±1°C.
(*vi*) Remote indication at a distance of about 60 m is quite satisfactory.

Limitations:

(*i*) Compensation is required in case of changes in ambient temperature and also in case long capillary tubes are used.
(*ii*) The transient response is primarily dependent on the bulb size and the thermal properties of the enclosed fluid.

8.4.1 Vapour Pressure Thermometer

Figure 8.4, shows a schematic diagram of a vapour pressure thermometer.

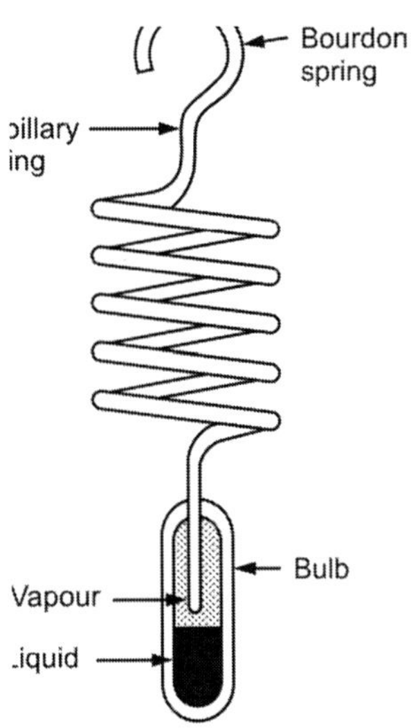

Fig. 8.4 Vapour pressure thermometer.

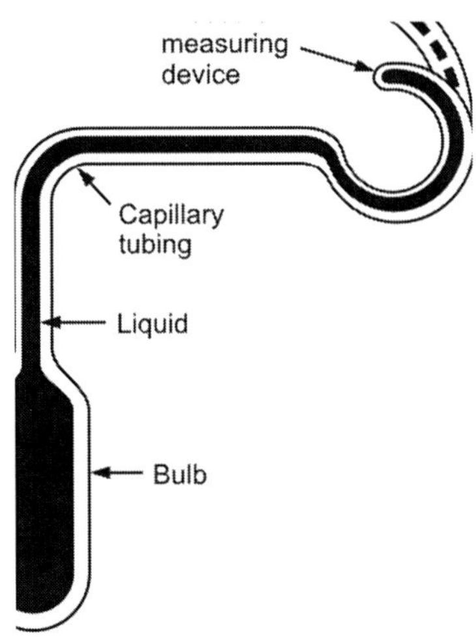

Fig. 8.5 Liquid-filled thermometer.

When the bulb containing the fluid is installed in the region whose temperature is required, some of the fluid vaporizes and increases the vapour pressure. This change of pressure is indicated on the Bourdon tube.

The relation between temperature and vapour pressure of a volatile liquid is of the *exponential form*. Therefore, the scale of a vapour pressure thermometer will not be linear.

8.4.2 Liquid-filled Thermometer

A liquid-filled thermometer is shown in Figure 8.5.

In this type of thermometer, the *expansion of the liquid causes* the pointer to *move in the dial*. Therefore, liquids having high coefficient of expansion should be used.

Mercury is the most widely used liquid because of the wide temperature range between its freezing and boilings points. However, when higher sensitivity is required, organic liquids like *toluene, and ethyl alcohol are used as their coefficients of volumetric expansion are approximately six times that of mercury.*

In such systems the pressure inside the system *must exceed* the vapour pressure of fill liquids to prevent vaporization. The minimum temperature of the system should be kept above the freezing point of the fill liquid to prevent plastic deformation of receiving element.

- The operating pressure varies from 3 to 100 bar.
- The bulb size varies inversely with expansivity of the fill liquid and inversely with span. Minimum range spans, determined by practical bulb size are around 15°C. The maximum span can be as high on 120°C to 230°C with range limits near −90°C and 310°C.

  - In actual design, the internal diameter of the capillary tube and Bourdon tube is made much smaller than that of the bulb. This is because the capillary tube is subjected to a temperature which is quite different from that of the bulb. Therefore, to minimise the effect of variation in temperature to which the capillary tube is subjected, the *volume of the bulb is made as large as possible as compared with the volume of the capillary*. However, large volume of bulb tends to increase time lag, therefore, a *compensating device is usually built into the recording or indicating mechanism, which compensates the variations in temperature of the capillary and Bourdon tubes.*
  - These thermometers could be used for a temperature upto 650°C in which mercury could be used as the liquid.

8.4.3 Gas-filled Thermometers

The temperature range for gas thermometer is practically the same as that of liquid-filled thermometer. The gases used in the gas thermometers are *nitrogen* and *helium*. Both these gases are chemically inert, have good values for their co-efficient of expansion and have low specific heats.

The construction of this type of thermometer is more or less the same as mercury-thermometer in which Bourdon spring is used. The errors are also compensated likewise. The *only difference in this case is that bulb is made much larger than used in liquid-filled thermometers. For good performance the volume of the bulb should be made at least 8 times than that of the rest of the system.*

These thermometers are generally used for *pressures below 35 bar.*

8.5 THERMISTORS

A **thermistor** *(contraction of the term "Thermal Resistor") is a thermally sensitive variable resistor made of a ceramic like semiconducting material.*

Thermistors, unlike metals, respond *negatively* to temperature; as the *temperature rises, the thermal resistance decreases.*

Thermistors are often composed of oxides of manganese, nickel, and cobalt in formulations having resistivities of 100 to 450,000 Ω cm; in cryogenic applications, doped germanium and carbon-impregnated glass are used. The metallic oxides are milled, mixed in appropriate proportions, are pressed into the desired shape with appropriate binders and finally sintered. The electrical terminals are either embedded before sintering or backed afterwards.

The electrical characteristics of thermistors are controlled by varying the type of oxide used and physical size and configuration of the thermistor.

Thermistors may be shaped in the form of beads, disc, washer etc.

Thermistors may be used bare but are usually glass coated or positioned under a thin metal cap. The change in resistance is measured by using circuitry similar to that of metal conductors.

Advantages:

(*i*) Fairly less cost.

(*ii*) Availability in very small sizes.

(*iii*) Fast thermal response.

(*iv*) High sensitivity.

(*v*) Easy adaptability to electrical readout devices.

- Owing to inherently *high sensitivity* possessed by thermistors :
  - *Very simple electrical circuitry can be used to measure the temperature.*
  - Ordinary ohmmeters may be used within the limits of accuracy of the meter itself.
  - More often one of the various forms of resistance bridge is used, either in the null-balance form or as a deflection bridge.
  - Simple ballast circuits are also usable.
  - In some cases, special linearizing circuits are used to obtain an output voltage that varies linearly with temperature.

8.6 THERMOCOUPLE THERMOMETERS

For higher range of temperatures, *i.e.,* above 650°C, filled thermometers are unsuitable. For higher range of temperatures, *thermocouples* and *pyrometers* are used.

8.6.1 Thermoelectric Effects

When direct current passes through a pair of thermocouples with junction maintained at different temperatures, the following *five* effects are observed:

1. Seebeck effect.
2. Peltier effect.
3. Thomson effect.
4. Joulean effect.
5. Conduction effect.

Seebeck effect. *When the two junctions of a pair of dissimilar metals are maintained at different temperatures, e.m.f. (electromotive force) is generated. The phenomenon of generation of e.m.f. is called* ***Seebeck effect****.*

8.6.2 Thermocouple and Thermoelectric Pyrometer

A thermocouple, in its simplest form, consists of *two dissimilar metals or alloys which develop e.m.f. when the reference and measuring junctions are at different temperatures.* The reference junction or cold junction is usually maintained at some constant temperature, such as 0°C.

Figure 8.6 shows a simple circuit of a *thermocouple and temperature measuring device.* In many industrial installations the instruments are equipped with *automatic compensating devices for temperature changes of the reference junction, thus eliminating the necessity of maintaining this junction at constant temperature.*

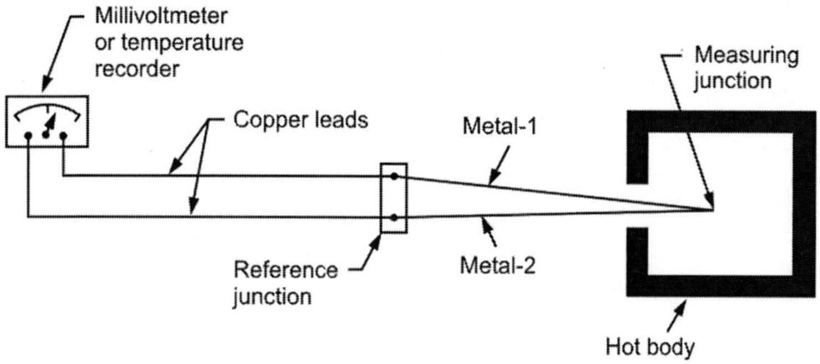

Fig. 8.6 Thermocouple.

Figure 8.7, shows the schematic diagram of a thermoelectric pyrometer.

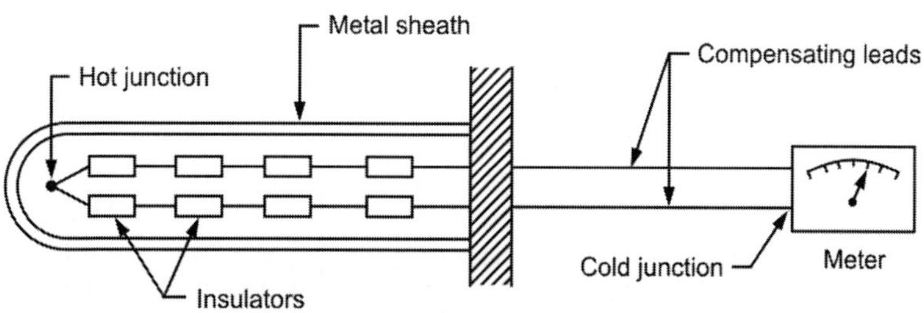

Fig. 8.7 Schematic diagram of a thermoelectric pyrometer.

Compensating leads. These are a pair of leads wires made of such material which when connected to a thermocouple behave such that the effective reference junction or junctions are removed to the other end of the leads. Its material should be such that it has the *same e.m.f. temperature characteristics* as the thermocouple over the range of the ambient temperature and is at the same time*cheap* also. The *resistance* of the compensating leads *should be as low as possible.* Usually they are covered with some protective material suitable for the installation.

Connectors or extension leads. These are generally copper conductors having a *negligible resistance* and are used to connect the compensating leads further to the indicators.

- *In order to prevent the unwanted e.m.f. or unwanted ground currents, the thermocouple leads and extension leads must be properly insulated.*

8.6.3 Thermocouple Materials

The thermocouple materials should possess the following *properties* :

1. The temperature-e.m.f. relationship should be reasonably linear and reproducible.
2. Adequate thermo e.m.f. per degree of temperature change to facilitate detection and measurement.
3. The materials should be physically strong enough to withstand high temperatures, rapid temperature changes and effects of corrosive atmosphere.
4. The thermocouple should maintain its calibration without drift over a long period of time.
5. Cost should be reasonable.
6. The thermocouple should have long life.

TABLE 8.1 Composition, Useful Temperature Range and e.m.f. produced for Some Thermocouples

S.No.	*Thermocouple*	*Composition*	*Temperature (°C)*		*Thermoelectric power*		*Remarks*
			Useful range	*Max.*	*°C*	*Millivolt*	
1.	*Platinum vs platinum-rhodium*	Pure platinum vs Pt + 10 or 13% Rh	400 to 1450	1700	0 500 1000 1500	0.0 4.219 9.569 15.498	Used for high temperature measurements.
2	*Chromel vs alumel*	90% Ni + 10% Cr *vs* 95% Ni + 5% (Al + Sn) Ma	– 200 to 1200	1450	– 200 0 300 600 900 1200	– 5.75 0.0 12.21 24.90 37.36 48.85	High resistance to oxidation.
3.	*Iron vs constantan*	Pure iron vs 45-60% Cu + 55-40% Ni	– 200 to 750	1000	– 200 0 300 600 900	– 8.27 0.0 16.59 33.27 52.29	—
4.	*Copper vs constantan*	Pure copper vs Cu-Ni constantan	– 200 to 350	600	– 200 0 200 400	– 5.539 0.0 9.285 20.865	Not suitable in air due to excessive oxidation.

Series combination (Thermopile):

In a series arrangement, the total output from n thermocouples *equals the sum of the individual e.m.fs.*

- The series arrangement provides *considerably more sensitivity* and is frequently used where it is desired to obtain a *substantially large e.m.f. for measurement of small temperature differences between the two junctions.*

Parallel Combination:

- This arrangement may be used to *obtain an average temperature of a number of points, as in determining the average temperature of gas flowing through a large area passage.*
- The total potential indicated by the potentiometer will be an *average of the junction potentials.*

8.6.4 Advantages and Disadvantages of Thermocouples

The following are the *advantages* and *disadvantages* of the thermocouples:

Advantages:

1. Thermocouples have a *better response.*
2. They have a *higher range of temperature measurements.*
3. The sensing element of a thermocouple can be *easily installed.*
4. *Cheaper* than resistance thermometers.
5. *Very convenient* for measuring the temperature at *one particular point* in a piece of apparatus.

Disadvantages:

1. *Lower accuracy,* and as such they cannot be used for precision work.
2. In order to ensure long life, they need to be *amply protected.* For preventing contamination of thermocouple, when precious metals like platinum or its alloys are being used, the protecting tube has to be made chemically inert and vacuum tight.
3. The *circuitry* for thermocouple is *very complex.*

8.7 RADIATION PYROMETERS

The principle, that the radiant energy emitted from a body increases with temperature, is used in measuring temperatures particularly in the *higher ranges.*

Radiation pyrometers are used in the following *situations :*

(*i*) High temperatures.
(*ii*) Where it is not possible to contact the material whose temperature is to be measured.
(*iii*) Where furnace atmosphere is detrimental to thermocouples and causes erratic measurements and short life.

Following two *"principles"* are used for construction of radiation temperature measuring devices:

(*i*) *Total radiation pyrometer.* In this case the *total radiant energy* from a heated body is measured.
(*ii*) *Selective (or partial) radiation pyrometer.* In this case, we measure the spectral radiant intensity of the radiant energy from the heated body at a *given wavelength.*

Since it is difficult to build a practical pyrometer that is responsive to radiations of all wavelengths or to spectral radiation of one wavelength, therefore, *practical radiation pyrometers response to a wide band of radiation of approximately 0.1 to 8.0 microns width in the visible and infrared band of radiation spectrum;* the actual width of this band, however, depends entirely on the *physical construction of the radiation receiver.*

Total radiation pyrometer:

A device which measures the total intensity of radiation emitted from a body is called total radiation pyrometer.

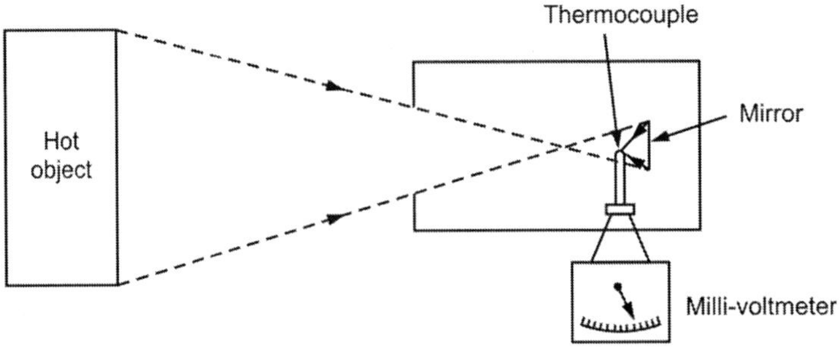

Fig. 8.8 A schematic diagram of radiation pyrometer.

The elements of a *total radiation pyrometer* are illustrated in Fig. 8.8. It collects the radiation from an object (hot body) whose temperature is required. A mirror is used to focus this radiation on a thermocouple. This energy which is concentrated on the thermocouple raises its temperature and in turn generates an e.m.f. This e.m.f. is then measured either by the galvanometer or potentiometer method. *Thus rise of temperature is a function of the amount of radiation emitted from the object.*

Advantages and Disadvantages of Radiation Pyrometers:

Following are the *advantages and disadvantages of radiation pyrometers:*

Advantages:

(*i*) Can be used to measure the temperature of an object which may be either stationary or moving, and so adaptable to continuous industrial processing,

(*ii*) High speed of response.

(*iii*) Accuracy ±2% of scale,

(*iv*) Primarily used to measure temperatures in the range 700-2000°C, where resistance thermometers and thermocouples cannot be used.

(*v*) Direct contact is not necessary with the object whose temperature is to be measured,

(*vi*) Suitable for measuring temperatures where the atmospheric or other environmental conditions prevent satisfactory operation of other temperature sensing devices.

(*vii*) Relatively independent of the distance between the measuring element and the heated body.

Disadvantages:

(*i*) In the situations where the temperature may be high due to operating conditions, *cooling is required* to protect the instrument from overheating.

(*ii*) The presence of dust and dirt on the mirrors or lens causes the instrument to read too low.

8.8 OPTICAL PYROMETERS

An **optical pyrometer** *works on the principle that matters glow above 480°C and the colour of visible* *radiation is proportional to the temperature of the glowing matter.* The amount of light radiated from the glowing matter (solid or liquid) is measured and employed to determine the temperature.

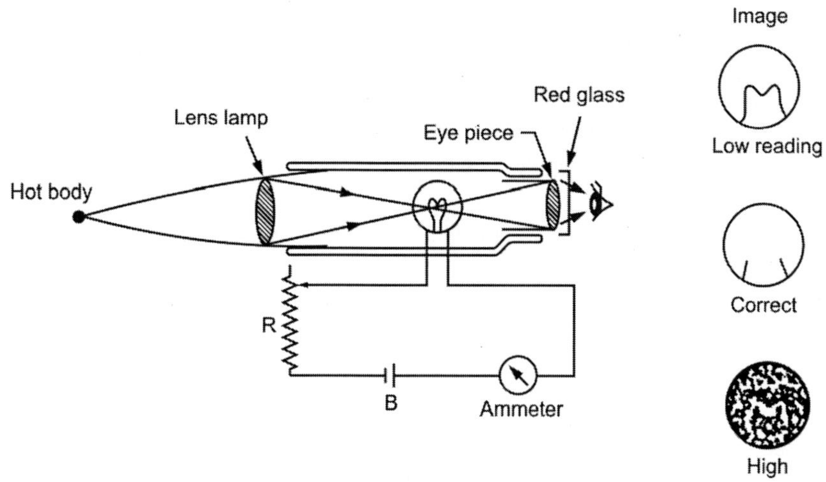

Fig. 8.9 Optical pyrometer.

Figure 8.9, shows a *disappearing filament pyrometer.*

Operation:

- The optical pyrometer is sighted at the hot body and focused.
- In the beginning filament will appear dark as compared to the background which is bright (being hot).
- By varying the resistance (R) in the filament circuit more and more current is fed into it, till filament becomes equally bright as the background and hence disappears.
- The current flowing in the filament at this stage is measured with the help of an ammeter which is calibrated directly in terms of temperature.
- If the filament current is further increased, the filament appears brighter as compared to the background which then looks dark.
- An optical pyrometer can measure temperature ranging from 700 to 4000°C.

Uses:

The optical pyrometer is widely used for *accurate measurement of temperature of:.*

- furnaces;
- molten metals;
- other heated materials.
- *This pyrometer has been accepted as the standard means for determining temperatures on the International Temperature Scale from the gold point and upwards.*

Advantages:

(*i*) Excellent accuracy within ±5°C for the operating range 700–3000°C.

(*ii*) No direct contact is necessary with the object whose temperature is to the measured. Thus this type of pyrometer can be used in situations where the measuring target is remote and inaccessible such as furnace interiors, molten metals etc.

(*iii*) Measurement is independent of the distance between the target and measuring instrument.
(*iv*) The skill in operating the thermometer can be acquired readily.

Disadvantages:

(*i*) The lower measuring temperature is limited to 700°C (The eye is insensitive to wavelength characteristics below this temperature).
(*ii*) Owing to the manual null-balance operation of this pyrometer it is not suitable for continuous reading or automatic control applications.

8.9 SOURCES OF ERRORS AND PRECAUTIONS IN MEASUREMENT OF TEMPERATURE

Sources of Errors:

The *errors in the measurement of temperature* may arise due to following *reasons:*

1. Error in sensing.
2. Error in lead wires.
3. Error in indicating instruments.
4. Dynamic errors.

1. Error in sensing. Error in sensing may be caused due to:

- Stratification and poor mixing of fluids.
- Incomplete immersion.
- Inhomogeneity of thermocouple wires.
- Radiation and convection losses from the measuring element surface.

2. Errors in lead wires:

- Parasitic thermoelectric.
- Parasitic voltage.
- Insulation breakdown.
- Errors due to lead wire junction at different temperatures.

3. Errors in indicating instruments:

- Drop in voltage due to circuit resistance and leakage current.

4. Dynamic errors:

- The instrument may not respond to instantaneous temperature changes because of thermal lag of the system.

Precautions in Measurement of Temperature

Following are the general requirements that should be met for correct measurement of temperature:

I. ***For contact methods:***

(*i*) *Satisfactory contact* between the testing body and the hot body.
(*ii*) The heat capacity (*i.e.*, mass) of the testing body should be small, so that temperature conditions are not disturbed.

TABLE 8.2 Summary of Temperature Range of Different Instruments

Type of Instrument	Liquid in glass thermometer	Bimetallic thermometer	Pressure thermometers			Resistance thermometers			Thermocouples				Radiation pyrometers	Optical pyrometers
	Mercury		Vapour Pressure	Liquid filled	Gas filled	Copper	Nickel	Platinum	Iron Constantan	Copper Constantan	Chromel Alumel	Platinum Rhodium	—	—
Low temperature limit	– 39°C	– 75°C	– 250°C	– 87°C	– 269°C	– 140°C	– 184°C	– 184°C	– 196°C	– 250°C	– 184°C	0°C	about room temp:	760°C
High temperature limit	600°C	540°C	316°C	650°C	540°C	120°C	316°C	760°C	760°C	350°C	1260°C	1540°C	as high as desired.	as high as desired.
Remarks	Low price, fairly accurate, easily broken	Easier to read than liquid-in-glass thermometers	Normally scale non-linear	To be compensated for variation in ambient temperatures	Requires large bulb, linear scale	Very sensitive. Suitable for narrow range Instruments.			Widely used in industry	Constantan is 60% Cu, 40% Ni	Chromel is 20% Cr, 80% Ni Alumel is 2% Al, 98% Ni	Expensive	—	—

(*iii*) The testing body *should not react chemically* with the hot body, otherwise some heat may be either liberated or absorbed as a result of chemical reaction.

(*iv*) The condensation of vapour on the testing body should be avoided, as far as possible; since the latent heat of condensation or wet bulb effect may cause serious errors in the readings.

II. ***For non-contact methods:***

(*i*) The testing body should be placed in an enclosure at sensibly the same temperature as itself.

(*ii*) The testing body should not have any abnormal fluorescent or lumiscent effects.

(*iii*) If there are intervening gas (*e.g.*, in the case of total radiation pyrometers), the presence of CO_2 or water vapour in the intervening gas will seriously affect the readings and in such cases, special precautions must be taken.

(*iv*) The presence of smoke or dirty windows having appreciable absorption coefficients would require special calibration of the instrument.

While measuring temperatures in order to achieve required degree of accuracy, the following *precautions* should be observed:

(*i*) Non-withdrawal of thermometer from the temperature field during reading.

(*ii*) Sufficient stirring and mixing of the fluid for average temperature measurements.

(*iii*) Proper inspection of the filament in a mercury thermometer with regard to breakage and separation.

(*iv*) Proper selection of thermal wells; exposed parts are to be insulated.

(*v*) Use of protecting shields around the sensitive element to minimise the radiation losses.

(*vi*) Thermocouple corrections for lead resistance and for cold junction temperatures.

(*vii*) Use of fine wire thermocouples to have a minimum lag, high sensitivity and low radiation error.

(*viii*) Where there is fast movement of the gas or liquid the thermometer bulb or thermometer well should be installed.

Essential considerations to obtain accurate temperature readings with minimum dynamic errors:

(*i*) Minimum thermal resistance at the bulb or well; keep the bulb clean and avoid accumulation of insulating film.

(*ii*) There should not be any formation of entrained air or gas around the bulb.

(*iii*) Ensure proper contact between the sensor and the system whose temperature is being measured.

(*iv*) There should be no heat gain to or heat loss from the sensing elements to the surroundings.

(*v*) Make sure that there is flow of gas or liquid over the bulb, or good exposure to radiant energy.

8.10 FACTORS INFLUENCING THE RESPONSE OF A TEMPERATURE SENSING DEVICE

The response of a temperature sensing device is influenced by the following *factors:*

1. **Thermal capacity and thermal conductivity of the element:** *Large thermal capacity* and *lower thermal conductivity* of the element result in *slow response* to attain a given percentage of temperature change. Evidently, for such elements the *lag coefficient is higher.*
2. **Surface area of the element per unit mass (of the element):** With the increase in surface area, the rate of heat flow increases and consequently the response rate is augmented.

3. **Velocity, thermal capacity and thermal conductivity of the film surrounding the element:** Higher the velocity, the lower the lag coefficient (At low velocity, a film is formed around the sensing element which retards the heat flow and decreases the response rate).
4. **Film coefficient of heat transfer:** With higher film coefficients of heat transfer, the heat flow rate increases and that results into a quick response.

8.11 CALIBRATION OF TEMPERATURE MEASURING DEVICES

Methods employed to calibrate temperature-measuring systems can be categorized as follows:

1. *Comparison with primary standards or the fixed temperature points or specified by the International Temperature Scale (Refer to Art. 8.2).*
2. *Comparison with reliably calibrated secondary standards.*

- The primary standard, basically consists of the fixed physical conditions *(melting points, triple points* etc.)
- The intermediate points are established by specified interpolation procedures. Therefore, for *primary calibration,* the problems are those of technique attending reproduction of these fundamental points and reproduction of interpolation methods.
- Besides primary fixed points, numerous *secondary fixed points* have been tabulated; *examples:* The sublimation of carbon dioxide or dry ice (– 78.5°C), the boiling point of water (100°C), and the boiling point of naphthalene (218°C).
- For many fixed points commercial "standards" are available. These standards consist of.a sealed container enclosing the reference materials. Glass is used for the container material for the lower temperatures and graphite is used for the higher temperatures. Integral heating coils are used.

— *To use the standard, the element to be calibrated is placed in a well extending into the centre of the container. The heater is then turned on, and the temperature carried above the melting point of the reference substance and held until melting is completed. It is then permitted to cool and when the freezing point is reached, the temperature stabilizes and remains constant at the specified value as long as liquid and solid are both present (several minutes).* It is claimed that accuracies of approximately *0.1°C* may be easily attained and that 0.01°C may be attained by exercising a *more care.*

HIGHLIGHTS

1. The *temperature* is a thermal state of a body which distinguishes a hot body from a cold body.
2. *Electrical sensors* are widely used particularly when automatic or remote recording is desired.
3. In a *bimetallic thermometer* differential expansion of bimetallic strips is used to indicate the temperature.
4. Gas filled themometers are generally used for pressures below 35 bar.
5. The sources of static errors in filled systems are:
   (*i*) The ambient temperature effect. (*ii*) The head effect.
   (*iii*) The barometric effect. (*iv*) The immersion effect.
6. A *thermistor* is a thermally sensitive variable resistor made of a ceramic-like semiconducting material.
7. A *thermocouple* in its simplest form, consists of two dissimilar metals or alloys which develop e.m.f. when the reference and measuring junctions are at different temperatures.
8. A device which measures the total intensity of radiation emitted from a body is called total *radiation pyrometer.*
9. An *optical pyrometer* works on the principle that matters glow above 480°C and the colour of visible radiation is proportional to the temperature of the glowing matter.

THEORETICAL QUESTIONS

1. Define the term "temperature".
2. How are temperature measuring instruments classified?
3. Explain briefly the following expansion thermometers.
   (*i*) Liquid-in-glass thermometers.
   (*ii*) Bimetallic thermometers.
4. State the advantages, disadvantages and applications of bimetallic thermometers.
5. Discuss briefly any two of the following pressure thermometers:
   (*i*) Vapour pressure thermometer.
   (*ii*) Liquid-filled thermometer.
   (*iii*) Gas-filled thermometer.
6. What are thermistors? What are their advantages?
7. "Give a comparison between "thermistors" and "metal resistors".
8. List the factors which should be considered while selecting resistance thermometers.
9. What is a "Seebeck effect"?
10. Explain with a neat diagram the construction and working of a thermoelectric pyrometer.
11. What are the desirable properties of thermoelectric materials?
12. State the advantages and disadvantages of thermocouples.
13. Explain briefly the working of a radiation pyrometer.
14. List the advantages and disadvantages of radiation pyrometers.
15. Describe with a neat sketch the working of an optical pyrometer.
16. What are the advantages and disadvantages of an optical pyrometer?
17. State the precautions which should be taken while making temperature measurements.
18. Explain the factors which influence the response of a temperature sensing device.

❑❑❑

UNIT II

METROLOGY AND INSPECTION

Introduction to Metrology and Inspection

Chapter 9

9.1 MEANING OF METROLOGY

Metrology, in literary sense, *means the pure science of measurements.* But for engineering purposes, it is restricted to measurements of length and angles and other quantities which are expressed in linear or angular terms.

- *Metrology is mainly concerned with:*
  (*i*) Establishing the units of measurements, reproducing these units in the form of standards and ensuring the uniformity of measurements;
  (*ii*) Developing methods of measurement;
  (*iii*) Analysing the accuracy of methods of measurement, researching into the causes of measuring errors and eliminating these.
- In the broader sense, metrology is not limited to length measurement but is also concerned with the *industrial inspection* and its various techniques. Inspection is carried out with gauges and the metrologist is intimately concerned with the design, manufacturing and testing of gauges of all types.

Legal metrology. *'Legal metrology' is that part of metrology which treats units of measurements, methods of measurements and the measuring instruments, in relation to the technical and legal requirements.*

The activities of the service of 'Legal Metrology' are:

(*i*) Control (testing, verification, standardisation) of measuring instruments;
(*ii*) Testing of prototypes/models of measuring instruments;
(*iii*) Examination of a measuring instrument to verify its conformity to the statutory requirements etc.

Dynamic metrology. *'Dynamic metrology' is the technique of measuring small variations of a continuous nature.* The technique has proved very valuable, and a record of continuous measurement, over a surface, for instance, has obvious advantages over individual measurements of an isolated character.

Deterministic metrology. Deterministic metrology is a new philosophy in which *part measurement is replaced by process measurement.* The new techniques such as 3D error compensation by CNC (Computer Numerical Control) systems and expert systems are applied, leading to fully adaptive control. This technology is *used for very high precision manufacturing machinery and control systems to achieve microtechnology and nanotechnology accuracies.*

9.2 OBJECTIVES OF METROLOGY

Although the basic objective of a measurement is to provide the *required accuracy at a minimum cost,* metrology has further *objectives* in a modern engineering plant with different shapes which are:

1. Complete evaluation of newly developed products.
2. Determination of the process capabilities and ensure that these are better than the relevant component tolerances.
3. Determination of the measuring instrument capabilities and ensure that they are quite sufficient for their respective measurements.
4. Minimising the cost of inspection by effective and efficient use of available facilities.
5. Reducing the cost of rejects and rework through application of *"Statistical quality control techniques".*
6. To standardise the measuring methods.
7. To maintain the accuracies of measurement.
8. To prepare designs for all gauges and special inspection fixtures.

9.3 STANDARDS OF MEASUREMENT

These days only *two standard systems of linear measurement, English* (yard) and *Metric* (metre) are in general use throughout the world.

The *metric system* was *originated in France* and is now being used in many countries in the world. The British system of linear measurement is based on one arbitrarily unit known as *yard.*

For linear measurements the various standards known are:

1. Line standard
2. End standard
3. Wavelength standard.

9.3.1 Line Standard

A yard or metre *is defined as the distance between scribed lines on a bar of metal under certain conditions of temperature and support.* These are legal standards and Act of Parliament authorises their use.

- **The *Metre*** is defined as 1650763.73 wavelengths of the orange radiation in vacuum of krypton-86 isotype.
- **The *Yard*** is defined as 0.9144 metre. This is equivalent to 1509458.35 wavelengths of the same radiation.

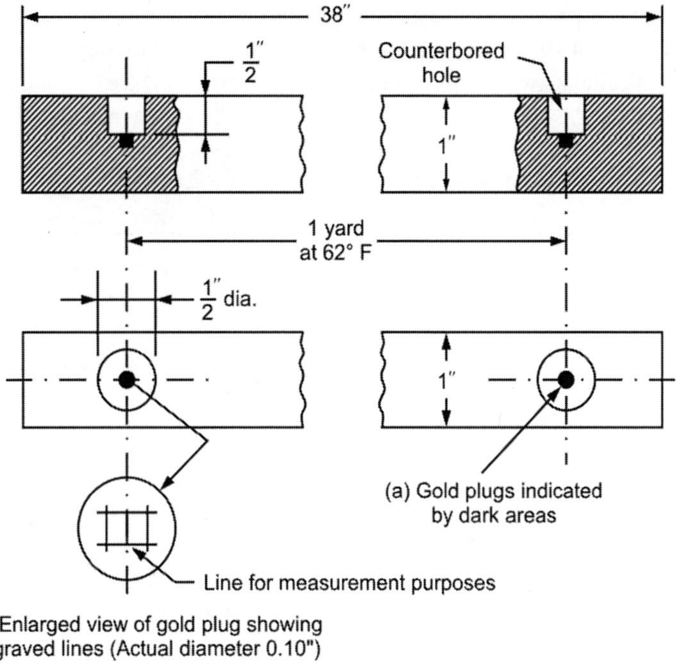

Fig. 9.1 Imperial standard yard.

9.3.1.1 Yard

A Yard was formerly known as the Imperial Standard Yard. Figure 9.1 are shows a diagram indicating its essential features. It consists of a bronze bar made from an alloy known as Baily's metal, consisting of 16 parts copper, $2\frac{1}{2}$ parts tin and 1 part zinc. The bar, 1 sq. in. in cross-section has an overall length of 36". Two counter bored holes, $\frac{1}{2}''$ diameter by $\frac{1}{2}''$ deep, at 36″ centres (1″ from each end of the bar) provide sighting holes for two gold plugs inserted in the holes at the base of each counter bore. The faces of the gold plugs are flush with bases of counter bores and, therefore, lie in the neutral plane of the bar where bending effects are minimised when the bar is resting on supports. These plugs are 0.01 in. diameter, and five lines are ruled on the upper polished face of each; three lines at right angles to the length of the bar and two parallel to the bar as shown in Fig. 9.1(*b*). The *length of the yard is defined as the distance between the two central transverse lines on the plugs when the temperature of the bar is constant at 62°F, and when the bar is supported on rollers, in a specified manner, to prevent flexure.*

- The original procedure, when intercomparisons were made between the standard and its copies, was to float the bars in the mercury; but proof that the bars could be effectively supported on rollers, while maintaining the previous accuracy, was provided by Airy. His method is shown in Fig. 9.2. The standard was directly supported on eight equally spaced rollers in conjunction with a special frame.

  The distance between the rollers was proved by Airy to be equal to $\frac{1}{\sqrt{n^2-1}}$ where, n and l represent number of rollers and length of bar respectively. This method of support was used for the purpose of inter-comparisons until 1922 when two supports at the Airy points were introduced. When a bar is supported specifically at two points symmetrically about its centre, a condition can be produced when the bar ends lie in a horizontal plane. With this condition the bar deflects at its centre, but the effective error in the length of bar is negligible. Applying Airy's formula the specific distance between the supports is equal to 0.577 *l*.

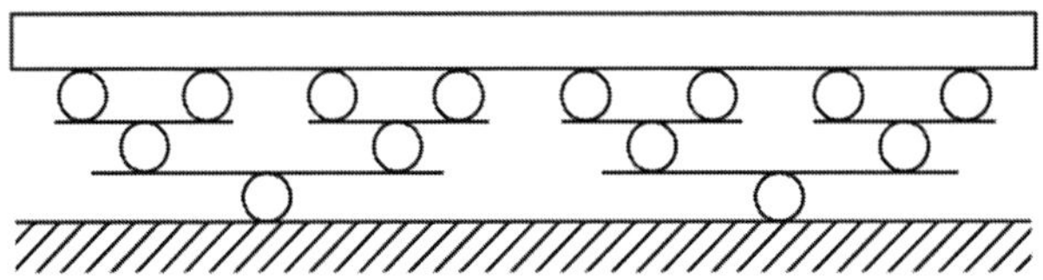

Fig. 9.2 8-point supporting system for imperial yard till 1922.

9.3.1.2 Metre

The *length of the metre is defined as the distance, at 0°C between the centre portions of pure platinum—irridium alloy (10% irridium) of 102 cm total length—and having a cross-section as shown in Fig. 9.3.* The graduations are on the upper surface of web which coincides with the neutral axis of the section.

9.3.1.3 Sub-division of Standards

The imperial standard yard and international prototype metre defined previously are considered to be perfect or master standards and cannot usually be used for general purposes. Thus depending upon the importance of accuracy required for the work, standards are subdivided into *four grades.*

1. **Primary standards.** To ensure that standard unit of length, *i.e.,* yard or metre does not change its value and it is strictly followed and precisely defined, there should be one and only one material standard preserved under most careful conditions. This has no direct application to a measuring problem encountered in engineering. *These are used only at rare intervals and solely for comparison with secondary standards.*
2. **Secondary standards.** Secondary standards are made as nearly as possible to the primary standard with which they are compared at intervals. Any error existing in these bars is recorded by comparison with primary standards after long intervals. *These standards are distributed to a number of places for safe custody and used in their turn for occasional comparison with tertiary standards. These standards also act as safeguard against the loss or destruction of primary standards.*

   Materials for secondary standards:

   1. Invar—an alloy of nickel and steel
   2. Fuse silica
   3. Elinvar—an alloy of nickel and chromium.

   All the above materials have *usually very low coefficient of linear expansion.*

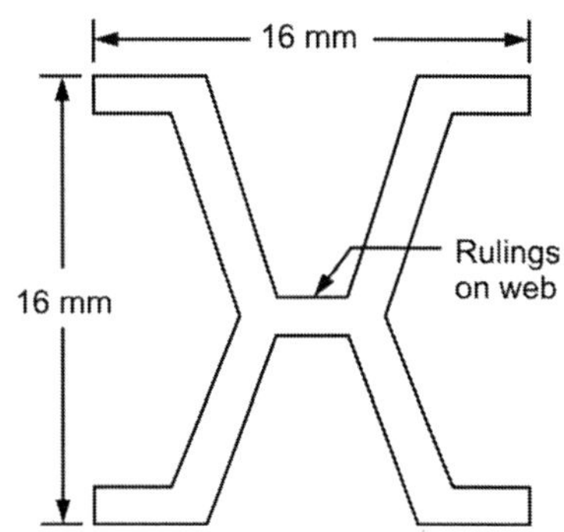

Fig. 9.3 International prototype metre (Cross-section).

3. **Tertiary standards.** Tertiary standards are *line first standards to be used for reference purposes in laboratories and workshops.* These should also be maintained as a *reference for comparison at intervals with working standards.*
4. **Working standards.** These standards are necessary for use in metrology laboratories and similar institutions. These are derived from fundamental standards.

   Sometimes standards are classified as:

   - **Reference standards** (used for reference purposes)
   - **Calibration standards** (used for calibration of inspection and working standards)
   - **Inspection standards** (used by inspectors)
   - **Working standards** (used by operators).

9.3.1.4 Characteristics of Line Standards

The characteristics of line standard are given below:

1. Accurate engraving on the scales can be done but it is difficult to take full advantage of this accuracy. For example, a steel rule can be read to about ±0.2 mm of true dimension.
2. It is easier and quicker to use a scale over a wide range.
3. The scale markings are not subject to wear although significant wear on leading end *leads to undersizing.*
4. There is no *''built in'* datum in a scale which would allow easy scale alignment with the axis of measurement, this again *leads to undersizing.*
5. Scales are subjected to the parallax effect, a source of both positive and negative reading errors.
6. For close tolerance length measurement (except in conjunction with microscopes) scales are not convenient to be used.

9.3.2 End Standard

***End standards**, in the form of the bars and slip gauges, are in general use in precision engineering as well as in standard laboratories such as the N.P.L. (National Physical Laboratory). Except for applications where microscopes can be used, scales are not generally convenient for the direct measurement of engineering products, whereas slip gauges are in everyday use in tool-rooms, workshops, and inspection departments throughout the world.*

A *modern end standard consists fundamentally of a block or bar of steel generally hardened whose end faces are lapped flat and parallel to within a few millionth of a cm.* By the process of lapping, its size too can be controlled very accurately. Although, from time to time, various types of end bar have been constructed, some having flat and some spherical faces, *the flat, parallel faced bar is firmly established as the most practical method of end measurement.*

Characteristics of end standards:

1. Highly accurate and well suited to close tolerance measurements.
2. Time-consuming in use.
3. Dimensional tolerance as small as 0.0005 mm can be obtained.
4. Subjected to wear on their measuring faces.
5. To provide a given size, the groups of blocks are *"wrung"* together. Faulty wringing leads to damage.

6. There is a *"built-in"* datum in end standards, because their measuring faces are flat and parallel and can be positively located on a datum surface.
7. As their use depends on *"feel"* they are not subject to the parallax effect.

End bars. Primary end standards usually consist *of bars of carbon steel* about 20 mm in diameter and made in sizes varying from 10 mm to 1200 mm. These are hardened at the ends only. They are used for the measurement of work of larger sizes.

Slip gauges. *Slip gauges* are used as standards of measurement in practically *every precision engineering* works in the world. These were invented by C.E. Johansom of Sweden early in the present century. These are made *of high-grade cast steel and are hardened throughout.* With the set of slip gauges, combination of slip gauge enables measurements to be made in the range of 0.0025 to 100 mm but in combinations with end/length bars measurement range upto 1200 mm is possible.

Note. The accuracy of line and end standards is affected by temperature changes and both are originally calibrated at 20 ± ½° C. Also care is taken in manufacture to ensure that change of shape with time is reduced to negligible proportions.

9.3.3 Wavelength Standard

In 1829, Jacqnes Babinet, a French philosopher, suggested that wavelengths of monochromatic light might be used as natural and invariable units of length. It was nearly a century later that the Seventh General Conference of Weights and Measures in Paris approved the definition of a standard of length relative to the metre in terms of the wavelength of the red radiations of cadmium. Although this was not the establishment of a new legal standard of length, it set the seal on work which kept on going for a number of years.

- *Material standards are liable to destruction and their dimensions change slightly with time. But with the monochromatic light we have the advantage of constant wavelength and since the wavelength is not a physical one, it need not be preserved. This is reproducible standard of length, and the error of reproduction can be of the order of* 1 *part in* 100 *millions. It is because of this reason that International standard measures the metre in terms of wavelength of krypton* 86 *(Kr* 86).
- Light wavelength standard, for sometime, had to be objected because of the impossibility of producing pure monochromatic light as wavelength depends upon the amount of isotope impurity in the elements. But now with the rapid development in atomic energy industry, pure isotopes of natural elements have been produced. *Krypton* 85, *Mercury* 198 *and Cadmium* 114 are possible sources of radiation of wavelength suitable as natural standard of length.

9.3.3.1 Advantages of Wavelength Standards

The following are the advantages of using wavelength standard as basic unit to define primary standards:

1. It is not influenced by effects of variation of environmental temperature, pressure, humidity and ageing because it is not a material standard.
2. There is no need to store it under security and thus there is no fear of its being destroyed as in the case of yard and metre.
3. It is easily available to all standardising houses, laboratories and industries.
4. It can be easily transferred to other standards.

5. This standard can be used for making comparative statement of a much higher accuracy.
6. It is easily reproducible.

9.3.4 Classification of Standards

To maintain accuracy and interchangeability it is necessary that the standards be traceable to a single source, usually the National Standards of the country, which are further linked to International Standards. The accuracy of National Standards is transferred to working standards through a chain of intermediate standards in a manner given below:

National Standards
↓
National Reference Standards
↓
Working Standards
↓
Plant Laboratory Reference Standards
↓
Plant Laboratory Working Standards
↓
Shop Floor Standards

Evidently, there is degradation of accuracy in passing from the defining standards to the shop floor standards. The *accuracy of a particular standard depends on a combination of the number of times it has been compared with a standard in a higher echelon, the frequency of such comparisons, the care with which it was done, and the stability of the particular standard itself.*

Standardising organisations

Taking into view the role of standardisation in the development of industry, organisations to handle the complexities of standardisation have been evolved in each of the chief industrial countries. In *India, Indian Standard Institution* (ISI) is responsible for evolving standards on metrological instruments etc. In Great Britain, British Standards Institution play very similar role to ISI. In Europe, the International Federation of National Standardising Associations, known as I.S.A., co-ordinate the work of the continental countries.

National Physical Laboratories (NPL) carry out lot of research work in various fields; are responsible for defining standards, and also issue certification marks for quality instruments.

International Organisations of Weight and Measures. It was established in 1975 under "International Metre Convention" in Paris with the object of maintaining uniformity of measurements throughout the world. It comprises of:

- General Conference of Weight and Measures.
- International Committee of Weights and Measures.
- International Organisation of Legal Metrology.

9.3.5 Relative Characteristics of Line and End Standards

TABLE 9.1

The relative characteristics of *line* and *end standards* are given in the tabular form below:

S.No.	Aspect	Line standard	End standard
1.	*Manufacture and cost of equipment*	Simple and low.	Complex process and high.
2.	*Accuracy in measurement*	Limited to ±0.2 mm. In order to achieve high accuracy, scales have to be used in conjunction with microscopes.	Very accurate for measurement of close tolerances upto ±0.001 mm.
3.	*Time of measurement*	Quick and easy.	Time consuming.
4.	*Effect of use*	Scale markings not subject to wear but the end of scale is worn. Thus it may be difficult to assume zero of scale as datum.	Measuring faces get worn out. To take care of this end pieces can be hardened, protecting type. Built-in datum is provided.
5.	*Other errors*	There can be parallax error.	Errors may get introduced due to improper wringing of slip gauges. Some errors may be caused due to change in laboratory temperature.

9.4 LIMITS, FITS AND TOLERANCE

9.4.1 General Aspects

- In the design and manufacture of engineering products a great deal of attention has to be paid to the *mating, assembly and fitting of various components.* In the early days of mechanical engineering during the nineteenth century, the majority of such components were actually mated together, their dimensions being adjusted until the required type of fit was obtained. These methods demanded craftsmanship of a high order and a great deal of very fine work was produced.
- Present day standards of quantity production, interchangeability, and continuous assembly of many complex compounds, could not exist under such a system, neither could many of the exacting design requirements of modern machines be fulfilled without the knowledge that certain dimensions can be reproduced with precision on any number of components.
- *Modern mechanical production engineering is based on a system of limits and fits, which, while not only itself ensuring the necessary accuracies of manufacture, forms a schedule or specifications to which manufacturers can adhere.*

  In order that a system of limits and fits may be successful, following *conditions* must be fulfilled:

  (*i*) The range of sizes covered by the system must be sufficient for most purposes.

  (*ii*) It must be based on some standards so that everybody understands alike and a given dimension has the same meaning at all places.

  (*iii*) For any basic size it must be possible to select from a carefully designed range to fit the most suitable one for a given application.

  (*iv*) Each basic size of hole and shaft must have a range of tolerance values for each of the different fits.

(*v*) The system must provide for both unilateral and bilateral methods of applying the tolerance.

(*vi*) It must be possible for a manufacturer to use the system to apply either a hole-based or a shaft-based system as his manufacturing requirements may need.

(*vii*) The system should cover work from high class tool and gauge work where very wide limits of sizes are permissible.

9.4.2 Nominal Size and Basic Dimensions

Nominal size. A *'nominal size' is the size which is used for purpose of general identification.* Thus the nominal size of a hole and shaft assembly is 60 mm, even though the basic size of the hole may be 60 mm and the basic size of the shaft 59.5 mm.

Basic dimension. A *'basic dimension' is the dimension, as worked out by purely design considerations.* Since the ideal conditions of producing basic dimension, do not exist, the basic dimenions can be treated as theoretical or nominal size, and it has only to be approximated. A study of function of machine part would reveral that it is unnecessary to attain perfection because some variations in dimension, however small, can be tolerated on size of various parts. It is, thus, *general practice to specify a basic dimension and indicate by tolerances as to how much variation in the basic dimension can be tolerated without affecting the functioning of the assembly into which this part will be used.*

9.4.3 Definitions

The definitions given below are based on those given in IS : 919 *Recommendation for Limits and Fits for Engineering,* which is in line with the ISO recommendation.

Shaft. The term *shaft* refers not only to diameter of a circular shaft but to *any external dimension on a component.*

Hole. This term refers *not only* to the diameter of a circular hole but to *any internal dimension on a component.*

Actual size of the shaft. This is the measured dimensions of the part.

Basic size. Refer to Fig. 9.4. The *basic size is the standard size for the part and is the same for both the hole and its shaft.* Example : A 60 mm diameter hole and shaft.

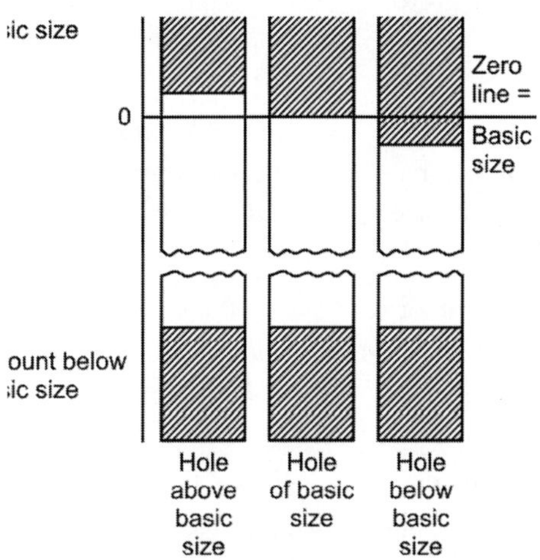

Fig. 9.4.

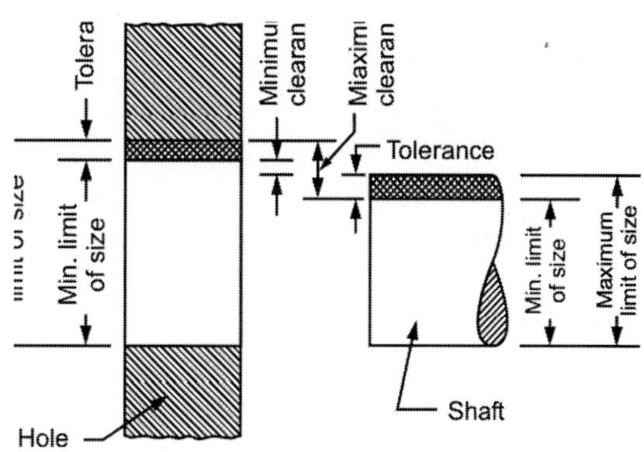

Fig. 9.5.

Zero line. Refer to Fig. 9.4. *This is the line which represents the basic size so that the deviation from the basic size is zero.*

Limits of size. These are *maximum and minimum permissible sizes of the part.*

Minimum limit of size. Refer to Fig. 9.5. The *minimum size permitted for the part.*

Maximum limit of size. Refer to Fig. 9.5. The *maximum size permitted for the part.*

Tolerance. Refer to Fig. 9.5. *The difference between the maximum and minimum limits of size.*

Tolerance size. This is the difference between the two limits of size.

Grade of tolerance. The tolerance grade *is an indication of the degree of accuracy of manufacture.* It is designated by the letter IT followed by a number. Tolerance grades are ITOI, ITO, upto IT16. The *larger the number the larger the tolerance.*

Standard tolerance unit. This is the unit used to calculate the various grades of tolerance for a given basic size.

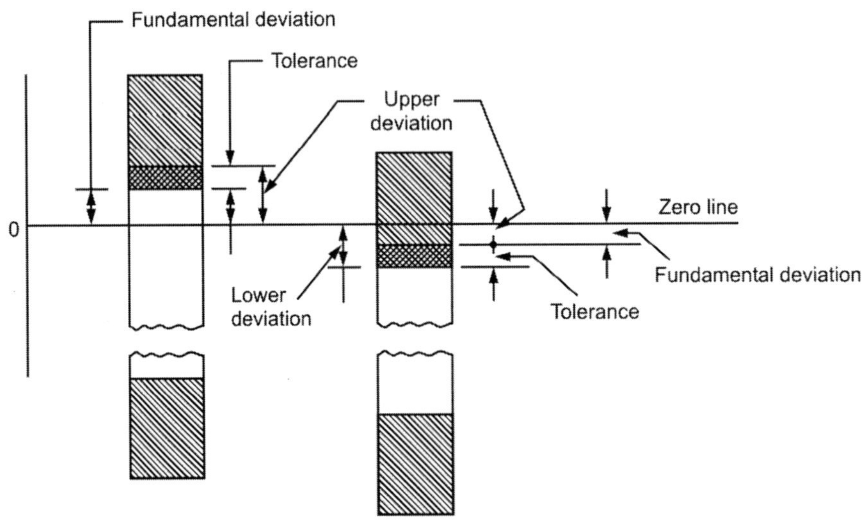

Fig. 9.6.

Upper deviation. Refer to Fig. 9.6. This is the *amount from the basic zero or zero line, on the maximum limit of size for either a hole or a shaft.* It is designated ES for a *hole* and *es* for a *shaft.* Upper deviation is a *positive quantity* when the maximum limit of size is *greater* than the basic size and *negative quantity* when the maximum limit of size is less than the basic size.

Lower deviation. Refer to Fig. 9.6. This is the *amount from basic size,* or *zero line,* to the *minimum limit of size.* It is designated EI for a *hole* and *ei* for a *shaft.* Lower deviation is a *positive quantity* when the minimum limits of *sizes* is *greater* than the basic size and a *negative quantity* when the minimum limit of size is *less* than the basic size.

Fundamental deviation. Refer to Fig. 9.6. This is the deviation, either the upper or lower deviation, which is the *nearest one to the zero line* for either a hole or a shaft. It fixes the position of the tolerance zone in relation to zero line.

Fit. Fit *means a degree of tightness or looseness between two mating parts to perform a definite function.*

Clearance. The *difference between the sizes of a hole and a shaft* which are to be assembled together when the *shaft is smaller than the hole.*

Interference. The *difference between the sizes of a hole and a shaft* which are to be assembled together when the *shaft is larger than the hole.*

Classes of fit: Refer to Fig. 9.7.

1. **Clearance fit:** A *clearance fit could be obtained by making the lower limit on the hole equal to or larger than the upper limit on the shaft.* Any hole and any shaft made to these tolerances would assemble with a clearance fit with certainty.
2. **Interference fit:** An *interference fit would be obtained with equal certainty by making the lower limit on the shaft equal to or larger than the upper limit on the hole.*
3. **Transition fit:** Between these two conditions lies a *range of fits known as transition fit.* These are obtained when the upper limit on the shaft is larger than the lower limit on the hole, and the lower limit on the shaft is smaller than the upper limit on the hole. It must be realised that transition fits exist only as a class; any actual hole and shaft must assemble with either a clearance or interference fit.

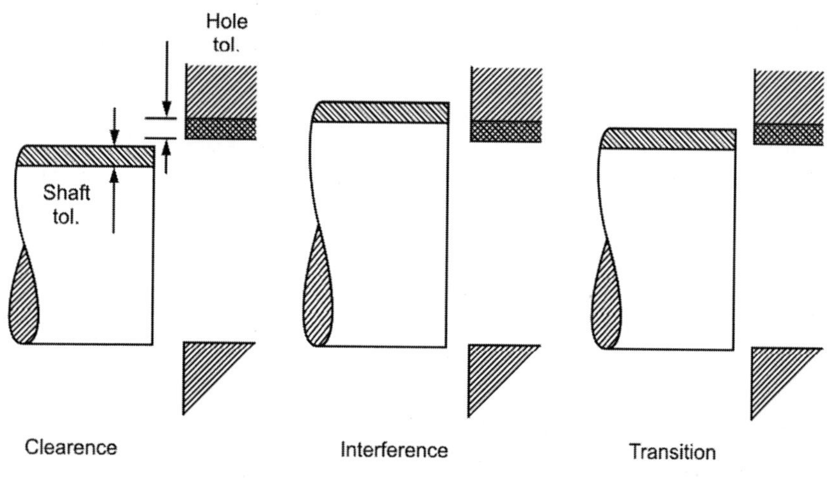

Fig. 9.7 Classes of fit.

Allowance: The *difference between the maximum shaft and minimum hole is known as allowance.* In a *clearance fit,* this is the *minimum clearance* and is and *positive allowance.* In an *interference fit,* it is the *maximum interference and is a negative allowance.*

9.4.4 Basis of Fit (or Limit) System

A fit or limit system consists of a series of tolerances arranged to suit a specific range of sizes and functions, so that limits of size may be selected and given to mating components to ensure specific classes of fit. This system may be arranged on the following basis:

1. Hole basis system
2. Shaft basis system.

Hole basis system: Refer to Fig. 9.8. *"**Hole basis system**" is one in which the limits on the hole are kept constant and the variations necessary to obtain the classes of fit are arranged by varying those on the shaft.*

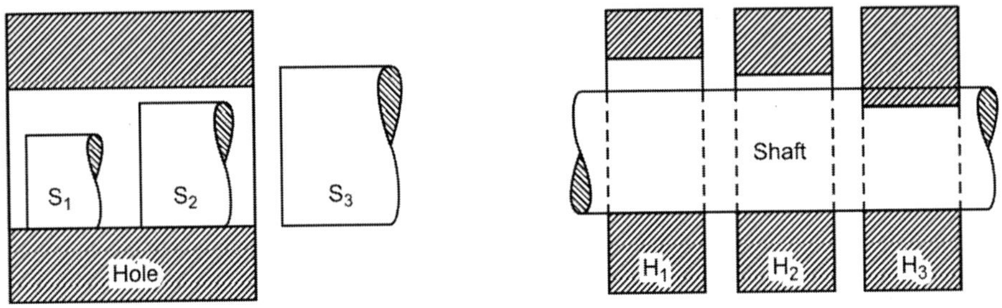

'S' denotes shaft to give various fits with hole
Hole Basic System

'H' denotes shaft to give various fits with shaft
Shaft Basic System

Fig. 9.8 Hole and shaft basis systems.

Shaft basis system: Refer to Fig. 9.8. *'**Shaft basis system**' is one in which the limits on the shaft are kept constant and the variations necessary to obtain the classes of fit are arranged by varying the limits on the holes.*

In present day industrial practice *hole basis system is used because a great many holes are produced by standard tooling,* for example, reamers drills, etc., whose size is not adjustable. Subsequently the shaft sizes are more readily variable about the basic size by means of turning or grinding operations. Thus the *hole basis system results in considerable reduction in reamers and other precision tools as compared to a shaft basis system because in shaft basis system due to non-adjustable nature of reamers, drills etc. great variety (of sizes) of these tools are required for producing different classes of holes for one class of shaft for obtaining different fits.*

9.4.5 Systems of Specifying Tolerances

The tolerance or the error permitted in manufacturing a particular dimension may be allowed to vary either on*one side* of the basic size or on *either side* of the basic size. Accordingly two systems of specifying tolerances exit.

Refer to Fig. 9.9.

1. Unilateral system
2. Bilateral system.

In the *unilateral system,* tolerance is applied only in *one direction.*

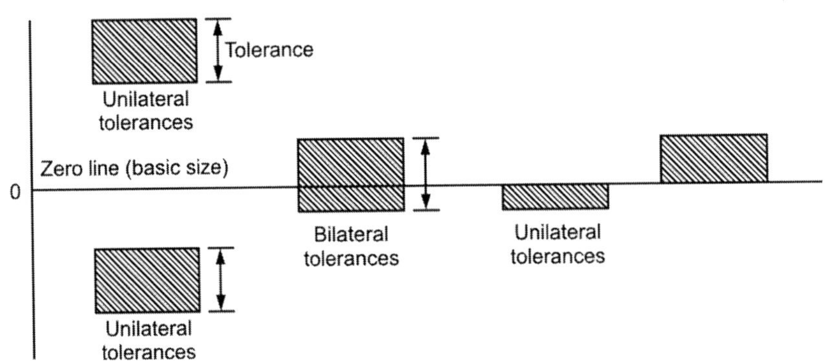

Fig. 9.9 Unilateral and bilateral tolerance.

Example: $40.0^{+0.04}_{+0.02}$, or, $40.0^{-0.02}_{-0.04}$

In the ***bilateral system*** of writing tolerances, a dimension is permitted to vary in *two directions.*

Example: $40.0^{+0.02}_{-0.04}$

- ***Unilateral system*** is more satisfactorily and realistically applied to certain machining processes where it is a common knowledge that dimensions will most likely deviate in one direction. Further, in this system the tolerance can be revised without affecting the allowance or clearance conditions between mating parts, *i.e.*, without changing the type of fit. This system is *most commonly used in interchangeable manufacture especially where precision fits are required.*
- It is not possible, in ***bilateral system,*** to retain the same fit when tolerance is varied. The basic size dimension of one or both of the mating parts will also have to be changed. This system clearly points out the theoretically desired size and indicates the possible and probable deviations that can be expected on each side of basic size. *Bilateral tolerances help in machine setting and are used in large scale manufacture.*

9.4.6 Designation of Holes, Shafts and Fits

A *hole* or *shaft is completely described if the basic size, followed by the appropriate letter and by the number of the tolerance grade, is given.*

Example:
- A 25 mm H-hole with the tolerance grade IT8 is given as :
  25 mm H8 or simply 25 H8
- A 25 mm *f*-shaft with the tolerance grade IT7 is given as :
  25 mm *f* 7 or simply 25 *f* 7,

A *'fit'* is indicated by combining the designations for both the hole and shaft with the hole designation written first, regardless of system *(i.e.,* hole-basis or shaft-basis).

Example: 26 H8—*f* 7 or

25 H8—*f* 7 or

$$25\frac{H8}{f7}$$

9.4.7 Commonly used Holes and Shafts

- In several engineering applications the fits required can be met by a quite small selection from the full range available in the standards. The holes and shafts commonly used are as follows:

  Holes *(commonly used):* H6, H7, H8, H9, H11.

  Shafts *(commonly used) : c11, d10, e9, f7, g6, h6, k6, n6, p6, s6.*

  IS : 919 gives the most commonly used holes and shafts upto 500 mm for the purpose of general engineering.

9.4.8 The Newall System

- The Newall system is the first standard evolved in Great Britain to standardise limits and fits and is still used to a certain extent although all the fits provided by this system can be obtained with approximately the same values by selection from 1916. This system provides a range of clearance, transition and interference fits for size upto 12". *It is a hole basis system,* which stipulates two grades of holes, specified with bilateral tolerances, together with 6 grades of shaft tolerances.
- This system is extremely simple and is earliest of all the systems. It specifies too few fits and those listed do not enforce to modern ideas as regards their basic deviations. Though this served a useful purpose in the past but is *not considered suitable for modern production.*
- Since it is based on hole basis system, therefore, *in this system provision is made in the size of the hole for error in workmanship, and the variation to obtain the quality of fit required is allowed for on the size of the shaft which has to enter the hole.*

9.4.9 ISO System of Limits and Fits

ISO system has presently been universally adopted and as a matter of fact IS : 919 is almost in line with this system. While ISO specifies 28 classes of holes designated A, B, C, CD, E, EF, F, FG, G, H, J, JS, K, M, N. P, R, S, T, U, V, X, Y, Z, ZA, ZB, ZC and 18 grades of tolerance exactly matching with those of ISO systems. Similarly ISO has 28 classes of shafts while IS : 919 specifies only 25 classes of shaft. Other characteristics such as fundamental deviation and tolerance unit etc. are same in both the systems.

9.4.10 Types of Fits

Some important types of fits are discussed below:

1. **Selective fit:** A *selective fit* may be a *transition* or an *interference fit.* This type of fit is required where the object is to make a shaft and hole with a finite and not a permissible range on it. It is customarily used for tight or interference fits where it is desired to avoid the extremes of maximum tightness or looseness. The ideal selective fit for the tightest class of fit would stress the hole just to its elastic limit, thereby giving the maximum holding power without overstressing or distorting the grain structure.
2. **Push *fit*:** *A push fit* is a *transition fit.* It is also known as 'sung fit' and represents the closest fit that permits assembling parts by hand. With a push fit. there should be no perceptible play between the mating parts.
3. **Driving fit:** A *driving fit* is an *interference fit.* When a plug or shaft is made slightly larger than the hole into which it is to be inserted and the allowance is such that the parts can be assembled by driving, this is known as a driving fit. *Such fits are employed when the parts are to remain in a fixed position relative to each other.* Before assembling parts with a driving fit, the bearing surfaces should be oiled. A hydraulic press is usually preferable for assembling.
4. **Forced or pressed fit:** A forced or pressed fit is an *interference fit.* It is the term used when a pin, shaft or other cylindrical part is forced into a hole of slightly smaller diameter, ordinarily by the use of hydraulic press or some other type of press capable of exerting a considerable pressure. *A force fit lias a larger allowance than a driving fit, and therefore requires greater pressure for assembling.* Forced fits are restricted to parts *of small* and *medium size, e.g., crankpins, car wheel axles,* and *similar parts* (which must be held very securely).

5. **Shrinkage fit:** A *shrinkage fit* is an *interference fit.* It is obtained by making the internal member slightly larger than the hole in the external member. In this type of fit, the pressure is not required for assembly but instead the external member is heated and expanded sufficiently to permit inserting the internal member easily. Then as the external part cools or is cooled by applying water or dry ice, it shrinks tightly around the internal part. In general practice, a smaller allowance for shrinkage fit is favoured.
6. **Freeze fit:** In a *freeze fit,* instead of heating the female member, the male member may be *contracted by cooling and subsequently allowed to expand into the female.* This process uses an industrial refrigerator giving a temperature of about – 50°C, or to obtain lower temperatures the component is cooled in liquid air (boiling point –190°C). Examples of this process are in the insertion of exhaust valve seats inserts in engine cylinder heads or blocks, or in the insertion of brass bushes in various assemblies.

This process is convenient only for small parts as otherwise the size of the refrigerator equipment is prohibitive or consumption of liquid air is excessive. However, on suitable parts the process is very convenient, since the temperature is controlled and is unlikely to damage the structure of the material in any way. A combination of freezing and heating is also used as this enables reasonable maximum temperatures to be used. A *convenient method of reaching –50/–60°C without expensive equipment, and suitable for occasional use is to cool the part in alcohol to which solid frozen dioxide (known as 'dry ice'* or *'dry cold') is added.*

9.5 CONCEPT OF INTERCHANGEABILITY

Interchangeability *refers to assembling a number of unit components taken at random from stock so as to build up a complete assembly without fitting or adjustment.*

A modern motor-car, for example, consists of many hundreds of separate components each of which is manufactured in large numbers. For complete interchangeability it should be possible simply to collect at random the constituent parts then to assemble the whole without the use of any cutting tools and for the assembly to function satisfactorily.

The contacts between the various parts constitute what are termed *fits.*

For correct functioning of parts the fits must be good within certain limits of accuracy. It would be possible so to choose these limits as to ensure absolute interchangeability, and this should always be done in the case of less-important fits such as bolts-fitting in bolt-holes and so on. Experience shows, however, that to do this in the case where the tolerance on the fit is very small may call for such fine limits that the cost is excessive.

In cases like this a process known as *selective assembly* is used. Thus if we have a shaft required to run in a close fitting bearing we can arrange, during inspection, to sort the shafts into say, three grades, those near the upper limit, those near the middle and those nearly at the bottom limit. The same selection is made and those nearly at the bottom limit. The same selection is made with the bearings. By arranging to mate the top-limit shafts with the top-limit holes, for example, we shall ensure a much better assembly than if the parts were chosen at random. We can, infact, increase the limits on the components and thereby very much reduce the cost of production.

One of the objects of interchangeability is to make it possible to replace a work part, such as a complete ball bearing, without making any adjustment to the old or new parts. Here, of course, selective assembly is difficult except by actual manufacturers of the components, and in such cases it is necessary that absolute interchangeability should be possible.

9.6 GAUGES

A **gauge** *is tool or instrument used to measure or compare a component.* It is here employed in the sense of an instrument which, having fixed dimension, is used to determine whether the size of some component exceeds or is less than the size of the gauge itself.

The true value of a gauge is measured by its accuracy and service life which, in turn, depends on the workmanship and materials used in its manufacture. Since all gauges are continually subjected to abrasive wear while in use the selection of the proper material is of great importance. *High carbon and alloy tool steels* have been the *principal materials* used for many years. *These materials can be accurately machined to shape,* and they *respond readily to heat treating operations which increase their hardness and abrasive resistance. Steel gauges* are subjected to *some distortion* because of the heat-treating operation and their *surface hardness is limited.* These objections are largely overcome by the use of *chrome plating or converted carbides as the surface material.* Chrome plating permits the use of steels having inert qualities, since wear resistance is obtained by the hard chromium surface. This process is also widely *used* in the *reclaiming of the worn gauges. Cemented carbides, applied on metal shanks by powder-metallurgy technique, provide the hardest wearing surface obtainable.*

9.6.1 Classification of Gauges

The following gauges represent those most commonly used in *production work.* The classification is principally according to the shape or purpose for which each is used. A complete classification would require more sub-divisions under the various headings.

1. Snap gauges
2. Plug gauges
3. Ring gauges
4. Length gauges
5. Form comparison gauges
6. Thickness gauges
7. Indicating gauges
8. Pneumatic gauges
9. Electric gauges
10. Electronic gauges
11. Projecting gauges
12. Multiple dimension gauges.

9.6.2 Description of Some Commonly Used Gauges

Snap gauges. A snap gauge, used in the measurement of *plain external dimensions,* consists of a U-shaped frame having jaws equipped with suitable gauging surfaces. A plain gauge has two parallel jaws or anvils which are made to some standard size and cannot be adjusted. This type of gauge is largely being replaced by adjustable gauges which provide means of changing tolerance settings or adjusting to wear. Most gauges are provided with the "Go" and *"Not Go"* feature in a single jaw, such a design being both satisfactory and rapid. The general design shown in Fig. 9.10 has been selected because it incorporates most of the advantages of similar gauges. It is *light in weight, sufficiently rigid, easy to adjust, provided with suitable locking means and is designed to permit interchangeability of as many of the parts as possible.*

The tolerance for settings on such gauges as shown in Fig. 9.10 must account for the total *gauge allowance* which is cusomarily taken as 10% of the tolerance of the part; that is, 5% of the part tolerance for each button and the *wear allowance* which is 5% of the part tolerance. The allowance for wear is usually made only for 'go' gauges, since *'no gauge'* have little wear.

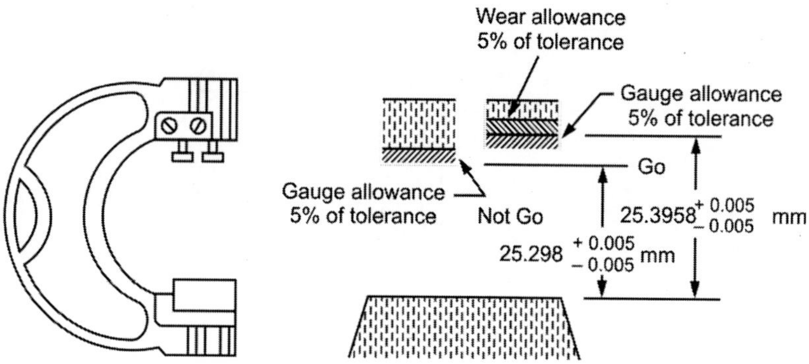

Fig. 9.10 Snap gauge.

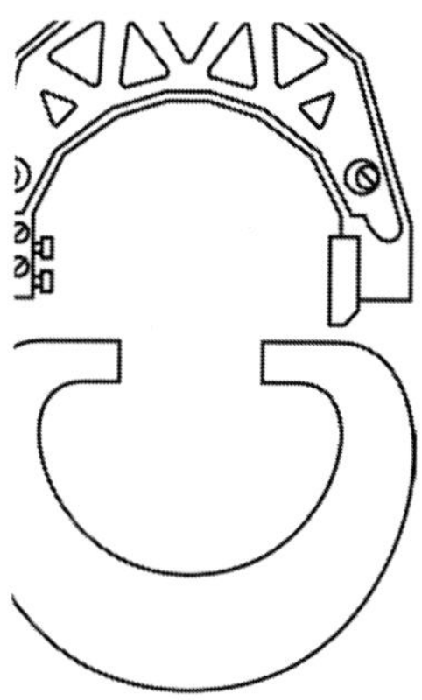

Fig. 9.11

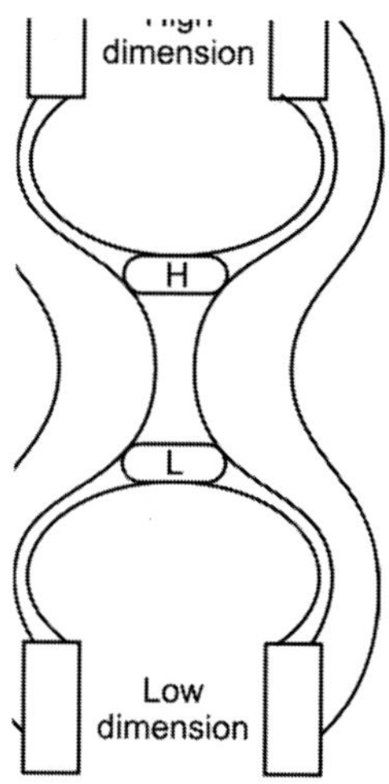

Fig. 9.12

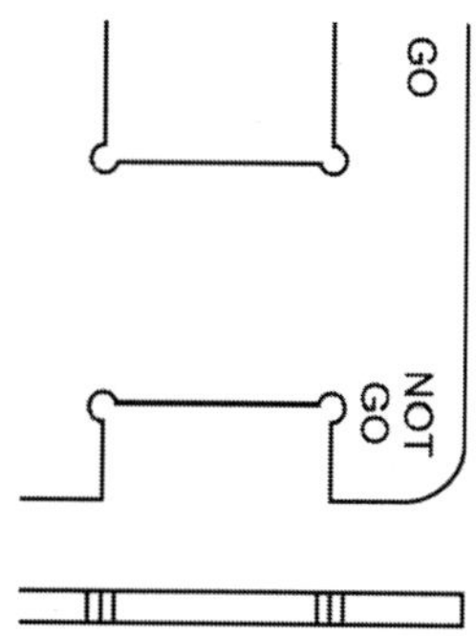

Fig. 9.13

The usual practice is to allocate both gauge tolerance and wear allowance entirely within the tolerance limits of the part to be inspected. An occasional part may be rejected though it has been made within its tolerance.

Figure 9.10 indicates the appropriate size for the "Go" and "Not Go" dimensions when the gauge is set to measure $25.40^{+0.00}_{-0.10}$ mm.

Other designs of snap gauges are shown in Fig. 9.11, 9.12, and 9.13.

Plug gauges. A plain plug gauge is an accurate cylinder used as an internal gauge for size *control of holes.* It is provided with a suitable handle for holding and is made in a variety of styles. These gauges may be either single or double ended. Double ended, plain gauges have "Go" and "Not Go" members assembled on *opposite ends,* whereas *progressive gauges have both gauging sections combined on one end.*

As for snap gauges, the allowance for manufacturing the gauge, as well as the allowance of the part to be inspected, must be considered in the design. Figure 9.14 shows a "Go" and "Not Go" gauge with the approximate dimensions for checking a hole $19.05^{+0.00}_{+0.10}$ mm.

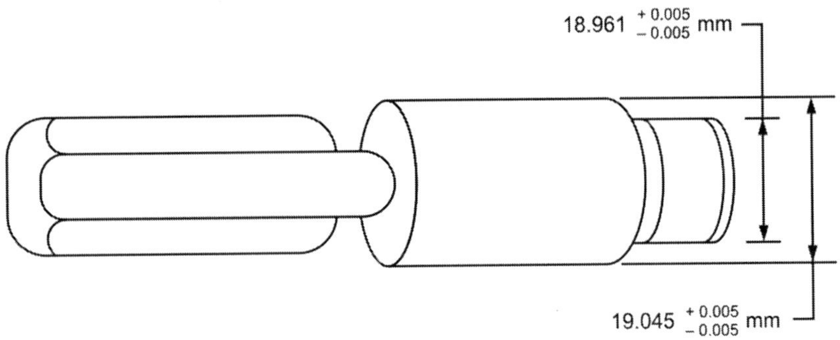

Fig. 9.14 Other types of gauges include ring, taper, thread, and thickness.

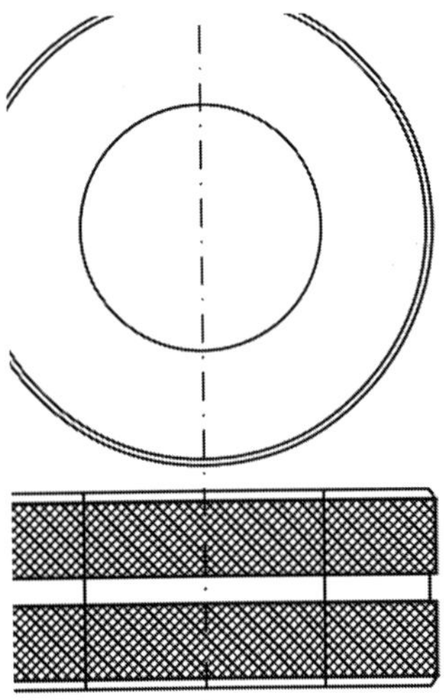

Fig. 9.15 Ring gauge.

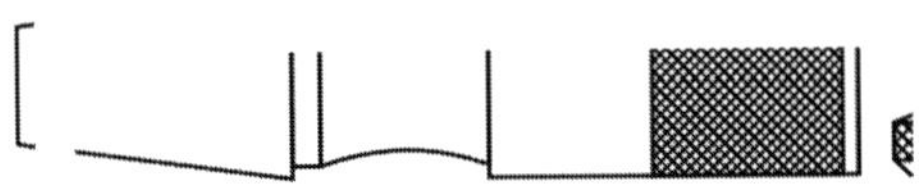

Fig. 9.16. Taper gauge for shafts.

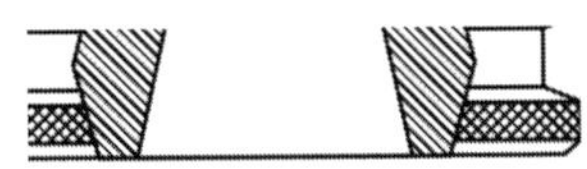

Fig. 9.17 Taper gauge for holes.

- **Ring gauges** (Fig. 9.15) for outside diameters, are used in pairs, a "Go" and "Not Go".
- **Taper gauges** are shown in Figs. 9.16 and 9.17. Taper gauges are not dimensional gauges but rather a means of checking in terms of degrees. Their use is a matter more of fitting rather than measuring.

A *thickness or feeler gauge* (Fig. 9.18) consists of a number of thin blades and is used in *checking clearances and for gauging in narrow places.*

Dial gauges. Dial gauges or dial test indicators are used for checking flatness of surfaces and parallelism of bars and rods. They are also used for testing the machine tools. They can also be used for measurement of linear dimensions of jobs which require easy readability and moderate precision. A dial gauge is shown in Fig. 9.19. It has two pointer arms which are actuated by a rack and pinion arrangement which acts as a mechanical amplifier. The rack is cut in a spindle. The spindle is made to come in contact with the workpiece. The linear displacement is converted into rotary movement of the pointers. The dial is divided into 100 equal divisions, each division represents a spindle movement of 0.01 mm. For 1 mm movement the bigger arm turns through one complete revolution. The smaller arm registers the number of full turns made by the bigger arm.

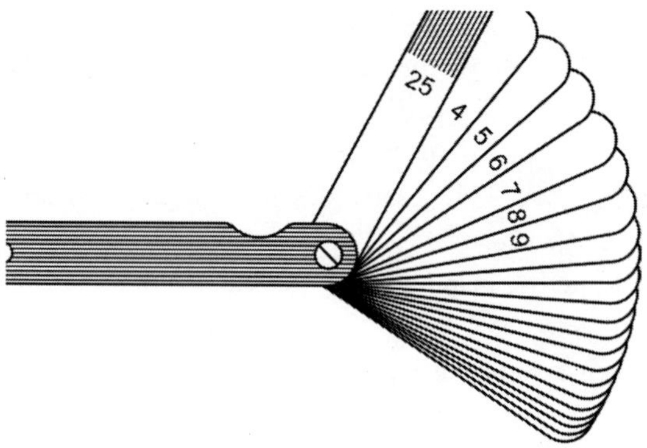

Fig. 9.18 Thickness or feeler gauge.

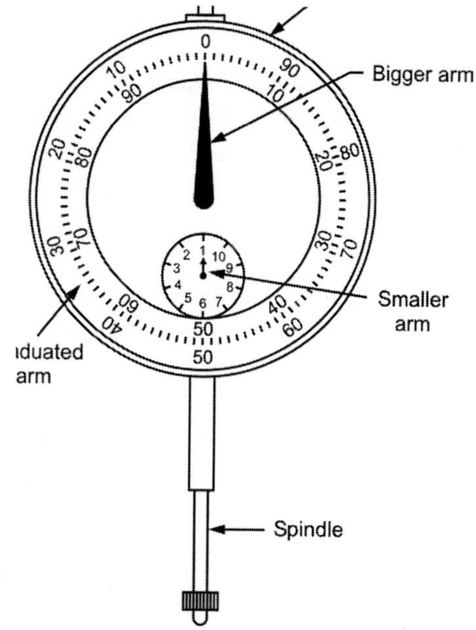

Fig. 9.19 Dial gauge.

Slip gauges. These are also known as *gauge blocks.* A gauge block is a *rectangular block mode up of hardened steel with two opposite faces separated by a defined distance.* The faces are ground and lapped to make them flat and parallel within fine limits.

- The blocks (slip gauges) are produced in sets with which it is possible to build a stack of any required height. The flatness and finish of the surfaces of the block is such that they are made to adhere to each other strongly sufficient enough to keep the stack together during normal use. *The process of making the blocks adhere is called* ***'wringing'.***

'Wringing' is done by first very *carefully cleaning* the faces, placing them in contact as in Fig. 9.20, *pressing them together lightly and twisting* as indicated by the arrows. Two true faces will adhere with very considerable force. It is not a question of air pressure, but a *molecular attraction; in fact if left too long in contact they may be damaged in separating them.* It is quite easy to wring ten or more slip gauges together and handle them as if they were one piece.

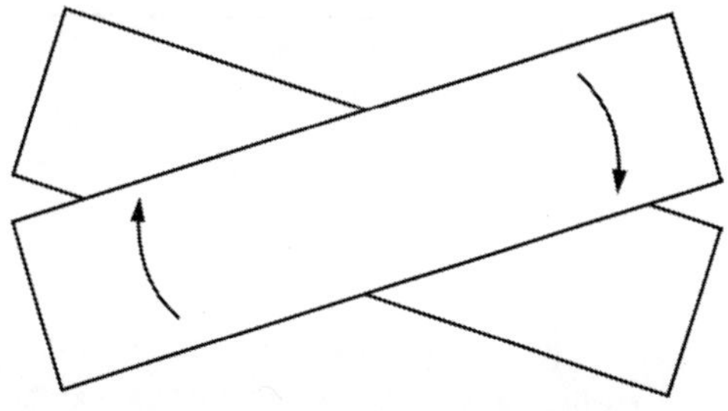

Fig. 9.20

- The rectangular blocks are 9 mm × 30 mm in size upto 10 mm length and 9 mm × 35 mm in size greater than 10 mm length. The lengths range upto 100 mm and the gauge blocks are graded depending upon their minimum accuracy of flatness, parallelism and face distance.

The gauge blocks are *classified* as follows:

1. **Reference grade.** Gauge blocks of reference grade are used as standards by manufactures of gauge blocks. They are also used in industries where work of similar nature is produced.
2. **Calibration grade.** Gauge blocks of calibration grade are used where highest level of accuracy is desired in normal engineering practice. They are used where tolerances are less than 2 mm. These gauge blocks are primarily used, in conjunction with suitable comparators, for calibration of other blocks.
3. **Grade 0, I and II.** These gauge blocks find extensive use in general engineering applications.

   Grade 0, Gauge blocks are used for jobs where high precision is required. These gauge blocks are of inspection grade.

   Grade I. These gauge blocks are used as general purpose manufacturing gauges in applications like gauge, tool and component production.

   Grade II. These gauge blocks are used for rough setting purposes and checking of components having wide tolerances.

The *Indian standard Specifications No.* 2984—1966 lists *two sets of gauge blocks.* They are given in the table below.

TABLE 9.2: Details of Gauge Blocks according to Indian Standard Specifications.

Normal Set

Size (mm)	*Increment (mm)*	*Number of pieces*
1.001 – 1.009	0.001	9
1.01 – 1.09	0.01	9
1.1 – 1.9	0.1	9
1 – 9	1	9
10 – 90	10	9
	Total	45

Special Set

Size (mm)	*Increment (mm)*	*Number of pieces*
1.001 – 1.009	0.001	9
1.01 – 14.9	0.01	49
0.5 – 9.5	0.5	19
10 – 90	10	9
	Total	86

9.7 LIMIT GAUGES

9.7.1 General Concepts

- It is not possible to make any part exactly to a given size or dimension. However carefully the part may be produced, some method of measurement could be found which can detect an error from the nominal size or variation in dimension. Therefore it is essential in engineering to allow a definite tolerance on every specified dimension, which can be a unilateral or a bilateral tolerance, within which the given dimensions is permitted to vary. The *limiting dimensions permitted by a given tolerance are called* **'limits'.**

- Present-day standards of quantity production, interchangeability, economical production and automatic assembly of complex components could not exist under a system in which the majority of such components are to be mated together and their dimensions are to be adjusted until the required type of fit is sustained. This can be fulfilled with the knowledge that certain dimensions can be reproduced with precision on any number of components based on a system of limits and fits.
- In mass production, it is necessary to check the dimensions of the manufactured components during production. Limit gauges are usually used. These gauges are of the "Go" and "Not Go" type. The *"Go" side of the limit gauses should enter the hole or just pass over the shaft under the weight of the gauge without using any force.* The *"Not Go" side of the gauge must not enter or pass.* The *basic or nominal size of the "Go" side is the high limit dimension of the shaft or the low limit dimension of the hole.* The basic size of the "Not-Go" side is low limit dimension of the shaft or the high limit dimension of the hole as shown in Fig. 9.21.

The three main grades of limit gauges used are:

1. **Workshop limit gauges** to be used on the machine for gauging the dimensions of the components during production.
2. **Inspection limit gauges** to be used in the inspection department for checking the component after production.
3. **Reference limit gauges** to be used in the metrology laboratory for reference proposes.

Grades 1 and 2 have the same production tolerances while the reference gauging are also provided with very narrow limits. Workshop limit gauges are also provided with wear allowance while the inspection gauges are provided with one-third of this wear allowance. This means that partly workshop gauges can be used as inspection gauges.

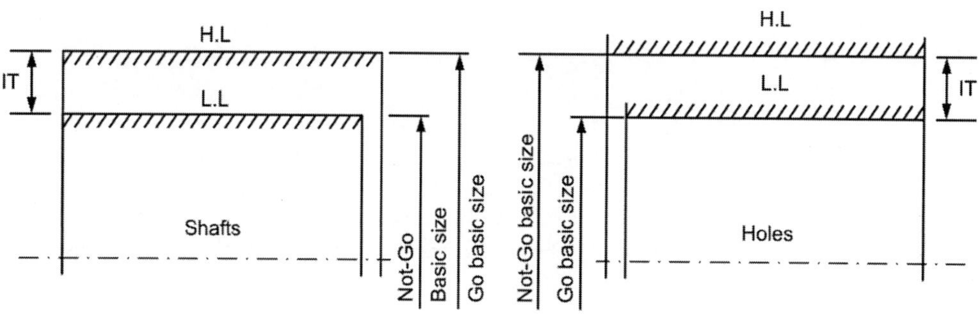

Fig. 9.21 Basic sizes for limit gauges.

- "Go" and "Not Go" gauges can be built up by using slip gauge blocks and their accessories as shown in Fig. 9.22. *These are usually used as reference gauges or when a small number of components have to be gauged.*

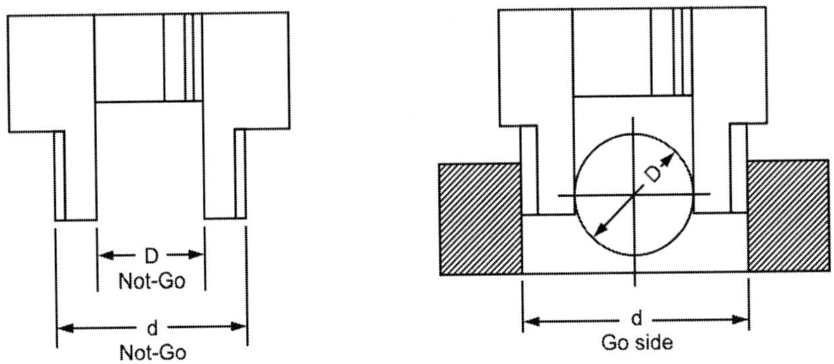

Fig. 9.22 Limit gauges built up by slip gauge blocks.

- For gauging holes, plug limit gauges are to be used. For gauging shafts, ring or shape limit gauges are used.

9.7.2 Taylor's Principle

"Go" and "Not-Go" gauges must fulfil *Taylor's principle* which states that:

"On the 'Go' code of a limit gauge, all related dimensions should be checked simultaneously, while on the 'Not-Go' side each dimension is to be checked separately."

This means that the *"Go"* gauge must have the ideal theoretical form of its mating part and it should be used *under the condition of maximum impassibility*. The *'Not Go"* gauge should ensure point contact with its mating part so that any geometrical inaccuracy that exists outside the limits of the dimensional tolerance can be detected. It must therefore, be used under the *condition of maximum possibility*.

THEORETICAL QUESTIONS

1. Define the following:
   (*i*) Metrology
   (*ii*) Legal metrology
   (*iii*) Dynamic metrology.
2. Enumerate the objectives of metrology.
3. What is the necessity and importance of metrology ?
4. State the need of precision measurement.
5. Discuss 'Metrology' as a means to achieve quality control.
6. What is actually understood by measurement ? What are the two standard systems of linear measurements?
7. Explain briefly the following:
   (*i*) Line standard (*ii*) End standard
   (*iii*) Wave length standard.
8. Disccus briefly the following:
   (*i*) Primary standards (*ii*) Secondary standards
   (*iii*) Tertiary standards (*iv*) Working standards.
9. State the charateristics of line standard.
10. What are the advantages of wavelength standards ?
11. Give classification of standards.

12. Give the relative characteristics of 'line' and 'end' standards.
13. Define the following:
    Nominal size, Basic dimension, Tolerance, Upper deviations, Fit, Allowance.
14. How do you classify fits?
15. Explain briefly basis of Fit (or limit) system.
16. What are the systems of specifying tolerances; which system is used most and why ?
17. How are holes; shafts and fits designated? Give examples.
18. Write a short note on 'Newell system'.
19. Explain briefly the following:
    (*i*) Selective fit
    (*ii*) Push fit
    (*iii*) Driving fit
    (*iv*) Forced or pressed fit
    (*v*) Shrinkage fit
    (*vi*) Freeze fit.
20. Write a short note on "Concept of Interchangeability'.
21. Explain any two of the following gauges:
    (*i*) Snap gauges.
    (*ii*) Plug gauges.
    (*iii*) Dial gauges.
22. What are "Limit gauge?
23. What is "Taylor's Principle" ? Explain.

❑❑❑

Linear and Angular Measurements, and Comparators

Chapter 10

10.1 LINEAR MEASUREMENTS

For linear measurements, the following instruments are used:

1. Engineer's Steel Rule
2. Calipers
3. Vernier Calipers
4. Vernier Height Gauge
5. Micrometers

These measuring instrument are discussed as follows:

10.1.1 Engineer's Steel Rule

An engineer's *steel rule* is also known as **Scale** and is a *line measuring device.* It is a precision measuring instrument and must be treated as such, and kept in a nicely polished condition.

It works on the basic measuring technique of *comparing an unknown length to the one previously calibrated.*

It consists of *strip of hardened steel* having line graduations etched or engraved at interval of fraction of a standard unit of length. Depending upon the interval at which the graduations are made, the scale can be manufactured in different sizes and styles. The scale is available in 150 mm, 300 mm, 600 mm and 1000 mm lengths.

Some scales are provided with some attachments and special features to make their use versatile, *e.g.*, very small scales may be provided with a handle to use it conveniently. They may be made in folded form so that they can be kept in pocket also. *Shrink rules* are the scales (used in foundry and pattern making shops) which take into account the shrinkage of materials after cooling.

Following are the *desirable qualities of a steel rule:*

(*i*) Good quality spring steel,
(*ii*) Clearly engraved lines,
(*iii*) Reputed make,
(*iv*) Metric on two edges, and
(*v*) Thickness should be minimum.

Use of scale. To get good results it is necessary that certain techniques must be followed while using a scale.

- The end of the scale must never be set with the edge of the part to be measured, because generally the scale is worn out at the ends and also it is very difficult to line up the end of the scale accurately with the part of the edge to be measured.
- The scale should never be laid flat on the part to be measured because it is difficult to read the correct dimension.

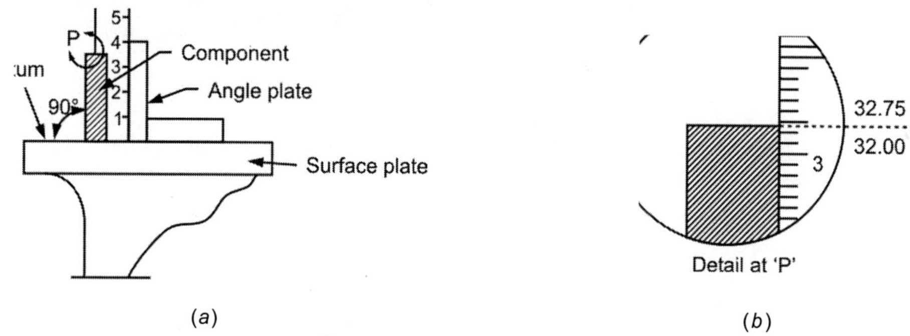

Fig. 10.1 Correct use of scale.

The principle of *common datums* should be employed while using a scale or rule. The principle is shown in Fig. 10.1(*a*), the set up indicating the correct method of measuring the length of a component. A surface plate is used as a datum face and its purpose is to provide a common location or position from which the measurement can be made. It may be noted that both the rule and key are at right angles to the working surface of the surface plate and the use of an angle plate simplifies the set-up.

The degree of accuracy which can be obtained while making measurements with a steel rule or scale depends upon:

(*i*) Quality of the rule, and

(*ii*) Skill of the user.

The correct technique of reading the scale is simply illustrated in Fig. 10.1(*b*).

10.1.2 Calipers

Caliper *is an instrument used for measuring distance between or over surfaces, or for comparing dimensions of workpieces with such standards as plug gauges, graduated rules etc.* In modern precision engineering they are not employed on finishing operations where high accuracy is essential, but in skilled hands they remain extremely useful. No one can prevent the spring of the legs affecting the measurement, and adjustment of *the firm-joint* type can be made only by tapping a leg or the head. Thus results obtained by using calipers depends very largely on the degree to which the user has developed a sense of touch.

Some firm-joint calipers (Fig. 10.2) have an adjusting screw which enables finer and more controlled adjustment than is possible by tapping methods. Thus at (*e*) in Fig. 10.2 is shown the Blacksmith's caliper made with firm joints and a long handle, the latter enabling the measurement of hot forgings without discomfort. The long arm is used for the greater, and the small arm for the smaller or furnished *size.*

At (*f*) is shown a *wide jawed caliper* used in rough measurement of the major diameters of threaded places. For measuring minor diameters a caliper with specially thinned points is sometimes used.

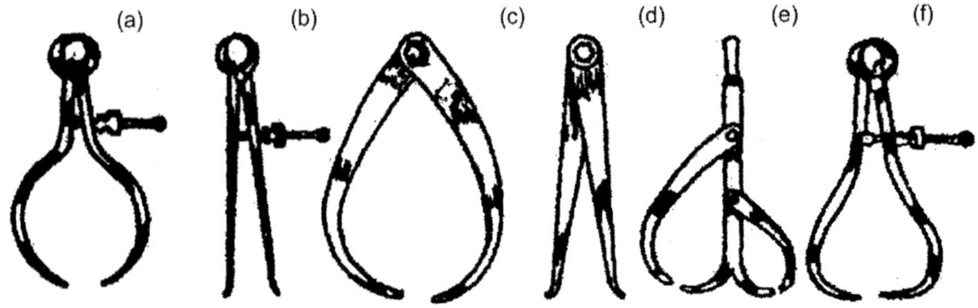

Fig. 10.2 Firm-joint calipers.

It is *unwise to use calipers on work revolving in a lathe.* If one contact point of the caliper touches revolving work, the other is likely to be sprung and drawn over it by friction.

10.1.3 Vernier Calipers

10.1.3.1 Introduction

The vernier instruments generally used in workshop and engineering metrology have comparatively *low accuracy*. The line of measurement of such instruments does not coincide with the line of scale. The accuracy therefore depends upon the straightness of the beam and the squareness of the sliding jaw with respect to the beam. To ensure the squareness, the sliding jaw must be clamped before taking the reading. The zero error must also be taken into consideration. Instruments are now available with a measuring range upto one metre with a scale value of 0.1 or 0.2 mm. They are made *of alloy steel,* hardened and tempered (to about 58 Rockwell C), and the contact surfaces are lap-finished. In some cases stainless steel is used.

10.1.3.2 The Vernier Principle

The principle of vernier is that when two scales or division slightly different in size are used, the difference between them can be utilised to enhance the accuracy of measurement.

Principle of 0.1 mm vernier. In the Fig. 10.3 is shown *the principle of 0.1 mm vernier.* The *main scale* is accurately graduated in 1 mm steps, and terminates in the form of a caliper jaw. There is a second scale which is movable, and is also fixed to the caliper jaw. The movable scale is equally divided into 10 parts but *its length* is only 9 mm; therefore one division on this scale is equivalent to 9/10 = 0.9 mm. This means the *difference between one graduation on the main scale and one graduation on the sliding or vernier scale* is 1.0 – 0.9 = 0.1 mm. Hence if the vernier caliper is initially closed and then opened so that the *first graduation on the sliding scale corresponds to the first graduation on the main scale a distance equal to 0.1 mm* has moved as shown in Fig. 10.4. Such a vernier scale is of limited use because measurements of greater accuracy are normally required in precision engineering work.

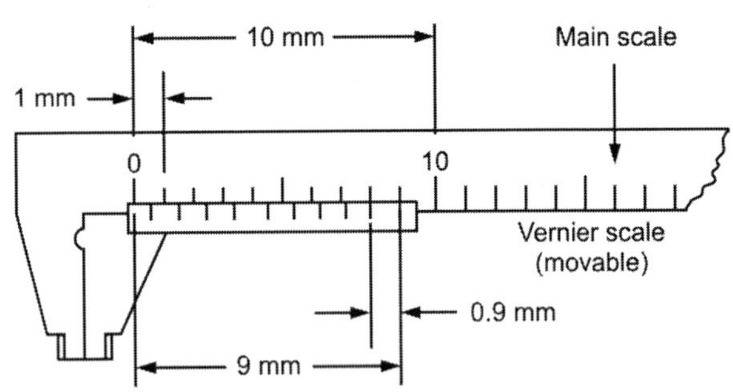

Fig. 10.3 Principle of 0.1 mm vernier.

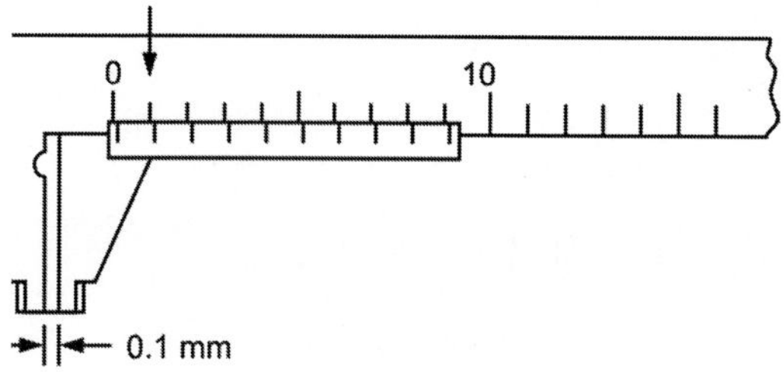

Fig. 10.4

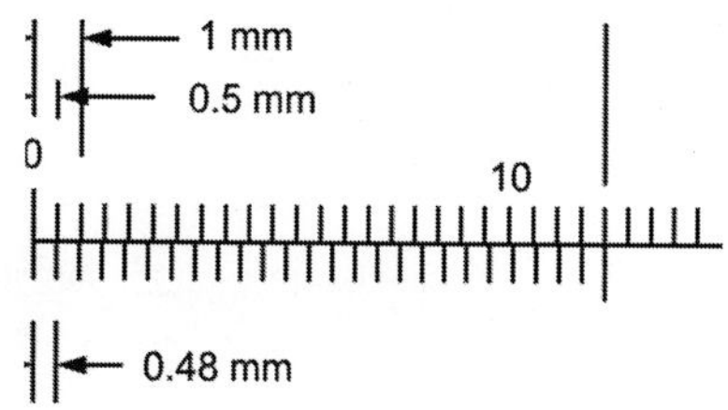

Fig. 10.5 Principle of 0.02 mm vernier.

Principle of 0.02 mm vernier. In Fig. 10.5 is shown the principle of a 0.02 *mm vernier.* The vernier scale has main scale graduation of 0.5 mm, whilst the vernier scale has 25 graduations equally spaced over 24 main scale graduations, or 12 mm. Hence each division on the vernier scale = 12/25 = 0.48 mm.

The difference between one division on the main scale and one division on the vernier scale

$$= 0.5 - 0.48 = 0.02 \text{ mm}$$

This type of vernier is read as follows:

- Note the number of millimetres and half millimetres on the main scale that are coincident with the zero on the vernier scale.
- Find the graduation on the vernier scale that coincides with a graduation on the main scale. This figure must be multiplied by 0.02 to give the reading in millimetres.
- Obtain the total reading by *adding the main scale reading to the vernier scale reading.*

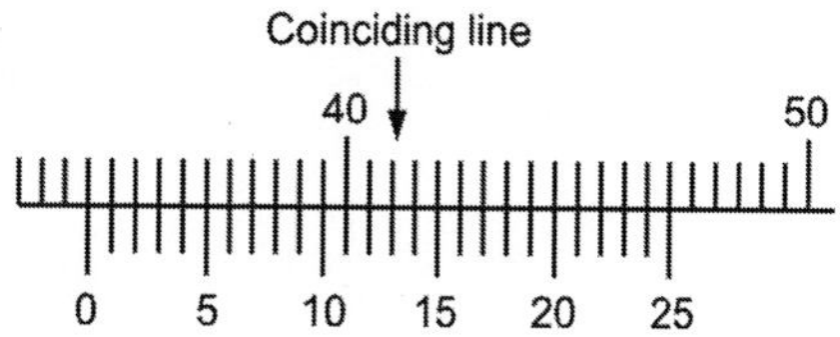

Fig. 10.6

Example: An example of a 0.02 mm vernier reading is given in Fig. 10.6.

Reading on the main scale upto zero of the vernier scale = 34.5 mm.

The number of graduation that coincides with the graduation on the main scale = 13th

This represents a distance of: $13 \times 0.02 = 0.26$ mm

$\therefore$ Total reading = 34.5 + 0.25 = 34.76 mm.

Note: While taking measurements using vernier calipers it is important to *set the caliper faces parallel* to *the surface across which measurements are to be made. Incorrect* reading will result if it is not done.

10.1.3.3 Types of Vernier Calipers

According to Indian Standard IS : 3651-1974, three types of vernier calipers have been specified to make external and internal measurements and are shown in Figs. 10.7, 10.8 and 10.9 respectively. All the three types are made with one scale on the front of the beam for direct reading.

- **Type A** (Fig. 10.7) vernier has jaws on both sides for *external and internal measurements, and a blade for depth measurement.*
- **Type B** (Fig. 10.8) is provided with jaws on one side for *external and internal measurements.*
- **Type C** (Fig. 10.9) has jaws on both sides for making the *measurements and for marking operations.*

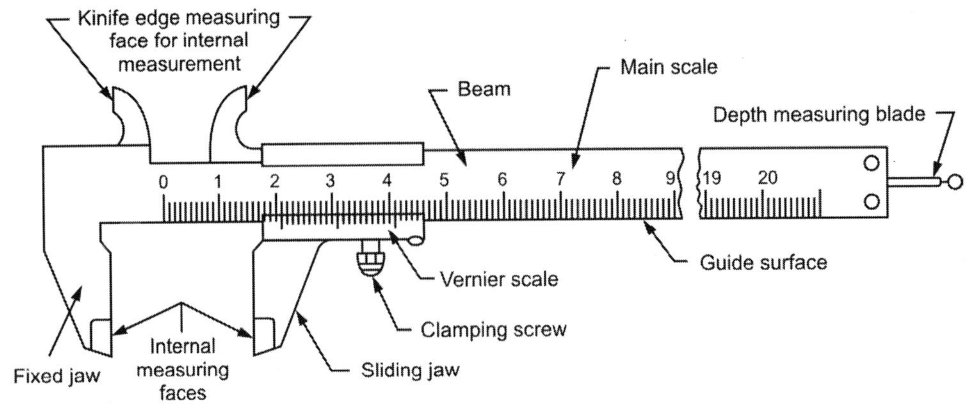

Fig. 10.7 Vernier caliper–Type A.

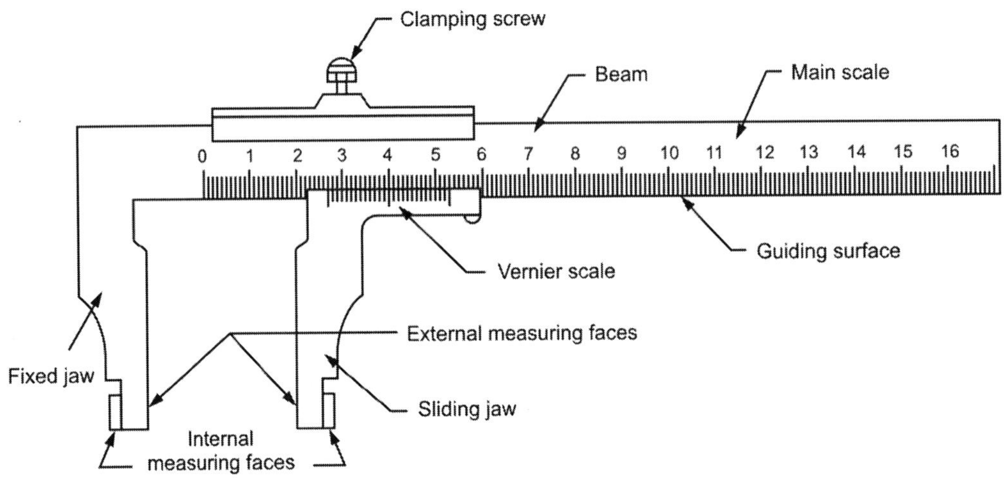

Fig. 10.8 Vernier caliper–Type B.

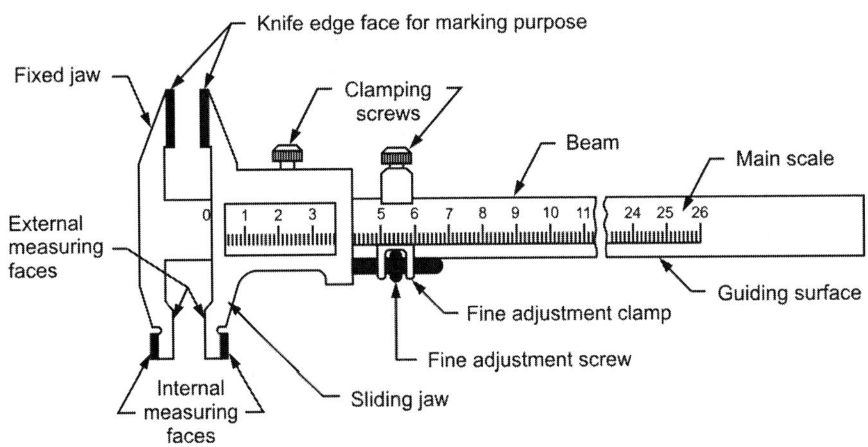

Fig. 10.9 Vernier caliper–Type C.

All parts of the vernier calipers should be of good quality steel and the measuring faces should possess a minimum hardness of 650 HV. The recommended measuring ranges (nominal size) of vernier calipers as per IS : 3651-1974 are:

0—125, 0—200, 0—250, 0—300, 0—500, 0—750, 0—1000, 750—1500 and 750—2000 mm.

10.1.4 Vernier Height Gauge

Refer to Fig. 10.10. The *vernier height gauge is mainly used in the inspection of parts and layout work.* It may be *used to measure and mark vertical distances above a reference surface, or an outside caliper.*

It consists of the following parts:

1. Base
2. Beam
3. Measuring jaw and scriber
4. Graduations
5. Slider.

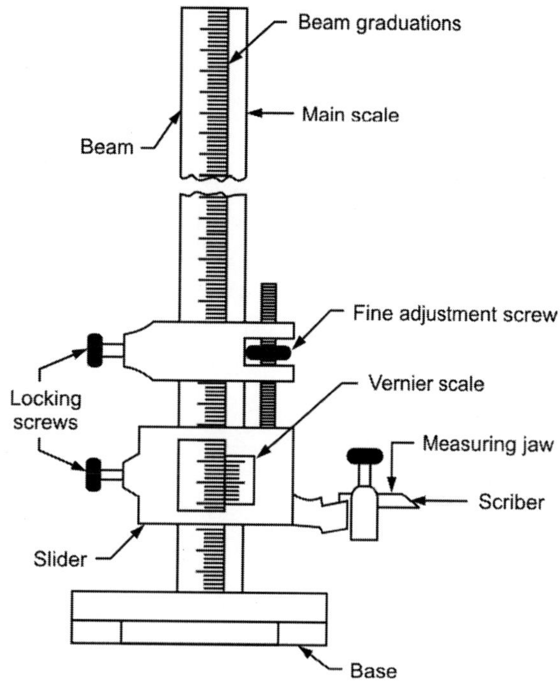

Fig. 10.10 Height gauge.

Base. It is made quite robust to ensure rigidity and stability of the instrument. The underside of the base is relieved leaving a surface round the outside edge of a least 7 mm width and an air gap is provided across the surface to connect the relieved part with the outside. The base is ground and lapped to an accuracy of 0.005 mm as measured over the total span of the surface considered.

Beam. The section of the beam is so chosen as to ensure rigidity during the use. The guiding edge of the beam should be perfectly flat within the tolerances of 0.02, 0.04, 0.06, 0.08 mm for measuring range of 250, 500, 750, 1000 mm respectively. The faces of the beam should also be flat within the tolerances of 0.04, 0.06. 0.10, 0.12 mm for vernier measuring heights of 250, 500, 750, 1000 mm respectively.

Measuring jaw and scriber. The clear projection of the measuring jaw from the edge of the beam should be atleast equal to the projection of the beam from the base. For all position of the slider, the upper and lower gauging surfaces of the measuring jaw should be flat and parallel to the base to within 0.008 mm. The measuring faces of the scriber should be flat and parallel to within 0.005 mm. The projection of the scriber beyond the jaw should be at least 25 mm. Vernier height gauges may also have an offset scriber and the scale on the beam is so positioned that when the scriber is co-planar with the base, the vernier is at zero position.

Graduations. The following *requirements* should be fulfilled in respect of graduations on scales :

- All graduations on the scale and vernier should be clearly engraved and the thickness of graduation both on scale and vernier should be identical and should be in between 0.05 mm and 0.1 mm.
- The visible length of the shortest graduation should be about 2 to 3 times the width of the interval between the adjacent lines.

- The perpendicular distance between the graduations on scale and the graduations on vernier should in no case be more than 0.01 mm.
- For easy reading, it is recommended that the surfaces of the beam and vernier should have dull finish and the graduations lines blackened in. Sometimes a magnifying lens is also provided to facilitate taking the readings.

Slider. The slider has a good sliding fit along the full working length of the beam. A suitable fitting is incorporated to give a fine adjustment of the slider and a suitable clamp provided so that the slider could be effectively clamped to the beam after the fine adjustment has been made.

An important feature of the height gauge is that a special attachment can be fitted to the part to which the scriber is normally fitted, to convert it, in effect, into a depth gauge.

Precautions:

1. When using any height gauge or surface gauge, care must be taken to ensure that the base is clean and free from burrs. It is essential, too, that final setting of vernier height gauges be made after the slider has been locked to the vertical column.
2. The height gauges are generally kept in their cases when not in use. Every care should be taken, particularly in case of long height gauges, to avoid its heating by warmth from the hands. The springing of measuring jaw should be always avoided.

10.1.5 Vernier Depth Gauge

A vernier depth gauge is used to *measure the depth of holes, recesses and distances from a plane surface to a projection.*

In Fig. 10.11 is shown a vernier depth gauge in use. The vernier scale is fixed to the main body of depth gauge, and is read in the same way as vernier calipers. Running through the depth gauge body is the main scale the end of which provides the datum surface from which the measurements are taken.

The depth gauge is carefully made so that the beam is perpendicular to the base in both directions. The end of the beam is square and flat, like the end of a steel rule and the base is flat and true, free from curves or waviness.

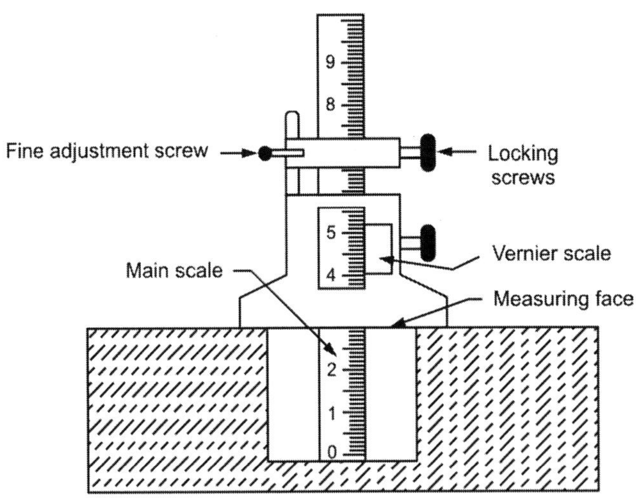

Fig. 10.11 Vernier depth gauge in use.

Use of vernier depth gauge:

- While using the vernier depth gauge, first of all, make sure that the reference surface, on which the depth gauge base is rested, is satisfactorily true, flat and square. Measuring depth is a little like measuring an inside diameter. The gauge itself is true and square but can be imperceptibly tipped or canted, because of the reference surface perhaps and offer erroneous reading.
- In using a depth gauge, press the base or anvil firmly on the reference surface and keep several kilogrammes hand pressure on it. Then, in manipulating the gauge beam to measure depth, be sure to apply only standard light, measuring pressure one to two kg—like making a light dot on paper with a pencil.

10.1.6 Micrometers

***Micrometers** are designed on the principle of **'Screw and Nut'**.*

10.1.6.1 Description of a micrometer

Figure 10.12 shows a 0-25 mm micrometer which is used for *quick, accurate measurements* to the *two-thousandths of a millimetre.* It consists of the following parts:

1. Frame
2. Anvil
3. Spindle
4. Thimble
5. Ratchet
6. Locknut.

The micrometer requires the use of an *accurate screw thread as* a means of obtaining a measurement. The screw is attached to a *spindle* and is turned by movement of a *thimble or ratchet* at the end. The barrel, which is attached to *the frame,* acts as a nut to engage the screw threads, which are accurately made with *a pitch of 0.05 mm.* Each revolution of the thimble advances the screw 0.05 mm. On the barrel a datum line is graduated with two sets of division marks. The set below the datum line reads in millimetres, and the set above the line reads in half millimetres. The thimble scale is marked in 50 equal divisions, figured in fives, so that each small division on the thimble represents 1/50 of 1/2 mm which is 1/100 mm on 0.01 mm.

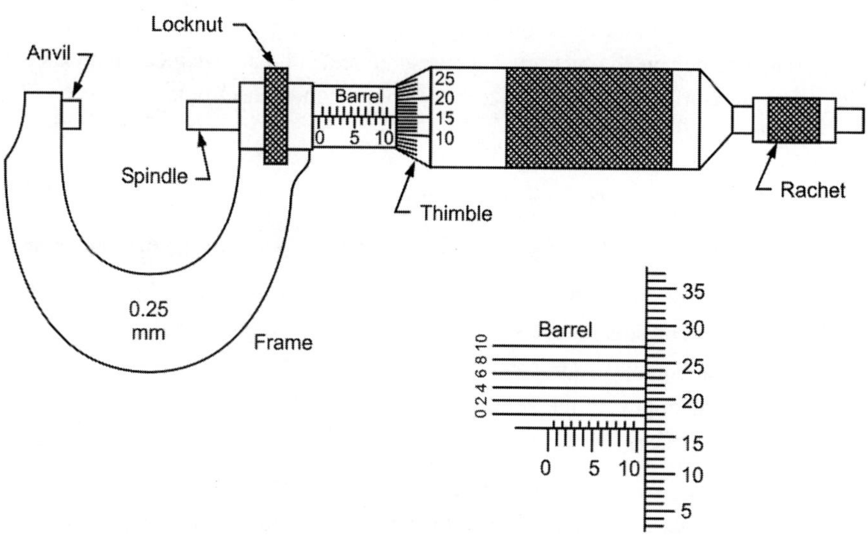

Fig. 10.12 0-25 mm micrometer.

- To *read the metric micrometer to 0.01 mm,* examine Fig. 10.12 and first note the whole number of major divisions on the barrel, then observe whether there is a half millimetre visible on the top of the datum line, and last read the thimble for hundredths. The *thimble reading is the line coinciding with the datum line.* The reading for Fig. 10.12 is as follows:

$$\begin{aligned} \text{Major divisions} &= 10 \times 1.00 \text{ mm} &&= 10.00 \text{ mm} \\ \text{Minor divisions} &= 1 \times 0.50 \text{ mm} &&= 0.50 \text{ mm} \\ \text{Thimble divisions} &= 16 \times 0.01 \text{mm} &&= 0.16 \text{ mm} \\ \hline & \text{Reading} &&= 10.66 \text{ mm} \end{aligned}$$

  Since a micrometer reads only over a 25-mm range, to *cover a wide range of dimensions, several sizes of micrometers are necessary.*

  The micrometer principle of measurement is also applied to inside measurements and depth reading and to the measurements of screw threads.

- To read the metric micrometer to 0.002 mm, vernier on the barrel is next considered. The vernier, shown rolled out in Fig. 10.12. has each vernier graduation represent two thousandths of a millimetre (0.002 mm), and each graduation is marked with a number 0, 2, 4, 6, 8 and 0 to help in the reading. To *read a metric vernier micrometer note the major, minor and thimble divisions. Next observe which vernier line coincides with a graduated line on the thimble.* This gives the number of two thousandths of a millimetre to be added to the hundredth's reading. For the cut out in Fig. 10.12, the reading is as follows:

$$\begin{aligned} \text{Major divisions} &= 10 \times 1.00 \text{ mm} &&= 10.00 \text{ mm} \\ \text{Minor divisions} &= 1 \times 0.50 \text{ mm} &&= 0.50 \text{ mm} \\ \text{Thimble divisions} &= 16 \times 0.01 \text{ mm} &&= 0.16 \text{ mm} \\ \text{Vernier divisions} &= 3 \times 0.002 \text{ mm} &&= 0.006 \text{ mm} \\ \hline & \text{Reading} &&= 10.666 \text{ mm} \end{aligned}$$

If the vernier *line coincident with the datum line is 0,* no thousandths of a millimetre are added to the reading.

Note. For shop measurements to 0.001 mm, a mechanical bench micrometer may be used. This machine is set to correct size by precision gauge blocks, and readings may be made directly from a dial on the head-stock. Constant pressure is maintained on all objects being measured and comparative measurements to 0.0005 mm are possible. Precision measuring machines utilizing a combination of electronic and mechanical principles are capable of an accuracy of 0.000 001 m.

10.1.6.2 Sources of errors in micrometers

Some possible *sources of errors* which may result in incorrect functioning of the instrument are:

(*i*) The anvils may not be truly flat.

(*ii*) Lack of parallelism and squareness of anvils at some, or all, parts of the scale.

(*iii*) Setting of zero reading may be inaccurate.

(*iv*) Inaccurate readings following the zero position.

(*v*) Inaccurate readings shown by fractional divisions on the thimble.

The parallelism is checked by measuring the diameter of a standard accurate ball across at least three different points on the anvil faces. The *squareness* of the anvils to the measuring axis is checked by using two standard balls whose diameters differ by an odd multiple of half a pitch which calls for turning the movable anvil at 180° with respect to fixed one. *Flatness* of the anvils is tested by the interference method using optical flats. The face must not show more than one complete interference band, *i.e.*, must be within 0.25 μm. Whenever tested at 20°C, the *total error should not exceed the following values*:

$$\text{For grade 1, total error} = \left(4 + \frac{L}{100}\right)\mu\text{m}$$

$$\text{For grade 2, total error} = \left(10 + \frac{L}{100}\right)\mu\text{m}$$

where, L = Upper limit of the measuring range in mm.

The micrometer must be so adjusted that the cumulative error at the lower and upper limits of the measuring range *does not exceed half the total error.*

10.1.6.3 Precautions in using the micrometer

The following *precautions* should be observed while using a micrometer :

1. Micrometer should be cleaned of any dust and spindle should move freely.
2. The part whose dimension is to be measured must be held in *left hand and the micrometer in right hand*.

   The way for holding the micrometer is to place the small finger and adjoining finger in the U-shaped frame. The forefinger and thumb are placed near the thimble to rotate it and the middle finger supports the micrometer holding it firmly. Then the micrometer dimension is set slightly larger than the size of the part and the part is slid over the contact surfaces of micrometer gently. After it, the thimble is turned till the measuring tip first touches the part and the final movement given by ratchet so that uniform measuring pressure is applied. In case of circular parts, the micrometer must be moved carefully over representative arc so as to note maximum dimension only.
3. Error in readings may occur due to lack of flatness of anvils, lack of parallelism of the anvils as part of the scale or throughout, inaccurate setting of zero reading etc. Various tests to ensure these conditions should be carried out from time-to-time.
4. The micrometers are available in various sizes and ranges, and the corresponding micrometer should be chosen depending upon the dimension.

10.1.6.4 Types of micrometers

Different types of micrometers are described below:

1. Depth micrometer. It is also known as *'micrometer depth gauge'*. Figure 10.13 illustrates a depth micrometer. The measurement is made between the end face of a measuring rod and a measuring face. Because the measurement increases as the measuring rod extends from the face, the readings on the barrel are *reversed* from the normal; *the start at a maximum* (when *the measuring rod is fully extended from the measuring face) and finish at zero* (when *end of the measuring rod is flush with the face).*

For example, the measurement on the depth micrometer as shown in Fig. 10.13 is: 16 + (18 × 0.01) mm

$$= 16 + 0.18 \text{ mm} = 16.18 \text{ mm}$$

Measuring rods in steps of 25 mm can be interchanged to give a wide measuring range. The thimble cap is unscrewed from the thimble which allows the rod to be withdrawn. The desired rod is then inserted and thimble cap replaced, so holding the rod firmly against a rigid face.

Figure 10.14 shows the applications of a depth micrometer.

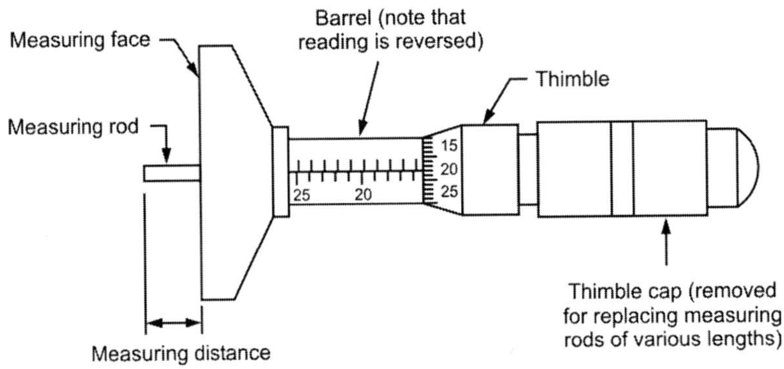

Fig. 10.13 Depth micrometer.

A depth micrometer is *tested for accuracy* as follows:

- In order to check the acuracy of a depth micrometer *unscrew the spindle* and *set the base of the micrometer on a flat surface like a surface plate or tool maker's flat.*
- Holding the base down firmly turn the thimble or screw in, or down, and when the tip of the micrometer depth stem contacts the flat firmly, with not more than one kg gauging pressure, read the barrel. *If the micrometer is accurate it should read zero.*
- Then rest the micrometer on a 25 mm slip gauge and screw the stem all the way down to contact with the flat. There it should register 25 mm.

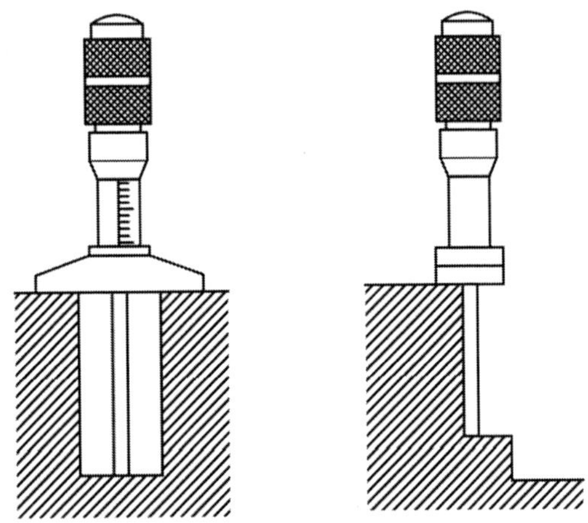

Fig. 10.14 Applications of a depth micrometer.

2. Height micrometer. Figure 10.15 shows a height micrometer. The same idea as discussed under depth micrometer is applied to the height micrometer.

3. Internal micrometers. These micrometers are used for measuring internal dimensions. The micrometer can be a rod provided with spherical anvils as shown in Fig. 10.16(*a*). The measuring range of this micrometer is from 25 to 37.5 mm, i.e., 12.5 mm. By means of exchangeable anvil rods, the measuring capacity can be increased in steps of 12.5 mm upto 1000 mm. Another type of internal micrometer is that shown in Fig. 10.16(*b*), in which the measuring anvils are inverted cantilevers. The measuring range of this micrometer is from 5 to 30 mm, i.e., 25 mm.

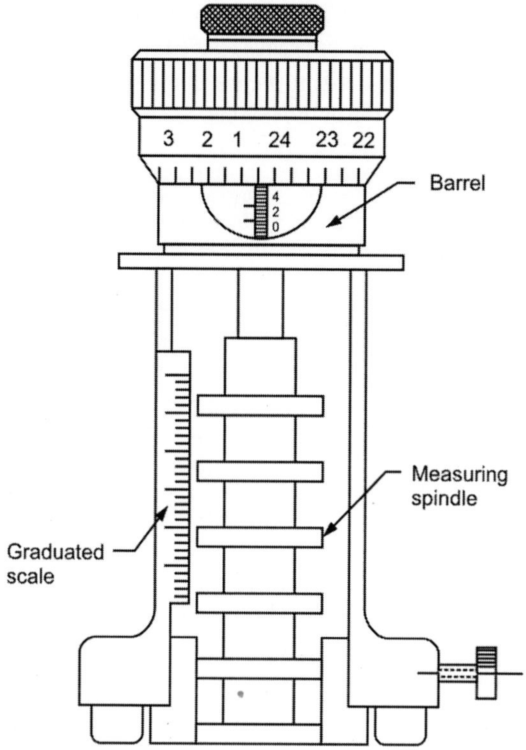

Fig. 10.15 Height micrometer.

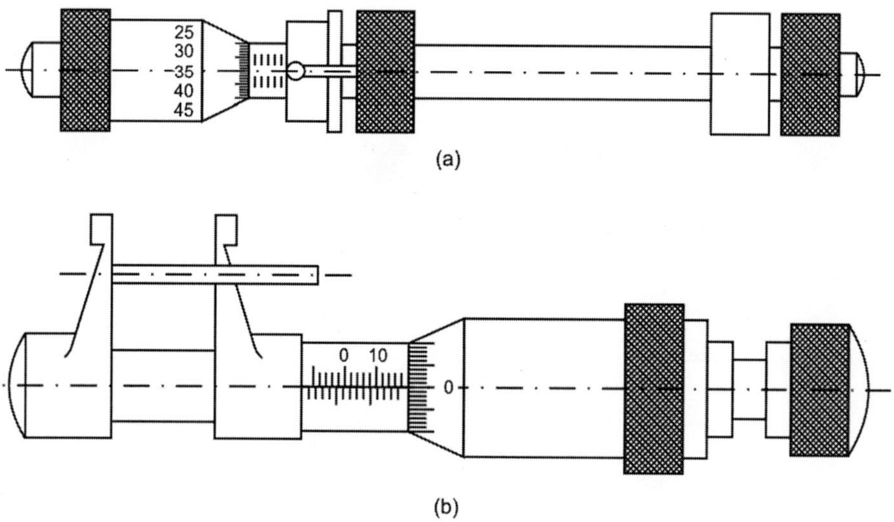

Fig. 10.16 Internal micrometer.

4. Differential micrometer. Refer Fig. 10.17. This type of micrometer is used to *increase the accuracy of the micrometers.* The right hand screws of different pitches p_1 (= 1.05 mm) and p_2(= 1 mm) are artranged such that due to rotation of the thimble, the thimble will move relative to the graduated barrel in one direction while the movable anvil, which is not fixed to the thimble but slides inside the barrel, moves in the other direction. The net result is that the movable anvil receives a total movement in one direction given by 1.05 – 1.0 = 0.05, *i.e.*, 1/20 mm per one revolution of the thimble. When the thimble scale is divided to 50 equal divisions, the scale value of the differential manometer will be $\frac{1}{20} \times \frac{1}{50} = 0.01$ mm.

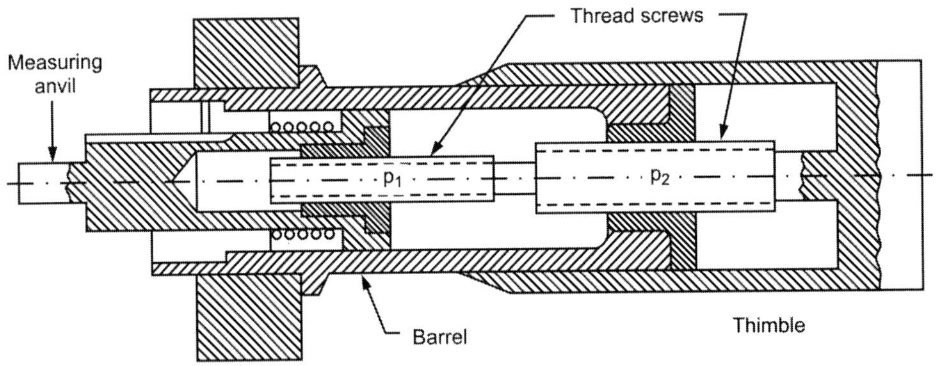

Fig. 10.17 Differential micrometer.

If a vernier scale is provided on the barrel, the micrometer would have a scale value of 0.1 μm. The measuring range, however, is comparatively small (about 5 mm).

5. Micrometer with dial gauge. Refer to Fig. 10.18. In order to enhance the accuracy of micrometers, different types are designed in which the fixed anvil is not merely a fixed one but moves axially to actuate a dial gauge through a lever mechanism. The micrometer can be used with the dial gauge anvil clamped as an ordinary micrometer for external measurement. Using the dial gauge, the micrometer works as comparator for checking similar components. The micrometer can be provided with a third anvil to improve and facilitate the mounting of the workpiece. Such a micrometer is called *snap dial gauge or snap dial micrometer.*

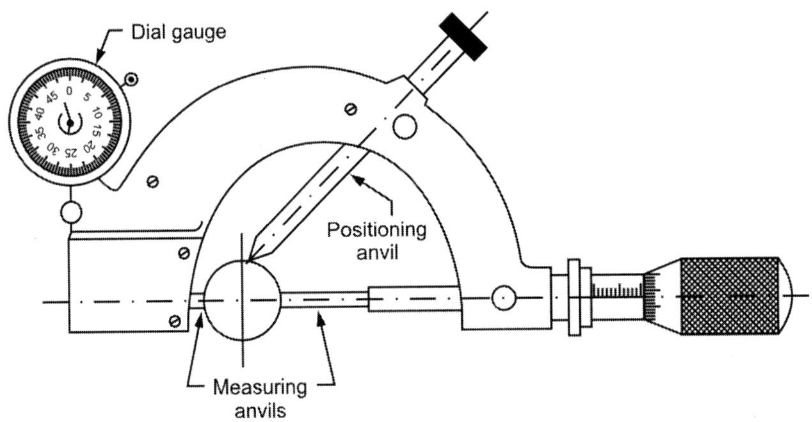

Fig. 10.18 Micrometer with dial-gauge.

Types of anvils: Figure. 10.19 shows different types of anvils in use:

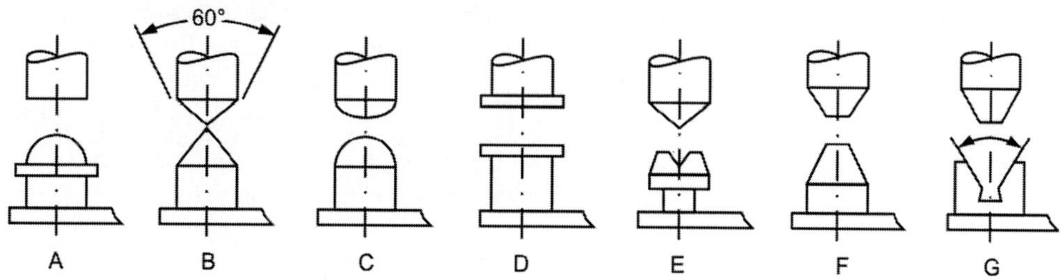

Fig. 10.19 Types of anvils.

A — These anvils are fitted with hardened, flat and spherical ends to enable *measuring tube thicknesses.* In such cases the radius of curvature of the fixed anvil must be less than that of the tube being measured.

B — These pointed anvils are used to measure *thicknesses of curved surfaces or depth of recesses.*

C — These spherical ended anvils are used for *measurements of thicknesses of curved sheets.*

D — These parallel flat anvils give greater area of contact and are employed for *measuring thickness of paper, rubber, cloth* or any soft materials.

E — This type of anvil is used in connection with the *measurement* of the *screw threads.*

F — These reduced diameter anvils are used for *measuring* the *depth of key ways.*

G — These anvils are used for measuring the diameter of odd fluted cylindrical parts such as reamers. In this case the V-block should have an angle of 60° and the side should intersect at a point on its bottom surface.

10.1.7 Advantages and Limitations of Commonly used Precision Instruments

Some of the advantages and limitations of verniers, micrometers and dial indicators are given below:

S.No.	Instruments	Advantages	Limitations
1.	*Vernier calipers*	Large measuring range on one instrument, upto 2000 mm, will measure external and internal dimensions.	Accuracy 0.02 mm. Point of measuring contact not in line with adjusting nut. Jaws can spring. Lack of 'feel'. Length of jaws limits measurement to short distance from end of component. No adjustment for wear.
2.	*Vernier height gauge*	Large range on one instrument, upto 1000 mm.	Accuracy 0.02 mm. Lack of 'feel'. No adjustment for wear.
3.	*Vernier depth gauge*	Large range on one instrument, upto 600 mm.	Accuracy 0.02 mm. Lack of 'feel'. No adjustment for wear.
4.	*External micrometer*	Accuracy 0.01 mm or, with vernier, 0.002 mm. Adjustable for wear. Ratchet or friction thimble available to aid constant 'feel'.	Micrometer head limited to 25 mm range. Separate instruments required in steps of 25 mm or by using interchangeable anvils.

S.No.	Instruments	Advantages	Limitations
5.	*Internal micrometer*	Accuracy 0.01 mm. Adjustable to wear. Can be used at various points along length of bore.	Micrometer head limited to 5 mm or 10 mm range. Extension rods and spacing collars required to extend range to 300 mm. Difficulty in obtaining feel.
6.	*Depth micrometer*	Accuracy 0.01 mm. Adjustable for wear. Ratchet or friction thimble available to aid constant 'feel'.	Micrometer head limited to 25 mm range. Interchangeable rods required to extend range to 300 mm.
7.	*Dial indicator*	Accuracy can be as high as 0.001 mm. Operating range upto 100 mm. Mechanism ensures constant 'feel'. Easy to read. Quick in use if only comparison is required.	Does not measure but will only indicate differences in size. Must be used with gauge blocks to determine measurement. Easily damaged if mishandled.

10.2 ANGULAR MEASUREMENTS

10.2.1 Introduction

The **angle** *is defined as the opening between two lines which meet at a point.* If one line is moved around a point in an arc, a complete circle can be formed and it is from this circle that units of angular measurement are derived. If a circle is divided into 360 parts, each part is called a degree (°). So *angle is such a thing which be just generated very easily requiring no absolute standard,* and it is the precision with which circle can be divided to get the correct measure of angle. Each degree is further divided into sixty parts called minutes (′), and each minute is further subdivided into 60 parts called seconds (″).

As in linear measurement, where two kinds of standards *(viz.,* end standard and line standard) are used; so also in angular measurement we have two types of tools. They are *angle gauges* corresponding to slip gauges and *dividing scales* corresponding to line standard.

The *difference in the linear and angular measurement is that no absolute standard is required for angular measurement.*

10.2.2 Instruments for Angular Measurement

The following instruments are used for angular measurement:

1. Protractors
   (*i*) Vernier bevel protractor
   (*ii*) Dial bevel protractor
   (*iii*) Optical bevel protractor
2. Sine bars
3. Sine tables
4. Sine centre
5. Angle gauges
6. Spirit level

7. Clinometers
8. Plain index centre
9. Optical instruments for angular measurement such as Auto-collimators.

10.2.2.1 Protractors

(*i*) Vernier bevel protractor. The angle between two faces of a component can be simply measured by means of a protractor. As shown in Fig. 10.20 the protractor has *two blades* (*viz.* fixed measuring blade and rotatable or movable blade) which can be set along the faces containing the angle. Its *body* contains a circular scale which is extended to form one of the blades. The second blade (rotatable blade) is sliding and can be locked in any position along its length to a *rotating turret* mounted on the body. Either the body or the turret carries the divided circular scale, while the other member carries a vernier or an index mark.

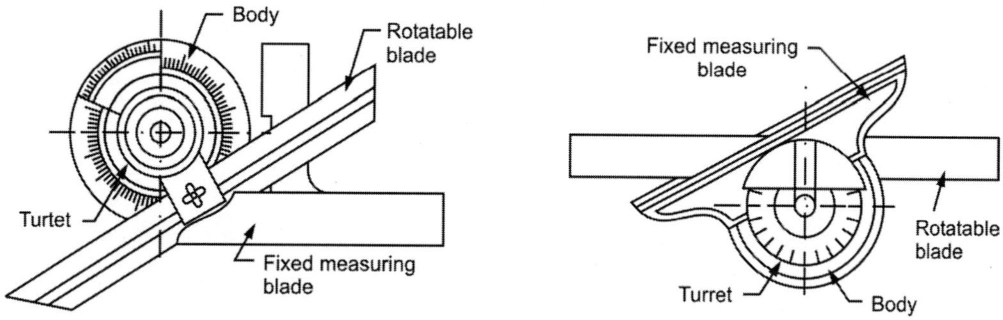

Fig. 10.20 Vernier bevel protractor.

Figure 10.20 shows an ordinary vernier bevel protractor which is a workshop instrument having scale value of 5' or 2.5'.

(*ii*) Dial bevel protractor. A dial bevel protractor is shown in Fig. 10.21. With this protractor the angles can be measured to within 5'.

In this type of protractor, the turret rotates a circular scale that serves as a gear actuating the dial gauge pointer. The magnification ratio is made so that the pointer rotates through one complete revolution of every 10° of the turret scale, *i.e., for* every 10° of blade angle movement. The scale of the dial gauge is divided into 10 equal divisions, each is subdivided into 12 so that a scale value of $\frac{1°}{12}$ or 5' is obtained. The scale of the dial gauge has an *opening* through which the angle in degrees is read against a fixed mark on the dial.

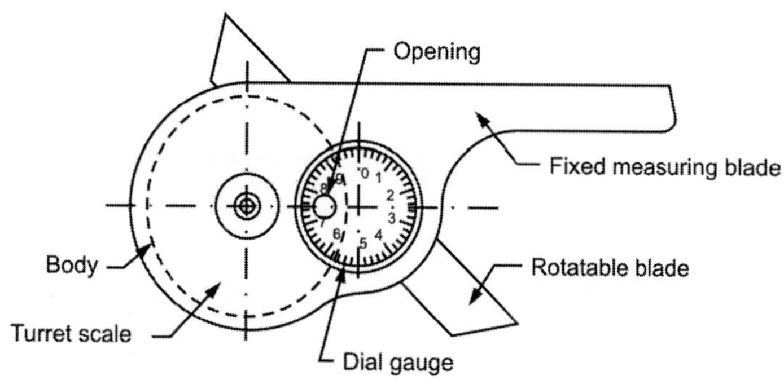

Fig. 10.21 Dial bevel protractor.

(*iii*) Optical bevel protractor. Refer to Fig. 10.22. A recent development of the vernier protractor is optical bevel protractor. In this device, a glass circle divided at 10 mm intervals throughout the circle is fitted inside the body. A small microscope is fitted throughout which the circle graduations can be viewed. The adjustable blade is clamped to a rotating member which carries its microscope. With the aid of microscope it is possible to read by estimation to about 2 min.

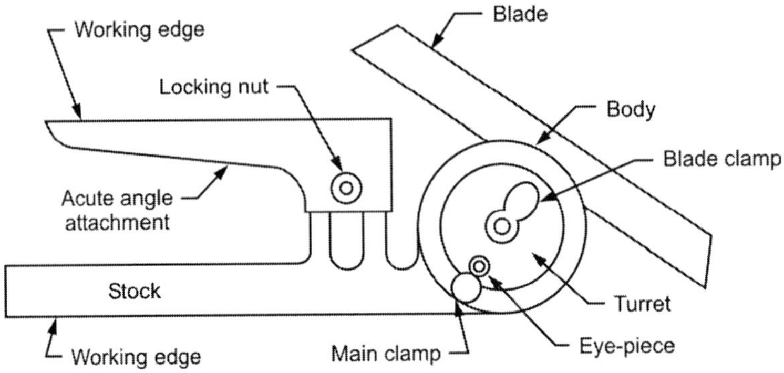

Fig. 10.22 Optical bevel protractor.

Figure 10.23 shows some applications of the optical protractor which can also be done by any other type of protractor.

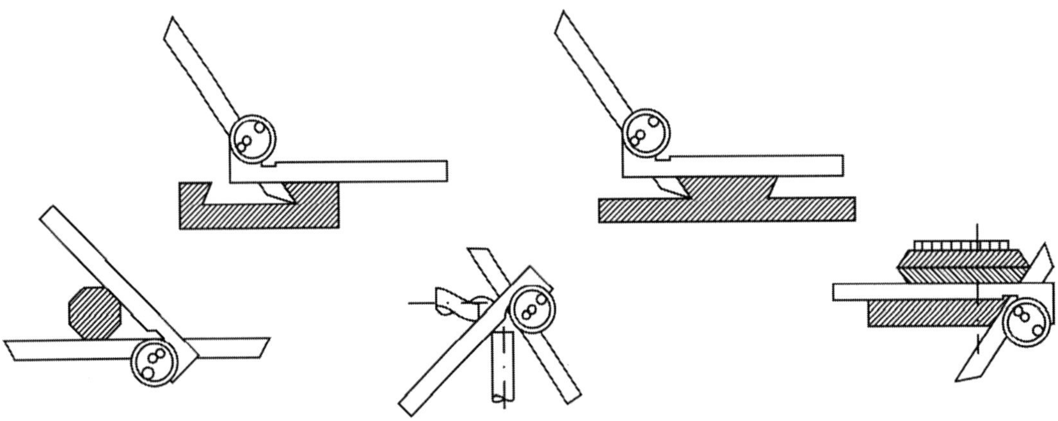

Fig. 10.23 Some applications of optical protractor.

10.2.2.2 Sine bars

Sine bar *is a tool used for accurate setting out of angles by arranging to convert angular measurements to linear ones.*

Figure 10.24 shows the *principle of sine bar.* If we wish to set out any angle θ, we could do so (Fig. 10.24) by setting off a horizontal line AB, next scribing an arc CD with a measured radius R and then setting out a vertical dimension EF, made so that $EF = R \sin \theta$. This is quite easily done by means of the sine bar.

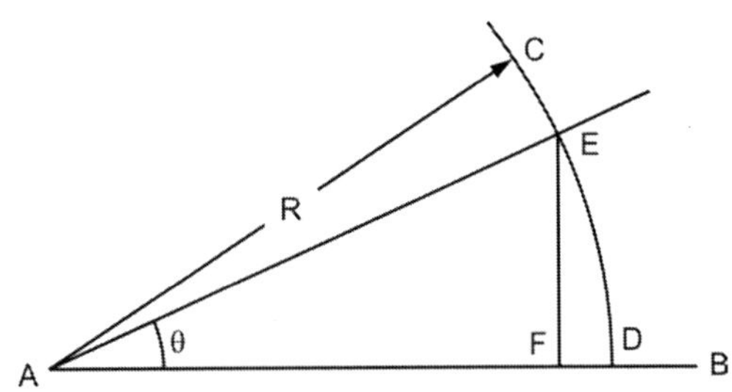

Fig. 10.24 Principle of sine bar.

- *Sine bars used in conjunction with slip gauges constitute a very good device for the precise measurement of angles.* Sine bars are used either to measure angles very accurately or for locating any work to a given angle within very close limits.

- *Sine bars are made from high carbon, high chromium, corrosion resistant steel, hardened, ground and stabilised.* Two cylinders of equal diameter are attached at the ends. The axes of these two cylinders are mutually parallel to each other and also parallel to and at exact distance from the upper surface of the sine bar. The distance between the axes of the two cylinders is exactly 5 inches or 10 inches in British system, and 100 and 200 and 300 mm. in metric system. The above recommendations are met and maintained by taking due care in the manufacture of all parts. The various parts are hardened and stabilised before grinding and lapping. All the working surfaces and the cylindrical surfaces of the rollers are finished to surface finish of 0.2 mm or better.
- Depending upon accuracy of the centre distance, sine bars are graded as of *A grade or B grade.* B grade of sine bars are guaranteed accuracy upto *0.02 mm/m of length and A grade sine bars are more accurate upto 0.01 mm/m length.*

Types of sine bars:

Different forms of sine bars are discussed below :

- Figure 10.25 shows a sine bar of simplest type. It consists of a lapped steel bar, at each end of which is attached an accurate cylinder, the axis of the cylinder being mutually parallel, and parallel to the upper surface of the bar. The axes are separated by a nominal distance, usually 100 mm or 300 mm for metric bars and 5 or 10 inches, for those in inch units. *The distance between the rollers forms the basis of the designation of sine-bars.* Thus a sine bar designated as 100 mm sine bar will have a distance of 100 mm between its rollers.

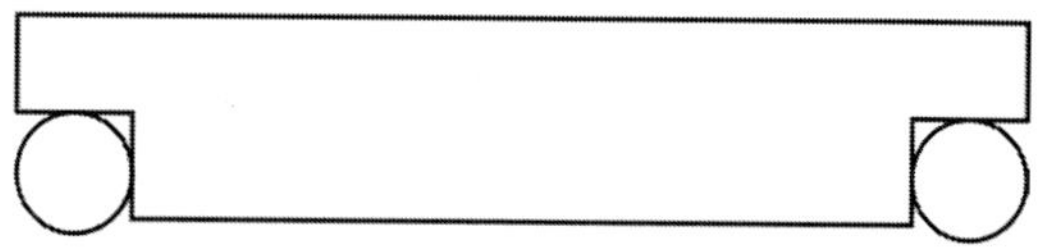

Fig. 10.25 Sine bar of simplest type.

- The form of the sine bar shown in Fig. 10.26 is most commonly used. Some holes are drilled in the body of the bar to *reduce the weight and to facilitate handling.*

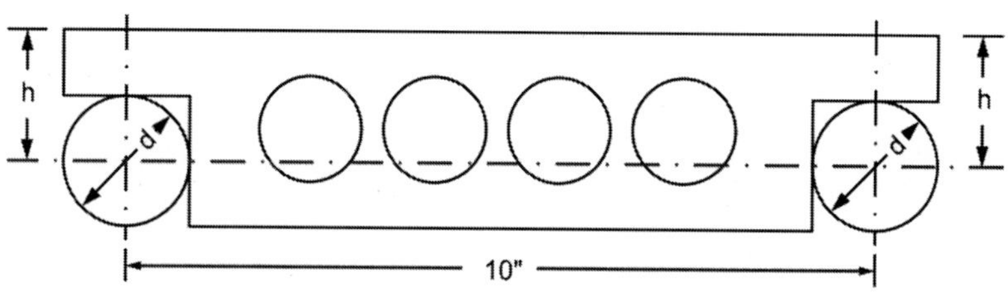

Fig. 10.26 Commonly used sine bar.

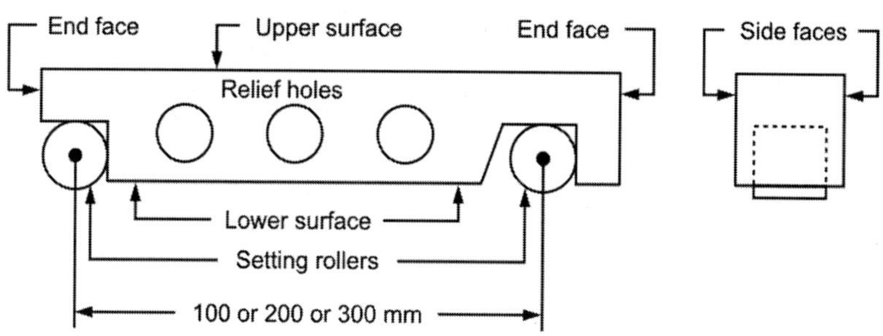

Fig. 10.27

- From the point of view of ease in manufacturing to ensure an exact distance between the cylinders, the form shown in Fig. 10.27 is preferable. Also this type of sine bar can be set to a steep angle without the slip gauges fouling the underside of the bar. But this point is immaterial as the accuracy of setting appreciably decreases with steep angles and from the point of stability also sine bars are not generally used for steep angles.
- In the Fig. 10.28 is shown a most commonly used form of sine bar in which the rollers are so arranged that their outer surfaces on one side are level with the plane to surface of the sine bar.
- In Fig. 10.29 is shown the form of sine bar which has hollow rollers whose *outside diameter is equal to the width of the sine bar* and is *used where the width* of the *sine bar enters into calculation of work height.*
- Figure 10.30 shows the form of the bar which is used where ordinary type cannot be used on the top surface due to interrupt.

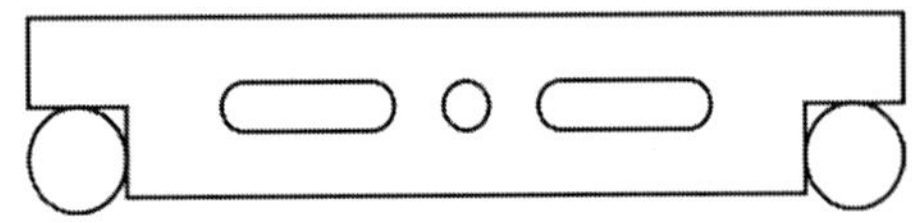

Fig. 10.28

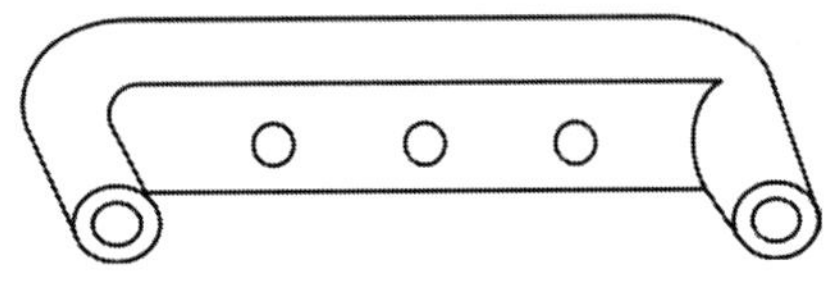

Fig. 10.29

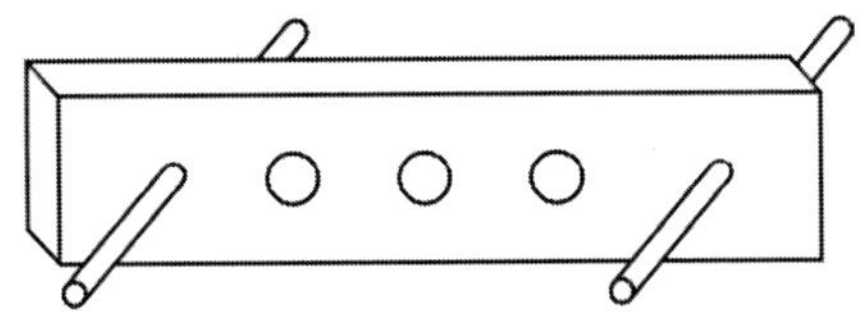

Fig. 10.30.

To get accurate results, in the use of a sine bar, it is essential that:

1. The contact rollers must be of equal diameter and true geometric cylinders.
2. The distance between the roller axes must be precise and known, and these axes must be mutually parallel.
3. The upper surface of the bar must be flat and parallel with the roller axes, and equidistance from each other.

Use of sine bar. A sine bar may be used in the following cases:

1. Measuring known angles or locating any work to a given angle.
2. Checking of unknown angles.
3. Checking of unknown angles of heavy components.

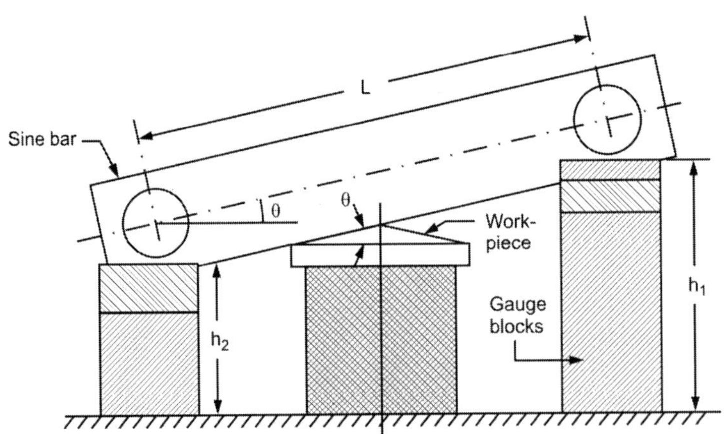

Fig. 10.31 Sine bars set up for measuring an angle of workpiece.

- Figure 10.31 shows sine bar set up on gauge blocks for measuring an angle on workpiece. In this case:

$$\sin\theta = \frac{h_1 - h_2}{L}$$

where, L = Known distance, such as 100 mm or 200 mm, depending on the size of bar used,
$h_1 - h_2$ = Heights built up to correct amounts with precision gauge blocks, and
θ = Angle being checked.

- Figure 10.32 shows how the sine bar is used to check small components that may be mounted upon it.
- In such cases where components are heavy and cannot be mounted on the sine bar, then sine bar is mounted on the component as shown in Fig. 10.33. In the example shown, the height over the rollers is measured by means of a vernier height gauge, using a dial test gauge. This is achieved by adjusting the height gauge until the dial gauge shows the same zero reading each time. The difference of the two readings being the rise of the sine bar as shown and,

$$\sin\theta = \frac{R_1 - R_2}{L}$$

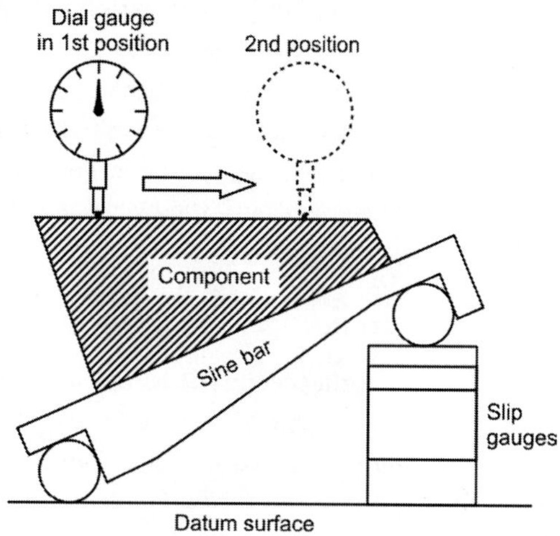

Fig. 10.32 Set up of sine bars for checking small components.

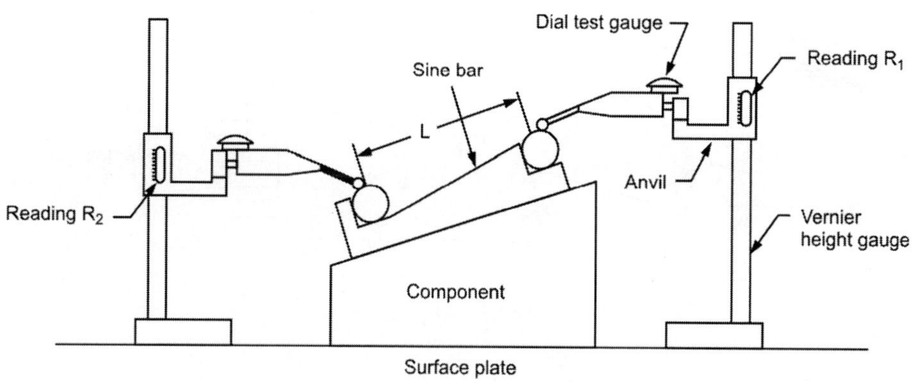

Fig. 10.33

Precautions in the use of sine bars.

The following *precautions* should be taken while using the sine bars:

1. As far as possible longer sine bar should be used since many errors are reduced by using longer sine bars.
2. The sine *bar should not be used for angle greater than 60°* because any possible error in construction is accentuated at this limit.

3. Accuracy of sine bar should be ensured.
4. A compound angle should not be formed by misaligning of workpiece with the sine bar. This can be avoided by attaching the sine bar and work against an angle plate.

10.2.2.3 Sine table

A **sine table** *is the development* of *the sine bar and the procedure of setting* it at any angle is *same for sine bars.* It has one hinged roller and the slots to facilitate work holding. A further development of this is the *compound sine table* having their axes of tilt set at right angles to each other are provided. These two tables are mounted on a common base and the table can be set at compound angle by resolving this compound angle into its individual angles in two planes at right angles to each other and setting each table accordingly.

Since a sine table is suitable for large work of greater height, so it is made a robust one. *Usually* the *sine tables are used to measure angles in two planes (i.e.,* compound angles) *but these are also used for linear as well as radial measurement.*

10.2.2.4 Sine centre

Refer Fig. 10.34. A **sine centre** is *basically a sine bar with block holding centres which can be adjusted and rigidly clamped in any position*. The *principle of setting is same as of sine table.*

The *sine centres are used for inspection of conical objects* (having male and female centres) *between centres.* These *are used upto inclination of 60°.* Rollers are clamped firmly to the body without any play. This is a very useful device for *testing the conical work centre at each end.*

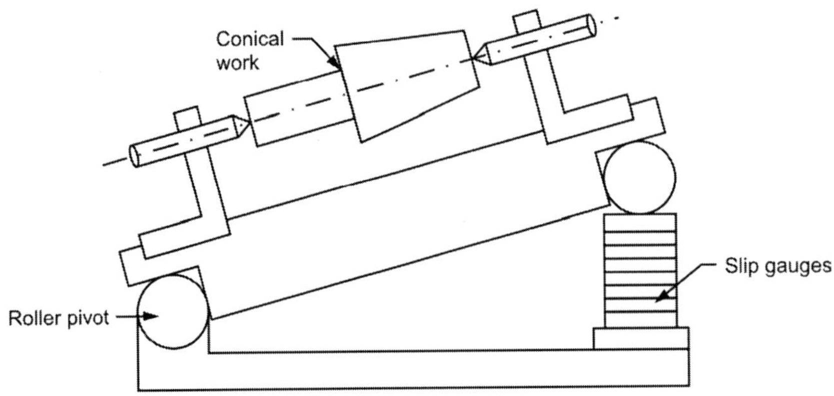

Fig. 10.34 Sine centre.

10.2.2.5 Angle gauges

An *angle gauge* (Fig. 10.35) *is a hardened steel block approximately 75 mm long, 16 mm wide which has two lapped flat working faces lying at a very precise angle to each other.* These are supplied in set, and can be *wrung together to form desired angles.* The angles of the gauges in thirteen piece set, which also includes a square block, are shown in the Table 10.1. The

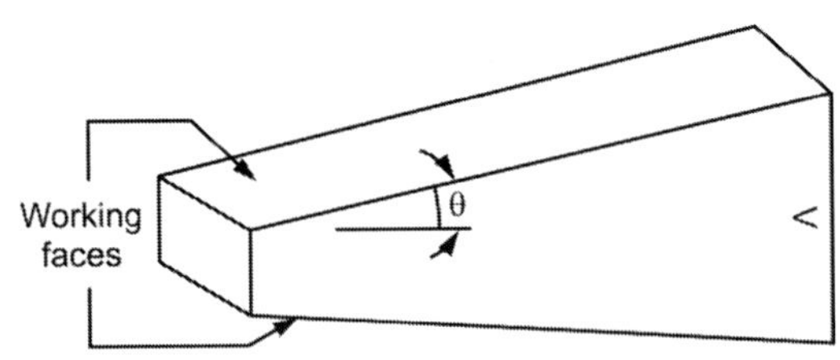

Fig. 10.35 Angle gauge.

arrangement allows angles in 3 seconds step to be obtained, thus no required angle expressed in whole seconds can be more than 1.5 seconds in nominal error.

TABLE 10.1 Nominal angles of combination angle gauges

Degrees (°)	**1**	**3**	**9**	**27**	**41**
Minutes (′)	1	3	9	27	
Fraction of minutes or seconds (″)	0.05 (3)	0.1 (6)	0.3 (18)	0.5 (30)	

An additional block having an angle of 9° is also available so that a 90° angle can be built up. Two different grades *A* and *B* are available. The only difference between these two types is that angle gauge of 0.05 minutes is not included in the set *B* and thus consists of only twelve gauges. For the visual inspection of angles, each set of angle gauges is provided with a straight-edge having parallel faces.

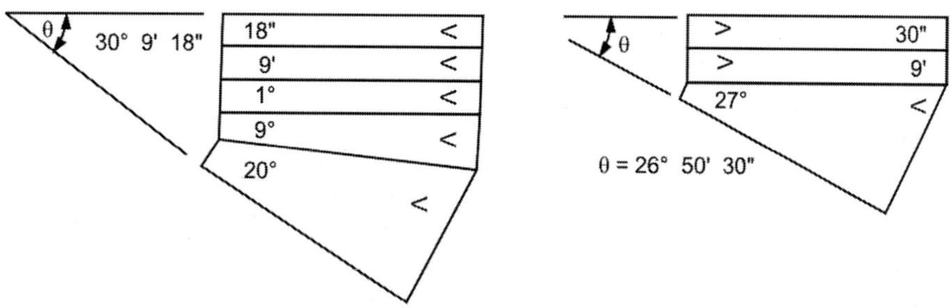

Fig. 10.36

Method of using angle gauges:

Figure 10.36 shows the correct use of angle gauges for making angles. The angles made are : 30° 9' 18" and 26° 50' 30".

It may be noted that each angle gauge is marked with engraved *V* which indicates the direction of included angle. *When the angles of individual gauges are to be added up then the* Vs *of all should be in line* and when any angle is to be *subtracted, its engraved V should be in other direction.*

Practical applications of angle gauges:

- In engineering industries the angle gauges have been widely used for the quick measurement of angles between the two surfaces. A frequent use of these gauges is to check whether the component is without its off angle tolerance.
- Where the angle to be measured between the two surfaces exceeds 90 degrees, the use of precision square becomes essential. Figure 10.37 shows a set up to test the angle of V-gauge whose included angle is 102°.

10.2.2.6 Spirit level

The spirit level is a simple form of mechanical measuring device. This instrument has been used in engineering for many years, mainly for static levelling of machinery and other equipment, yet a calibrated level can be, in fact, an angle measuring instrument of high precision with a wide variety of applications.

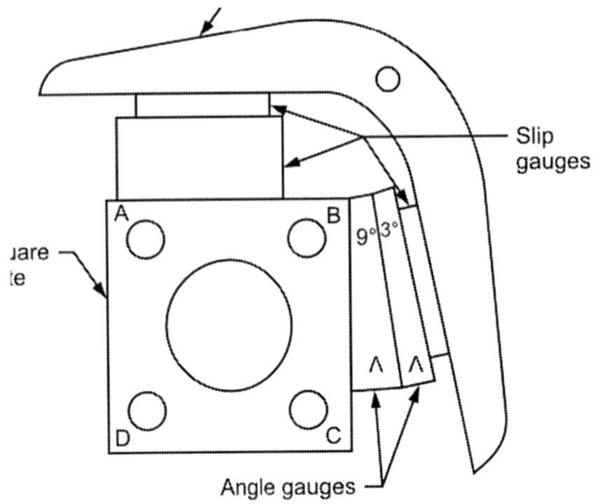

Fig. 10.37

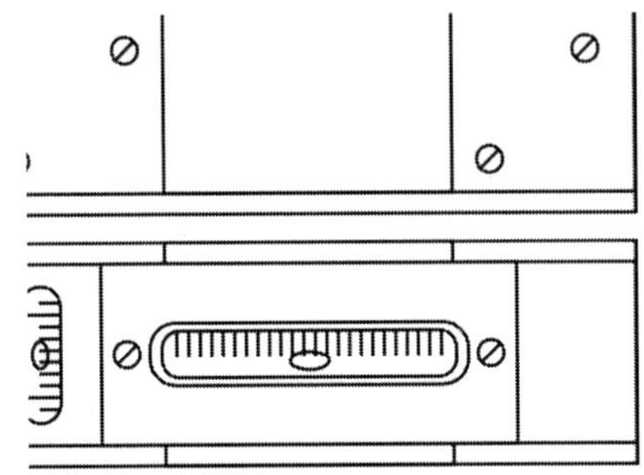

Fig. 10.38 Block spirit level.

Refer Fig. 10.38 (Block spirit level). The sensitive spirit level consists of a curved glass tube of uniform bore, nearly filled with a liquid so that an air bubble is left. The glass tube is fitted to a metal block in such a manner that when the lower surface of the block is horizontal, the air bubble will centre itself at the highest point of the tube, *i.e.*, in its centre. In order that the movement of the air bubble be proportional to the angle of tilt, it is necessary that the radius of curvature of the glass tube is made constant over its whole length. The internal cross-section of the tube should be uniform.

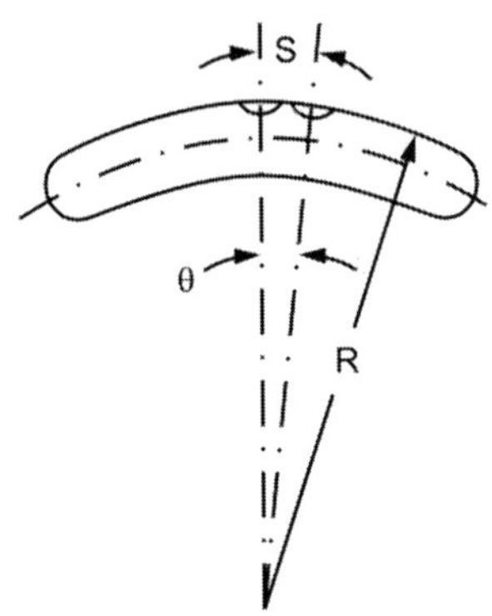

Fig. 10.39 Principle of spirit level.

The sensitivity of a spirit level is *expressed as the angle of tilt in seconds for which bubble will move by one division on the tube. One* division is generally about 2.5 mm in length.

Thus, $$\text{Sensitivity} = \frac{\text{angle in seconds}}{\text{1 division on the tube}}$$

Figure 10.39 shows the *principle of spirit level.*

If, R = Radius of curvature of glass tube,

S = Scale division, *i.e.*, the length of one division, and

α = Angle of tilt corresponding to 1 division movement of bubble.

Then, $$\alpha = \frac{S}{R}$$

Generally the graduations are at 2.5 mm intervals and these represent a tilt of 10 seconds of arc, *i.e.*, sensitivity of level desired is 10 sec. per 2.5 mm movement of the bubble.

Then, $$10 \text{ sec.} = 0.0000485 \text{ radian} = \frac{2.5 \text{ mm}}{R}$$

or, $$R = \frac{2.5}{0.0000485} = 51500 \text{ mm}$$

or, $$R = 51.5 \text{ m approximately.}$$

Thus for above sensitivity, radius of curvature of the tube must be about 51.5 m and it is obvious now that *sensitivity of the spirit level is governed solely by the radius of the tube and the base length of its mount.*

If the base length of any spirit level be about 250 mm, then the height h by which one end must be raised for 2.5 mm bubble movement is given by:

$$0.0000485 = \frac{h}{250}$$

or,
$$h = 0.0000485 \times 250 = 0.0121 \text{ mm.}$$

If the base length be reduced to 125 mm, then sensitivity is increased twice, and in this case each graduation represents 0.006 mm.

The sensitivity of a spirit level can also be expressed by tangent of the angle of tilt in mm/m. The *accuracy of a spirit level depends upon the setting of the tube relative to the base.*

10.2.2.7 Clinometers

A **clinometer** *is an instrument used for measuring angle relative to the horizontal plane.* By a simple calculation it is possible to check the angles between two parts of a job even if separated by considerable distances or on different levels. There are various makes and types of clinometers obtainable, but they are all designed on one of the two basic principles : One involves the *use of a spirit level* and the other makes use *of the plumb-bob (or pendulum) principle* and works from the vertical.

Clinometers are *used for checking angular faces, and relief angles on large cutting tools and milling cutter inserts.* These can also be *used for setting inclined table on jig boring machines and angular work on grinding machines etc.*

The different types of clinometers available are:

1. Vernier clinometer
2. Micrometer clinometer
3. Dial clinometer
4. Pendulum clinometer
5. Optical clinometer.

The most commonly used clinometer is of Hilger and Watt type. The circular glass scale is totally enclosed and is divided from 0° to 360° at 10' intervals. Sub-division of 10' is possible by the use of an optical micrometer. A coarse scale figured every 10 degrees is provided outside the body for coarse work and approximate angular reading. In some instruments worm and quadrant arrangement is provided so that reading upto 1' is possible.

In some clinometers, there is no bubble but a graduated circle is supported on accurate ball bearings and it is so designed that when released, it always takes up the position relative to the true vertical. The reading is taken against the circle to an accuracy of one second with the aid of vernier.

10.2.2.8 Plain index centre

Fundamentally index centres were meant for use on milling machines but they are now being used for inspection work also. This is specially suited to those problems which involve the measurement of a large number of angular dimensions about a common centre. The work is set between the centres and correct angular dimensions are set directly from the indexing plate.

To obtain wide range of angles the following types of indexing methods may be employed:

1. Single indexing
2. Differential indexing
3. Compound indexing.

10.2.2.9 Optical instruments for angular measurement

Autocollimator. An autocollimator is an instrument designed to measure *small angular deflections* and may be used in conjuction with a plane mirror or other reflecting surface. An *autocollimator is essentially an infinity telescope and a collimator combined into one instrument.*

An autocollimator is based on the *principle that a collimating lens can project and receive a parallel beam of light and that the reflected beam of light will change its direction by changing the angle of the surface reflecting the light.*

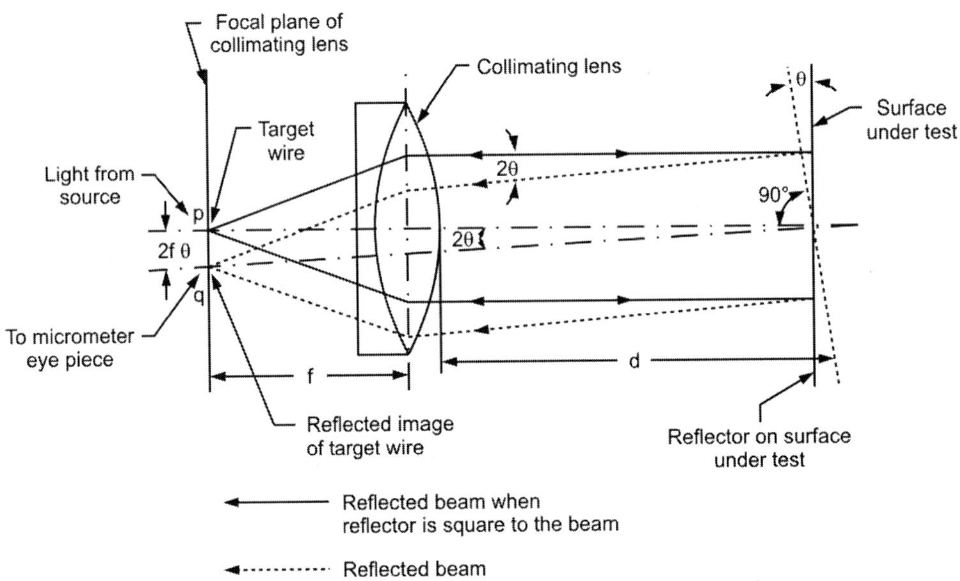

Fig. 10.40 Principle of autocollimator.

The autocollimator's principle is shown in Fig. 10.40. It shows the collimator lens that projects a parallel beam of light along the optical axis of the system when the light source (electric bulb) is placed at the focus of the collimator lens. A reflector, which can be a slip guage block, an optical flat or a mirror is attached to the surface under rest. If this reflection is accurately normal to the optical axis, the effected beam will return along its original incident path back to the lens and will be collected at a point '*p*' exactly at the length of the lens where cross-wires are placed. Therefore the image of the cross wires formed by the reflected light beam will, in this case coincide with the cross-wires at point '*p*'. If the reflection is slightly tilted at an angle θ, the reflected beam will change its path at twice the angle θ. The reflected beam will, therefore, form an image '*q*' at the cross-wires at distance '*pq*', the value of which is a measure of the angular deflection θ, where $pq = 2f\theta$.

- It may be noted that the position of the final image does not depend upon the distance of reflector from the lens, *i.e.*, separation '*d*' is independent of the position of reflector from the lens. But if reflector is moved too much back then reflected rays will completely miss the lens and no image will be formed. Thus, for full range of readings of instrument to be used, the maximum remoteness of the reflector is limited.

For high sensitivity, *i.e.,* for large value of '*d*' for small angular deviation θ, a long focal length is required.

Hilger angle dekkor

Figure 10.41 shows a typical auto-collimator-the Hilger Angle Dekkor, the description of which is given below:

- It consists of a *tubular body* containing the *collimating lens* at one end and viewing *eye piece* at the other end.
- At the focal plane of the collimating lens, there is a scale instead of cross-wires which is illuminated by means of an electric lamp mounted by the side.
- The *eyepiece* is provided with a *fixed scale* similar to and at right angles to the illuminated one.

  Both scales are divided from 0 to 40 and each division represents 1′ of arc.

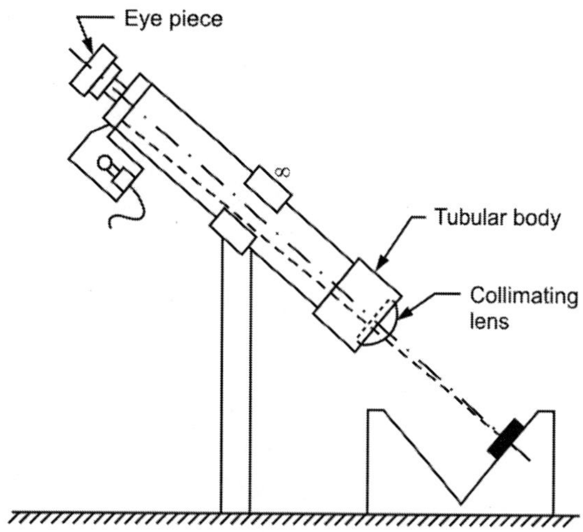

Fig. 10.41 Hilger angle dekkor.

- If the *reflector* is exactly normal to the optical axis of dekkor, the image of the illuminated scale will intersect the fixed scale such that the central reading of both scales will coincide. The central readings of the scales correspond to the point of intersection of cross-wires. If the reflector is tilted at an angle θ, this angle can be read directly from the position of the reflected scale with respect to the fixed one.

Figure 10.42 shows the method which can be used to measure the angle of inclination of one face of a V-block by the angle dekkor. The instrument should be first set at the theoretical angle by means of angle gauges and the variation of the theoretical and actual can be measured by the angle dekkor.

Micro-optic auto-collimator telescope

In the Fig. 10.42(*a*) is shown micro-optic auto-collimator telescope. It is a sensitive instrument similar to a dekkor. Its description is given below :

- The eye piece is provided with a *micrometer* whose drum is graduated either in 0.01 mm or in 1/2 second.
- The large cross-wires are illuminated by a *lamp* at the side.
- The reflected image of the wires, formed in the same plane as the wires themselves, is viewed through a 10 × microscope fitted with the micrometer eye-piece.
- Direct measurement of angles can be done by means of a horizontal circular table and micro-optic auto-collimator telescope [Fig. 10.42(b)]. The table is set so that the face of the component is accurately normal to the telescope and the telescope reads zero. In this position the table reading is observed. The table is then rotated until the face *Q* of the component becomes accurately normal to the telescope

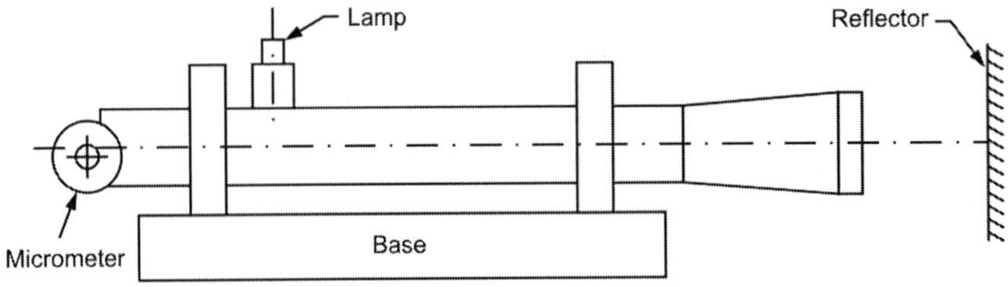

(a)

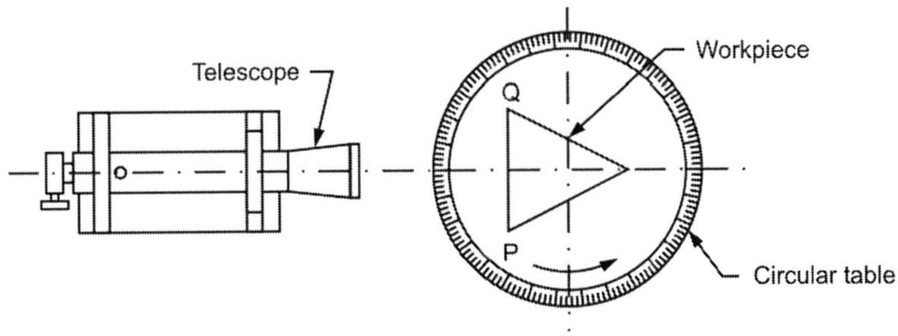

(b)

Fig. 10.42

whose reading remains zero. The difference between the initial and final readings of the table subtracted from 180° gives the value of the angle included between the two faces *P* and *Q*. In this case, the scale value of circular table will be the one which affects the measuring result.

- With the optical circular tables which are similar in principle to the optical clinometers, a scale value of 0.1' or 6" can be obtained.

Photo-electric micro-optic auto-collimator:

Photo-electric setting makes measuring and checking by auto-collimator far simpler and faster. Micrometer adjustment is provided for setting but coincidence of setting graticule and target image is *detectedphotoelectrically,* and shown on a meter as a null reading. This provides a high degree of sensitivity and repeatability, also reducing eye fatigue to a minimum. The eyepiece is normally only used to assist in initial setting up.

The photo-electric auto-collimator is particularly *suitable for calibrating polygons, for checking angular indexing and for checking small linear displacements.*

Application of auto-collimator. The auto-collimator can be used for measurement of angles of *surfaces which are of reflective nature.* Thus newly machined and polished surfaces will reflect. In case the *surface does not reflect then a parallel slip gauge is wrung to the surface.* Fig. 10.43 shows the use of auto-collimator to *compare* the angle of a precision angle plate with that between two faces of an eight-edged reference plane:

- The instrument is firmly fixed in position relative to the surface plate and the reading is taken with polygon in position.
- The polygon is then removed and replaced by an angle plate plus the slip gauges if necessary and a new reading is taken.
- The difference between the two sets of readings must be added or subtracted from the known angle between the particular pair of polygon faces, to give the angle of face plate.

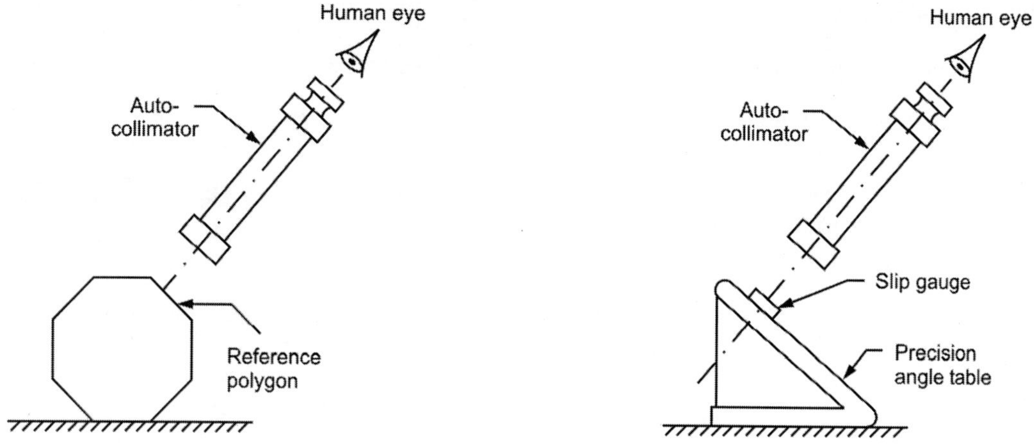

Fig. 10.43 Use of auto-collimator.

10.3 COMPARATORS

10.3.1 The Comparator

The **comparator** *is an instrument used for comparing the dimensions of a component with a standard of length.*

All comparators consist of following *essential parts* :

1. *A fixed surface* from which all measurement are taken.
2. *A very sensitive indicator* which will show the movement of a sliding piece usually terminating in an anvil with a *curved face.*
3. Some *means of setting the curved face* (referred to in 2) *at an adjustable distance from surface* (1).

This arrangement is used to measure the difference between the length or diameter of a component and a standard of length, usually made up of slip gauges.

Thus the *purpose of a comparator,* in general, *is to detect and display the small difference between the unknown linear dimension and length of the standard.* The difference in lengths is detected as a displacement of a sensing probe. The *important and essential function of the instrument is to 'Magnify'* or *'Amplify' the small input displacement so that it is displayed on an analogue scale.*

10.3.2 Uses of Comparators

Some of the *important uses of comparators* are given below:

1. To inspect newly purchased gauges.
2. In mass production, where components are to be checked at a very fast rate.
3. As laboratory standards from which working or inspection gauges are set and correlated.
4. In selective assembly of parts, where parts are graded in three groups depending upon their tolerance.
5. As working gauges, to prevent work spoilage and to maintain required tolerance at all important stages of manufacture.

10.3.3 Types of Comparators

The *most common commercially available comparators* can be classified into the following types:

1. Mechanical comparators.
2. Optical comparators.
3. Electrical and electronic comparators.
4. Fluid displacement comparators.
5. Pneumatic comparators.

A combination of magnifying principles has lead to the development of a few more types of comparators. These are:

6. Mechanical-optical comparators.
7. Electro-mechanical comparators.
8. Multi-check comparators etc.

10.3.4 Mechanical Comparators

The mechanical comparators utilize mechanical methods of magnifying the movement of the contact plunger; their manufacture requires a high degree of skill and accuracy. The usual magnifications of the mechanism range from about 250 to 1000. Some of the comparators which belong to this class are:

- Dial indicator
- Reed type comparator
- Sigma comparator
- Johansson Mikrokrator
- Eden Roll Millionth comparator

Whereas these comparators are used for higher precision work and are costly, the most frequently used one is 'dial indicator'.

10.3.4.1 Dial indicator

Figure 10.44 shows the main features of a plunger-type dial gauge indicator. The main scale is graduated into equal divisions corresponding to a 0.01 mm movement of the plunger. A second but small dial is set in the main dial face to indicate the number of completed revolutions turned through one revolution being equivalent to 1 mm of plunger movement. To enable the instrument to be zero for any convenient position, the main scale can be rotated and locked into place, using the scale locking screw indicated in Fig. 10.44.

The *principle of operation of the instrument* is shown in Fig. 10.45, where it can be seen that the plunger is attached to a rack. Meshing with a gear wheel, the straight or linear motion of the rack is converted into an angular or turning motion, the movement being magnified by using a large gear in mesh with a small gear wheel. It is the small gear wheel that is fitted to the main scale pointer shown in Fig. 10.45. The mechanism described above is *simple, reliable* and *very sensitive.* However this sensitivity means that *great care must be exercised* when using a dial gauge indicator.

Figure 10.46, shows a dial gauge indicator fitted to a pillar stand; the stand enables the indicator to be locked in any required position.

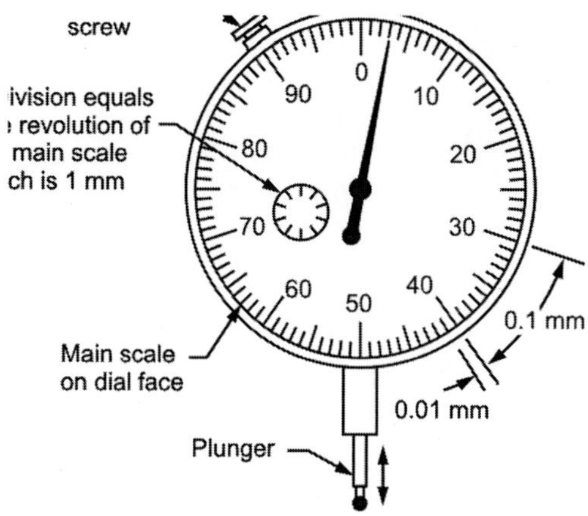

Fig. 10.44 Dial indicator.

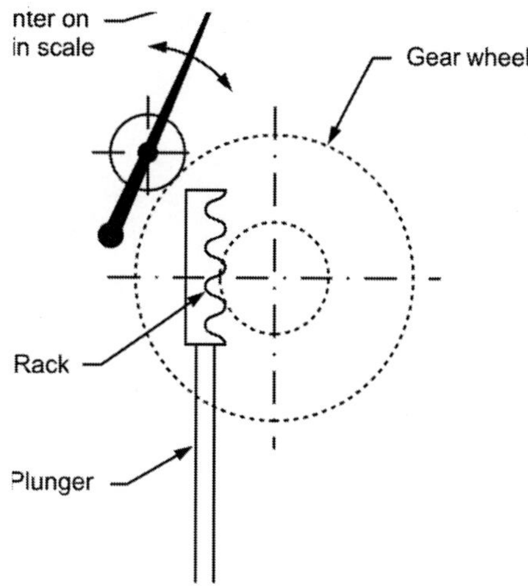

Fig. 10.45 Principle of operation of a dial indicator.

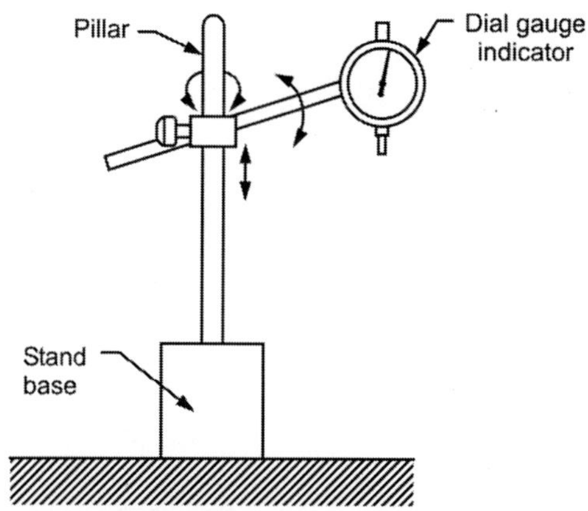

Fig. 10.46 Dial gauge indicator fitted to a pillar stand.

Fig. 10.47 shows a *workpiece being compared with a known standard,* which may be, for example, slip gauges. With the dial gauge indicator placed so that the *plunger is in contact with the reference face,* the dial scale can be *set to zero and locked.* Moving the pillar stand (with the dial gauge indicator fitted to it) across to the workpiece, which is positioned on the same flat surface, *any difference between the original setting and the face being checked can be read off the main scale.*

10.3.4.2 Read type mechanical comparator

Figure 10.48 shows a *Reed comparator* which is strictly a mechanical comparator because the linkages required scale for magnification are purely mechanical amplifiers.

- A comparator has a sensing probe, a spindle, attached to a movable *member A*.
- Member *A* moves through a diameter *x,* the input displacement, with respect to member B which is fixed.

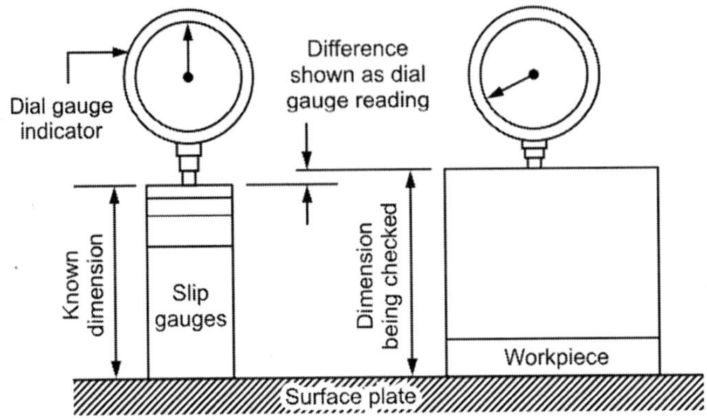

Fig. 10.47 Plunge type dial indicator in use.

- Member A is *constrained* by flexture strips or reeds C, to move to B.
- The pointer is attached to reeds D.
- A small input displacement x, produces a *large angular movement*, X, of the pointer on account of their orientation relative to the motion. The scale is calibrated by means of gauge blocks and indicates the difference in displacement of the fixed and the movable elements. There is no friction and the hysteresis effect is minimised by using suitable steel for the reeds.

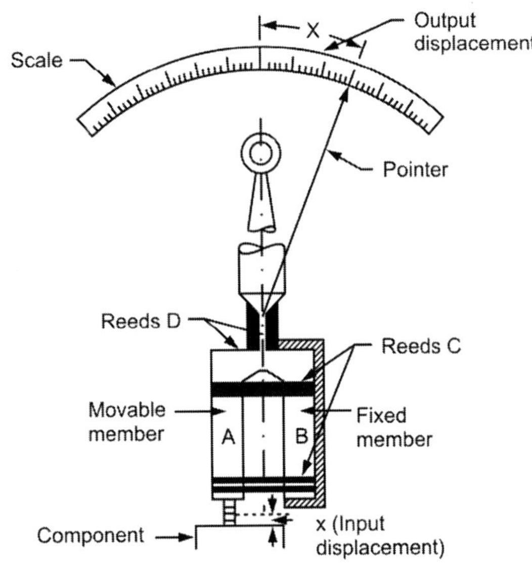

Fig. 10.48 Reed type mechanical comparator.

Comparators of this type have sensitivities of the order of 0.25×10^{-3} mm/scale division. The mechanical amplification is usually less than 100, but it is multiplied by the optical lens system. It is available in amplification ranges from × 500 to × 1000.

10.3.4.3 Sigma comparator

The sigma comparator is an excellent example of very successful instrument in which a high magnification is obtained entirely by mechanical means. Magnification ranges from 300 to 5000. The use of this instrument may be greatly extended by means of specially designed contacts and attachments to include internal diameter of screw threads.

The details and magnifying system of the comparator are shown in Fig. 10.49.

- The *plunger* is mounted on a pair of *slit diaphragm* to give a frictionless linear motion, has mounted upon the face of the moving member a *cross strip hinge* (Fig. 10.50). This hinge consists of the *moving component and a fixed member* connected by *flexible strips* alternately at right angles to each other. It can be shown that it would pivot as would a hinge about the line of intersection of the strips.

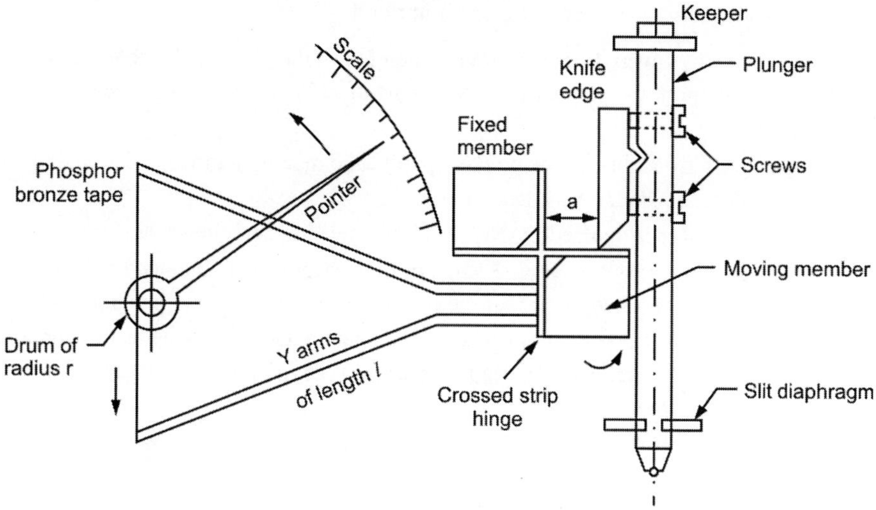

Fig. 10.49 Sigina comparator.

- An *arm* (which divides into *Y* form) is attached to the moving member. If the length of the arm is *l* and the distance from the hinge pivot to the knife edge is *a, then first stage of magnification is* $\frac{l}{a}$.
- To the extremities of the *Y* arm is attached a *phosphor bronze band* or *strip* which is passed around a *drum* of radius *r* attached to the pointer spindle. If the pointer is of length *R, the second stage of magnification* is $\frac{R}{r}$ and the total magnification,

$$M = \frac{1}{a} \times \frac{R}{r}$$

The magnification can be changed by tightening one and slackening the other screw attaching the knife edge to plunger and thus adjusting the distance *a*. Still another way to produce instruments of different magnification is to use drums of different radii *r* and suitable strip.

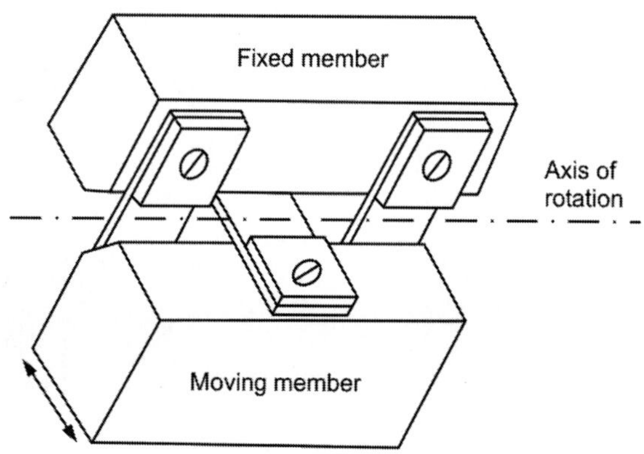

Fig. 10.50 Cross strip hinge.

Advantage:

It has got a bold scale and large indicator pointer.

Disadvantages:

1. Due to motion of the parts, there is wear in the moving parts.
2. It is not as sensitive as optical or other types of comparator due to friction being present in the moving parts.

Some of the *important features* of a Sigma comparator are:

1. As the knife edge moves away from the moving member of the hinge and is followed by it, therefore, if too robust movement of the plunger is made due to shock load, that will not be transmitted through the movement.
2. By mounting a non-ferrous disc on the pointer spindle and making it move in the field of a permanent magnet, dead beat readings can be obtained.
3. The error due to parallax is avoided by having a reflective strip on the scale.
4. The constant measuring pressure over the range of the instrument is obtained by the use of a magnet plunger on the frame and a keeper bar on the top of the plunger. As the plunger is raised the force required increases but the keeper bar approaches the magnet and the magnetic attraction between the two increases. Thus as the deflecting force increases, the assistance by the magnet increases and the total force remains constant.

10.3.4.4 Johansson 'Mikrokator'

Figure 10.51 shows diagrammatically the Johansson 'Mikrokator'.

- This instrument uses the simplest and most ingenious method for obtaining the mechanical magnification designed by H. Abramson which is called *Abramson movement.*
- *It works on the principle of a button spinning on loop of string.*

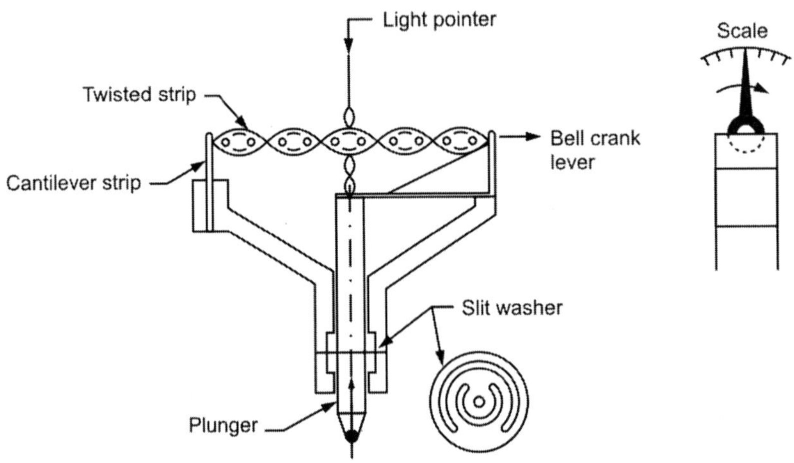

Fig. 10.51 Johanson mikrokator.

- A twisted thin metal strip carries at the centre of its length a very light pointer made of thin glass. The two halves of the strip from the centre are twisted in opposite directions so that any pull on the strip will cause the centre to rotate. One end of the strip is fixed to the adjustable cantilever strip and the other end is anchored to the spring below, one arm of which is carried on the measuring plunger. As the measuring plunger moves either upwards or downwards, the elbow acts as a bell crank lever and causes twisted strip to change its length thus making it further twist or untwist. Thus the pointer at the centre of the twisted strip rotates by an amount proportional to the change in length of strip and proportional to the plunger movement. The bell crank lever is formed of flexible strips with a diagonal which is relatively stiff. The length of cantilever can be varied to adjust the magnification of the instrument. Since the centre line of the strip is straight even when twisted, therefore, it is directly stretched by the tension applied to the strip. Thus in order to prevent excessive stress on the central portion, the strip is perforated along the centre line by perforations as shown in Fig. 10.51.

- It can be shown that the amplification of the instrument

$$\frac{d\theta}{dt} \propto \frac{1}{w^2 n}$$

where, θ = The twist of mid-point of strip with respect to the end,

l = The length of twisted strip measured along its neutral axis,

w = Width of the strip, and

n = Number of turns.

Thus to *increase the amplification of the instrument, a very thin rectangular strip must be used.* Further amplification can be adjusted by the cantilever strip.

THEORETICAL QUESTIONS

Linear Measurements

1. What is an Engineer's steel rule (or, scale) ?
2. What are the desirable qualities of a steel rule ?
3. State the techniques which must be used to get good results.
4. Write a short note on 'Steel rule'.
5. How should a steel rule or scale be taken 'care' of ?
6. What is the vernier principle ? Explain the following vernier principles:
   (*i*) Principle of 0.1 mm vernier.
   (*ii*) Principle of 0.02 mm vernier.
7. Name the different types of vernier calipers and draw their neat sketches.
8. State the probable errors in calipers.
9. Write down the precautions which should be taken while using a vernier caliper.
10. Explain the construction and working of a vernier height gauge with the help of a neat sketch.
11. State the field of application of a vernier height gauge.
12. What precautions should be taken while using a vernier height gauge ?
13. Draw a neat sketch of a vernier depth gauge and explain its construction and working.
14. State the 'principle' on which micrometers are designed.
15. Give the description of an outside micrometer. How is it read ?
16. Enumerate the sources of errors in micrometers.
17. What precautions should be observed while using the micrometer ?
18. Enumerate different types of micrometers and explain with neat sketches any two of them.
19. Explain briefly the construction and working of a depth micrometer.
20. Name different types of anvils, draw their sketches and state their uses.
21. State the advantages and limitations of following precision instruments:
   (*i*) Vernier caliper (*ii*) Vernier height gauge
   (*iii*) Vernier depth gauge (*iv*) External micrometer
   (*v*) Internal micrometer (*vi*) Depth micrometer.

Angular and Taper Measurements

22. Define an 'angle'.

23. Enumerate the instruments used for angular measurement.
24. Explain any two of the following instruments:
    (*i*) Vernier bevel protractor. (*ii*) Dial bevel protractor.
    (*iii*) Optical bevel protractor.
25. Differentiate between the following instruments:
    (i) Sine bar, (*ii*) Sine table
    (*iii*) Sine centre.
26. What is a sine bar ? Name the materials of which it is made up of.
27. Explain with the help of a diagram the principle of a sine bar.
28. Sketch various forms of sine bars and explain any one of them in detail.
29. Enumerate the uses of sine bars.
30. Write down the essential requirements in the use of a sine bar to get accurate results.
31. What precautions should be taken while using sine bars?
32. What is a sine table?
33. What is a sine centre? Is its principle of setting same as that of sine table?
34. What are angle gauges? How are they used?
35. State the practical applications of angle gauges.
36. Explain with a neat diagram block spirit level and its principle of working.
37. What is a clinometer? Name different types of commonly used clinometers.
38. Write a short note on 'Plain index centre'.
39. Give the description of an auto-collimator stating clearly its principle of working.
40. Explain with a neat sketch the construction and working of Hilger angle dekkor.
41. Explain any one of the following:
    (*i*) Micro-optic auto-collinator telescope.
    (*ii*) Photo-electic micro-optic auto-collimator.
42. Give applications of an auto-collimator.

Comparators

43. What is a comparator?
44. What are the essential parts of comparators?
45. What is the purpose of a comparator?
46. List the desirable features of a comparator.
47. Give the uses of comparators.
48. How are comparators classified?
49. Name important mechanical comparators.
50. Describe a 'dial indicator' with a neat sketch.
51. Give the description of a Reed type mechanical comparator.
52. Explain any one mechanical comparator.

❑❑❑

Measurement of Geometric Forms

Chapter 11

11.1 INTRODUCTION

In most of the methods the basis of measurement is either a straight line or a plane, and these two qualities are represented in precision engineering by the straight edge and the surface plate or table, either as separate units or as part of the construction of a machine tool or instrument. In angular measurement, the right angle assumes special importance, and there are, of course, methods for its production and measurement which are not applicable to angles in general.

The increased demand for product reliability and efficiency has placed a corresponding emphasis on the geometric integrity of components and their assembly. In engineering applications, one often comes across the problems of measurement of geometrical parameters such as alignment, *straightness, flatness, squareness etc.*

11.2 STRAIGHTNESS

- It is difficult to define exactly what is meant by straightness although a **straight line** *may be defined as the shortest distance between two points.* This is not satisfactory because any definition for metrological work must be such that measurements of departure from the ideal can be made. The ISI specifications for straight edges and surface plates give limits for the departures from an ideal straight line or mean true plane but do not define what is meant by these terms. It is generally understood that, for the shortest distances commonly met with in this type of work a ray of light is straight although in fact, its path is affected by the temperature, pressure, humidity of the air. For small areas, the surface of a liquid at rest is considered flat although in reality it is a small portion of a sphere.
- One method of testing long machine beds for straightness of the top surfaces is by *using the surface of a liquid* and another is by means of *rays of light.* Between these two methods there is a discrepancy of 0.025 mm for a distance of 30 m.

  Sometimes it is advisable to use the liquid surface method. For example, when laying the guide rails for testing models of ships in a long water tank, the rails must be parallel to the surface of water.

11.2.1 Definition of Straightness of Line in Two Planes

- A line is said to be *straight over a given length,* if the variation of the distance of its points from two planes perpendicular to each other and parallel to the general direction of the line remains within the specified tolerance limits; the reference planes being so chosen that their intersection is parallel to the straight line joining two points suitably located on the line to be tested and the two points being close to the ends of the lengths to be measured.

- *The tolerance on the straightness of a line is defined as the maximum deviation in relation to the reference straight line joining the two extremities of line under examination* as shown in Fig. 11.1

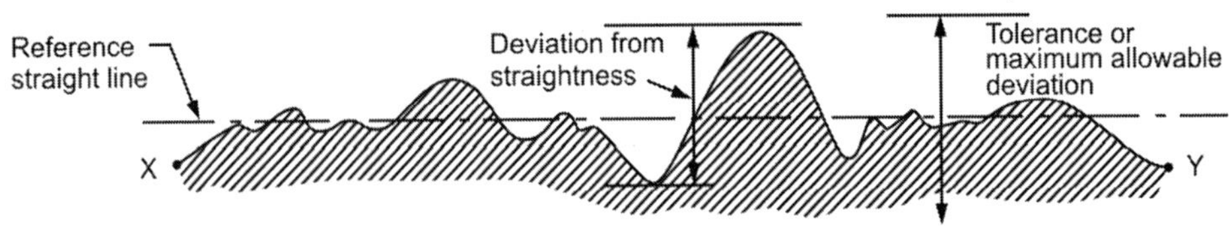

Fig. 11.1

- It is the usual practice to state the range of measurement, *i.e.,* the length to be checked; and the position of the tolerance in relation to the reference straight line. In most cases, the parts very close to the ends, which most often have local errors of no great importance may be neglected.

11.2.2 Straight Edge

- *A straight edge is a measuring tool which consists of a length of steel or other suitable material usually of narrow and deep section and vary in length from several millimetres to a few metres. The object of the deep and narrow section is to provide considerable resistance to bending in the plane of measurement without excessive weight.*
- For general use the *smaller sizes* are *unhardened* and may be bevelled at the measuring edge.
- The straight edges may be made in *either steel or iron.* The cast type, which is really a long narrow surface plate, is straightened with a deep web of parabolic shape, chosen so that the resistance to bending at each point is proportional to the bending moment due to the weight of the material.
- *Precision straight edges* are generally hardened and lapped on edge only to a small radius in cross-section, forming a blunt knife edge which is straight to a few thousandths of a millimetre. Figure 11.2 shows different types of straight edges.
- The accuracy of the straight edge should be high and permissible deviation of the measuring edge from the true straight line should not exceed.

$$\pm\left(0.001 + \frac{L}{500000}\right) \text{mm}$$

where L is the length in mm of the straight edge.

- Large straight edges should be supported at the points of minimum deflection, these points are usually engraved with arrows on the straight edge and lie symmetrically at a distance of 0.544 L, where L is the length of the straight edge.

11.2.3 Measurement of Straightness

The straightness can be measured by the following *methods:*

1. Straight edge method,
2. The wedge method.
3. The level method.
4. The autocollinator method.

11.2.3.1 Straight edge method

The most obvious method of testing straightness is to *use a straight edge whose accuracy is known.* For many applications it is sufficient to apply the straight edge to the surface to be tested, and gauge the continuity or degree of contact by *marking, feelers* or *light gap.*

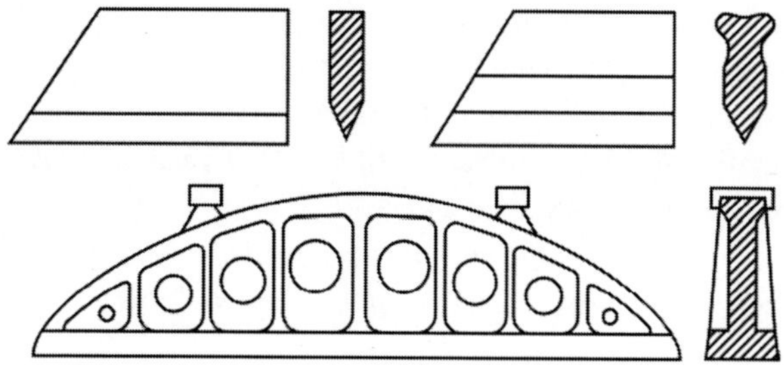

Fig. 11.2 Different types of straight edges.

(*i*) The '*light gap method*' is particularly applicable to the checking of lapped surfaces with a precision tool maker's straight edge. Viewed against a good light, a gap of less than 0.00025 cm can easily be seen. At about 0.000125 cm to 0.000175 cm the light appears red, whilst at 0.00075 cm it appears blue. These colour effects are due to interference caused by diffraction, and form a useful guide to the width of the gap in this region.

(*ii*) For coarser work, the 'blueing or feeler methods are quite useful, although here again, precautions must be taken against sagging of straight edges. It is very easy to assume that for example, that a large surface is flat when the straight edge is actually taking up the same contour as the surface.

(*iii*) *A more accurate method* is to support the straight edge on equal slip gauges at the correct points for minimum deflection, and measure the uniformity of the space under the straight edge with slips.

11.2.3.2 Wedge method

Figure 11.3 shows the testing of straight edge by *wedge method.* The straight edge is supported at the points for minimum deflection on two unequal piles of slip guages which rest on a surface which is straight or of which the errors in straightness are known. Greater certainty in measurement is assumed by having two unequal stacks of slip gauges for the purpose of supports. The difference between the slip values is made definite by 0.1 mm, and the distance between the supports is divided into a number of equal parts suitably marked on the straight edge. Figure 11.3 shows an arrangement where the distance between two supports differing by 0.1 mm is divided into 10 equal parts.

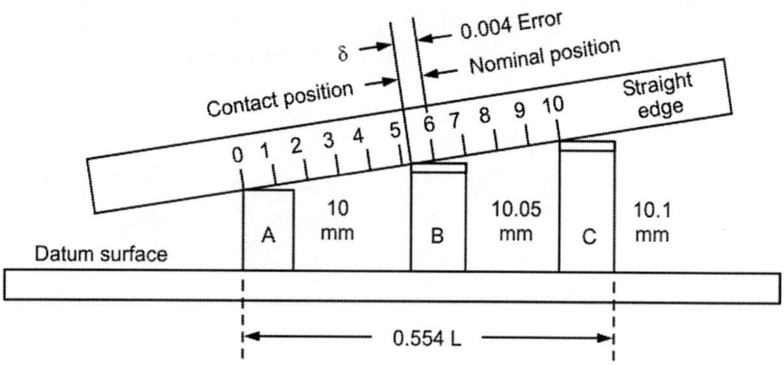

Fig. 11.3 Wedge method.

It is obvious that, if *both straight edge and surface are perfectly straight, the gap will vary by 0.01 mm at each point, making a wedge.* Consequently, *slip guage can be inserted corresponding to each position,* and *will make contact with surfaces exactly at the marked positions.* If, however, *there are errors in the straightness of the surface, the slip gauge may not fit exactly opposite their corresponding marked positions and will be displaced one way or the other along the straight edge by amounts proportional to errors.* For a separation of 200 mm between supports differing by 0.1 mm, each 0.01 mm variation corresponds to a displacement of 20 mm, and this provides a very sensitive method of measurement. Care must be taken to *see that slips do not wring to the surface, otherwise the whole sensitivity will be lost.* An *alternative method to measure the actual contact position of each nominal pile of slips is to alter the slip value until the contact is made at the nominal position.*

This may be noted that the *wedge method can be made as sensitive as desired by choosing a small wedge angle and a large number of measuring positions.*

11.2.3.3 The level method

In order to measure the straightness with a level (say a spirit level) the level has to be supported on two equal columns of slip gauges placed at suitable distance. The angular variation or the difference in height between the two supports can be measured. The results are obtained with reference to the horizontal datum plane and they will be influenced by the angle of inclination of the surface to the horizontal plane. This method *can only be used when the measured line lies approximately horizontal within the angular measuring range of the spirit level.*

11.2.3.4 The autocollimator method

The principle of this method is same as that of the level method. The *spirit levels can be used only for the measurement of straightness of horizontal surfaces while autocllimator method can be used on surfaces on any plane.*

In the autocollimator method, the instrument is placed at a distance of 0.5 to 0.75 metre from the surface to be tested on any rigid support which is independent of the surface to be tested. The parallel beam from the instrument is projected along the length of the surface to be tested. A block fixed on two feet and fitted with a plane vertical reflector is placed on the surface and the reflector face is facing the instrument (Fig. 11.4). The reflector and the instrument are set such that the image of the cross-wires of collimator appears nearer the centre of the field and for complete movement of reflector along the surface straight line, the image of cross-wires will appear in the field of eyepiece. The reflector is then moved to the other end of the surface in steps equal to the centre distance between the feet and tilt of the reflector is noted in seconds from the eye piece. The readings of the angle of tilt are converted into differences in height of the feet and the contour is plotted.

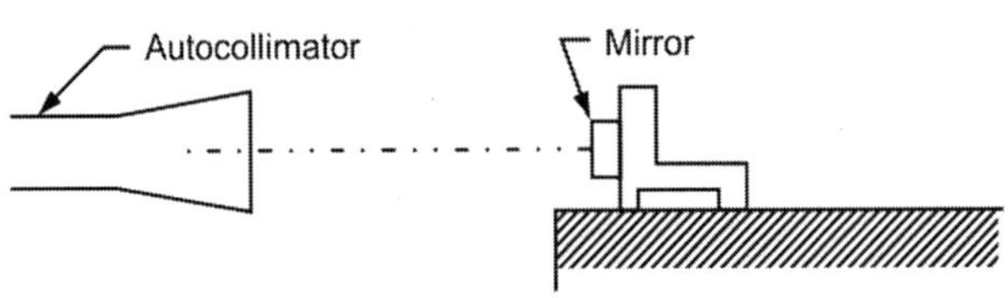

Fig. 11.4 Set up for autocollimator method.

11.3 FLATNESS

11.3.1 Definition

The **flatness** *may be defined as the minimum distance between two planes which coverall the irregularities of the surface under examination.*

According to Indian standard specification IS: 2063 a surface is deemed to be flat within a given range of measurement when the variation of the perpendicular distance of its points from a geometrical plane (this place should be exterior to the surface to be tested) parallel to the general trajectory of the place to be tested remains below a given value. The geometrical plane may be represented either by means of a surface plate or by a family of straight lines obtained by the displacement of a straight edge as a spirit level or a light beam.

11.3.2 Flatness Measurement

The following *methods* may be used for flatness measurement:

1. Direct comparison
2. Dial gauge
3. Level or autocollimator methods
4. Flatness comparators
5. Liquid method.
6. Interference methods.

11.3.2.1 Direct comparison

This method is restricted to relatively small area. For instance, the surface to be checked *can be compared with a surface plate or tool marker's flat.* One of the surfaces is marked with Prussian blue or some other suitable substance, and the surfaces are then rubbed together. The result of this is to *expose local departures from flatness, but their magnitude remains unknown.*

11.3.2.2 Dial gauge

In order to test the flatness of a surface plate, it has to be placed on a larger reference plate and supported by slip gauges or adjustable screws for the purpose of setting the two surfaces parallel. This can be done by means of a dial gauge with a stand that can be placed on the reference plate as shown in the Fig. 11.5.

To ensure parallelism, the reading of the dial gauge at each edge of the surface plate should be the same.

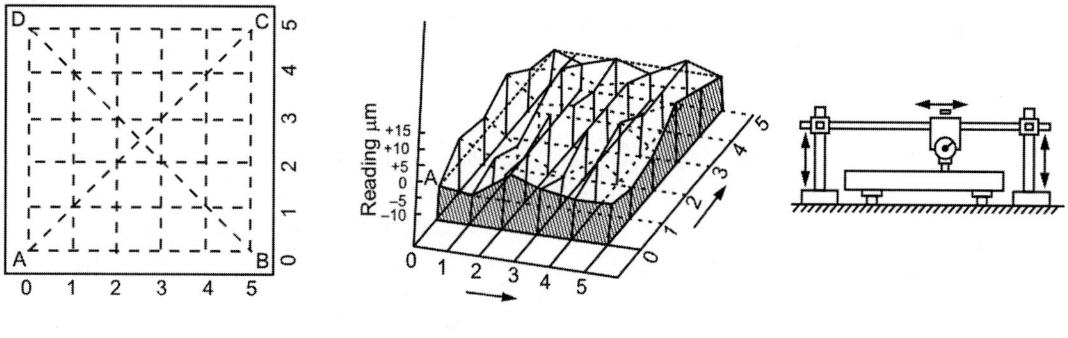

Fig. 11.5 Measuring the flatness of surface by a dial gauge.

The dial gauge should be *rigidity mounted* on the stand to prevent deformation and deflection of the overhang which can be quite large compared to errors of flatness. Therefore, it is recommended to mount the dial gauge on a bridge and dial gauge can slide on it over the surface plate.

- The method of measuring the flatness is to measure the straightness along a number of lines as shown in Fig. 11.5 so that the whole surface under test is covered with a grid of lines. The dial gauge is then moved along these lines and readings taken at each position or at the points of intersection of the lines have to be related to the first measuring point *'A'* that can be considered as a reference point. The results obtained can be drawn in a three co-ordinated diagram as shown, so that the flatness error can be evaluated. *The difference in height between every two successive points on the grid of lines drawn on the surface being tested can be measured by one of the instruments used for measuring the straightness, such as spirit level of the autocollimator.*

The important feature of this method is that it provides *reliable results* for determining the mean contour of the surface *particularly if the distance between the measuring points is selected as small as possible.*

11.3.2.3 Level or autocollimator method

- Tests for flatness using either a precision level or an autocollimator are similar to those for straightness, except that they are more elaborate.
- The method adopted for areas is to traverse the precision level or the reflector block over a network of lines. The measurements are related to a selected datum plane. In this respect it is important to ensure that the lines of test are linked by common positions of measurement. The actual number of positions occupied by the instrument concerned, in any particular instance, is determined by a *number of factors.* These include *the extent of area concerned, the possible spacing of the feet of the instrument and the degree of accuracy necessary for the required results. The level or reflector may be guided along the chosen lines, by using a straight edge or ordinary rule.* In the use of a *rectangular or square surface a general impression of its flatness may be obtained by checking along its two diagonals.*

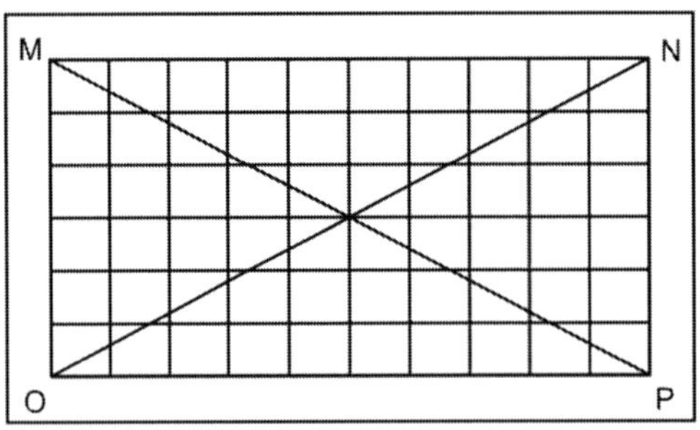

Fig. 11.6 Network of lines for checking flatness of. rectangular surface.

- When full examination is to be carried out the surface may be checked over a network of lines as shown in Fig. 11.6 which represents a regular surface. It will be seen that two series of lines, parallel to the sides of the work, are taken and also two diagonals. The controller of surface along each line is found, the test being taken in both forward and backward directions and the mean values used, as in normal straightness testing. The results can be plotted and the contour of the surface established. In this instance the lines of test could be linked by taking any three corners (say MON) as common positions of measurement and treating them as zero points.

11.3.2.4 Flatness comparators

The following two comparators are useful instruments for measuring flatness of a surface.

1. Beam comparator
2. Bar comparator.

Beam comparator

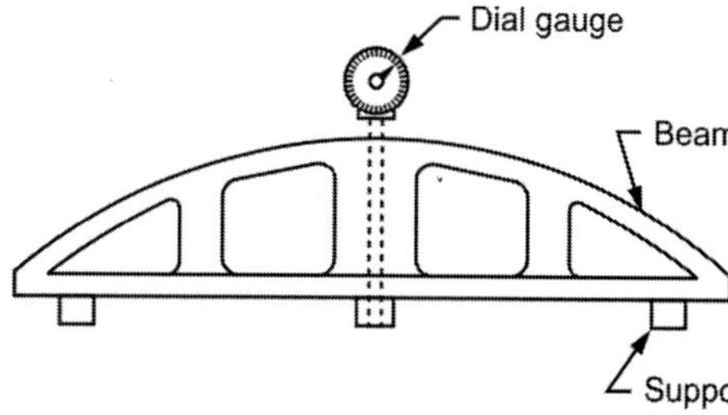

Fig. 11.7 Beam comparator.

- Refer to Fig. 11.7. The instrument is a *beam having four supports,* the two at the centre of the beam being offset from the line joining the other two supports. One of the middle supports is spring loaded so that it can be self-adjusted to lie in the plane passing through the other three fixed supports. This plane is considered as a reference true plane relative to which the flatness of the surface can be measured. For the purpose of mounting a suitable dial gauge the beam comparator is provided with a hole.
- By placing the beam comparator on a surface plate, the initial zero reading of the dial gauge can be adjusted. *In this position, the anvil of the dial gauge that contacts the surface plate will be lying on the reference true plane of measurement.*

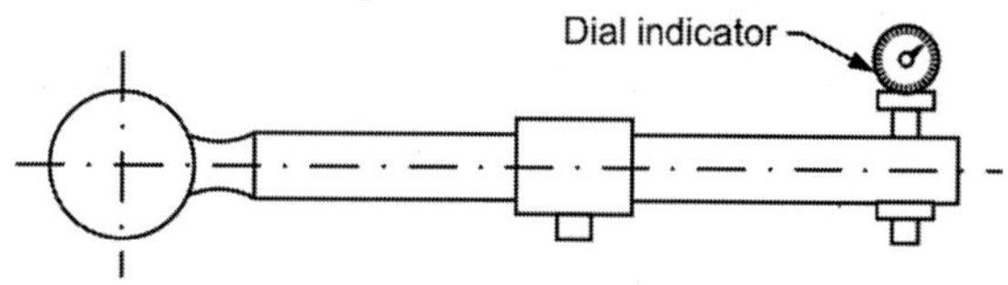

Fig. 11.8 Bar comparator.

- The instrument (beam comparator) is placed on the surface under test such that the *dial gauge measures the level of the point on the surface with respect to the reference plane.* The beam comparator has to be moved along the surface so that the level of each point on the marked grid of lines can be measured.
- *The beam comparator is used mainly for testing* surface plates having scraped finish.

Bar comparator

- This instrument can also be used in a similar manner and for the same purpose.
- Refer to Fig. 11.8. It consists of a bar supported on three fixed points, two at its centre and the third at one end. The measuring anvil of a micrometer or indicator, is arranged at the other end of the bar. The measuring anvil should lie on the reference plane containing the other three supporting points, when the micrometer or dial gauge read zero.

11.3.2.5 Liquid method

This method utilises a trough of liquid, such as mercury or soda solution, supported over the plate, and a form of height gauge with an overhanging micrometer head having a conical point to the spindle. The micrometer head spindle makes contact with the liquid, the position of the contact being determined with the help of an electrical indicator circuit. Variations in the surface are thus compared with the true plane of the liquid surface. Such methods, utilising liquids, are not new, and have appeared in various forms, but *on the whole they are not easy to apply,* and are *often of doubtful accuracy* unless sometime is devoted to practice and complete cleanliness is ensured. The type of instrument which consists of two jars of water, connected by a rubber tube, in which variations in water level between the jars is measured with micrometer dippers, is not *really suitable for accurate surface plate measurement, although it is quite useful for levelling of other types.*

11.3.2.6 Interference method

This method is *excellent for checking small lapped surfaces, such as micrometer anvils, block gauges and similar parts.*

- The *use of interference phenomena provides a very sensitive method of measuring flatness on a lapped and polished surface.* The flatness of the surface is directly compared with that of a glass plate, known as an *optical flat or proof plane.* The flatness of such a plate can be produced and measured to an accuracy of a few millionth of a cm. The optical flat is placed in contact with the surface, and the *interference patterns produced provide a highly accurate contour map of the variations in the flatness of the surface.*
- **Optical flats.** These are flat lenses with very accurately polished surfaces that have light-transmitting quality. Optical flats are usually made from natural quartz because of its hardness, low co-efficient of expansion, and resistance to corrosion. These are available from 35 to 300 mm in diameter with a thickness about 1/6 the diameter. It is not necessary that the two surfaces of a flat be absolutely parallel.

The optical flat is placed on the flat surface to be tested, and light is reflected both from the optical flat and the surface being tested through the very thin layer of air between the two surfaces. *When the light waves are in phase, there is a light band; when they are out of phase, a dark band is created.* If the thickness of the air layer measures of light or more, an interference effect occurs. The interference between the rays reflected from the bottom of the flat and from the top work causes dark bands (Newton's rings) or areas to appear.

If the surface is irregular, the appearance is similar to a contour map. The *position and number lines show the location and extent of irregularities. When the bands are straight, evenly spaced, and parallel to the line of contact, the surface is perfectly flat. If the bands were straight but not evenly spaced or if the bands curved, the surface would not be flat.* If the wavelength of the light source is known, any deviation from this pattern indicates an error in the surface, the amount of which can be measured.

Figure 11.9 shows the use of flats to determine curvature.

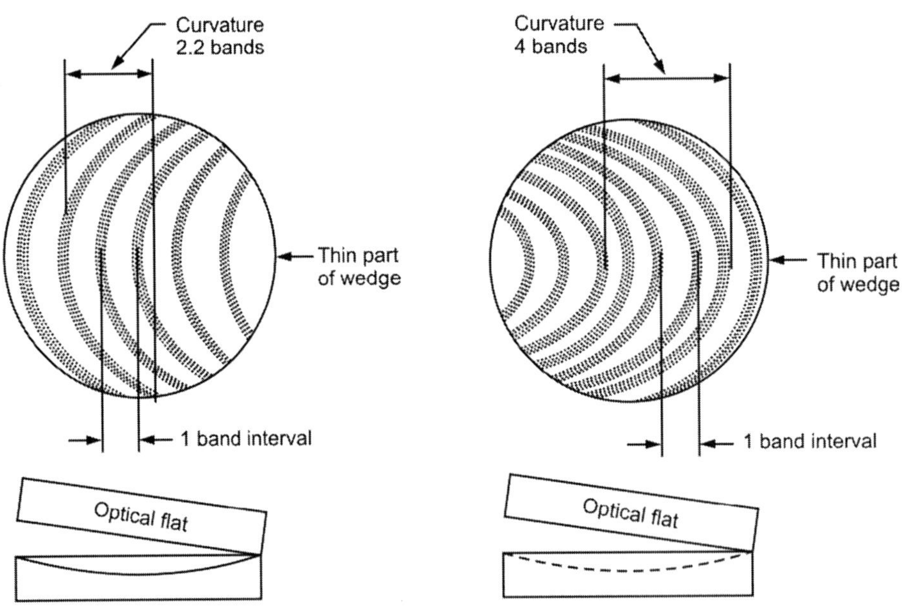

Fig. 11.9 Use of flats to determine curvature.

Refer to Fig. 11.10. The principle applied to a flatness interferometer is same as that of optical flat. The optical flat is supported above the surface of the gauge to be tested, and is not in contact with it. The flat is adjustable for inclination in two planes, so that the pitch and direction of the interference fringes may be set to the best position.

Figure 11.11 shows the method of testing a surface plate.

- The surface is surrounded by a brass ring containing paraffin liquid of about 10 mm thickness. The upper surface of the liquid forms the first reflecting surface, corresponding to the underside of an optical flat.
- From the *source of light (monochromatic)* the light rays are reflected at *two reflectors* after passing through a double *concave lens, a filter* and *an aperture* placed at the focus of the double concave lens. The light rays reach the *collimator lens* whose focus coincides with that of the *double concave lens* at the aperture. From the collimating lens, parallel rays of light fall on the surface plate after passing through the layer of paraffin liquid.
- *When the surface plate is tilted with respect to the horizontal plane through a small angle, interference takes place between the incident rays on the surface of the paraffin liquid and those refracted and then reflected at the surface plate.* The *reflected light returns back through another aperture to be delivered in a reflecting prism to* the ocular through which the interference fringes can be observed.

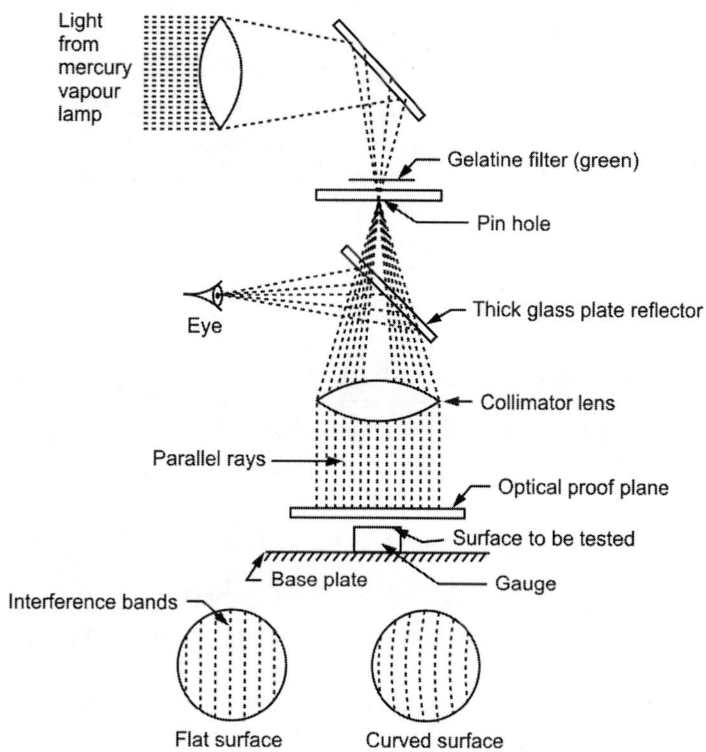

Fig. 11.10

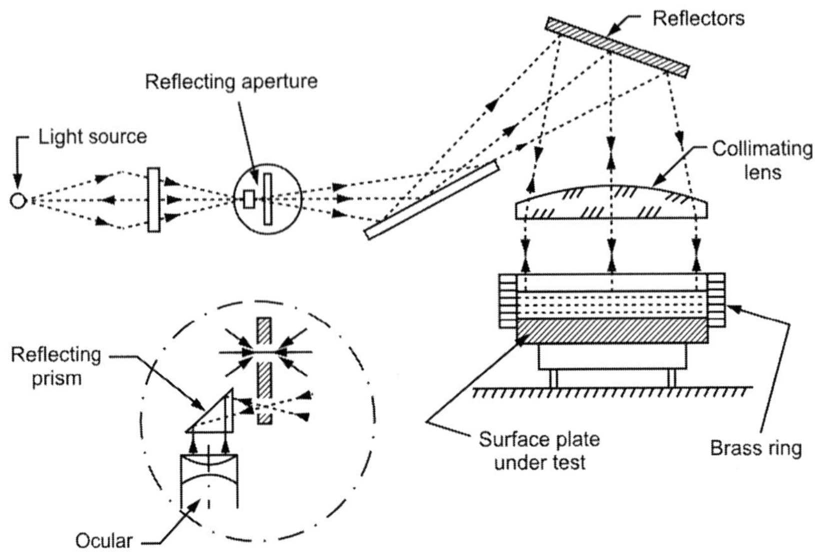

Fig. 11.11 Optical system of liquid surface interferometer.

11.4 CIRCULARITY AND ROUNDNESS

The terms roundness and circularity are often used interchangeably. **Roundness** *is defined as a condition of a surface of revolution* (like cylinder, core or sphere) *where all points of the surface intersected by any plane perpendicular to a common axis in case of cylinder and cone* (or passing through a common centre in case of sphere) *are equidistant from the axis* (or centre). Since the axis and centre do not exis physically, measurements have to be made with reference to surface of the figures of revolution is the circular contour. It may be understood that *while roundness expresses a particular geometric form of a body of revolution in all the three dimensions, the* ***circular contour*** *is the characteristic form of the entire periphery of a plane figure. For measuring roundness, it is only the circularity of the contour which is determined.*

11.4.1 Types of Irregularities of a Circular Part

The errors of circularity at a cross-section can be of the following nature :

1. Ovality
2. Lobbing
3. Irregularities of no specific form.

1. **Ovality.** Figure 11.12 shows difference between the major and minor axes.
2. **Lobbing.** During certain machining operations, especially all forms of grinding and occasionally turning, all too frequently, it is possible to produce a deviation from the geometric form of a components. One of the most common imperfections is *lobed* surface (Refer Fig. 11.13). In this case the *diameters at any two opposite points are constant, but still it is not circular form.* The error in form may often be attributed to possible irregular conditions of original stock, or to the component holding mechanism during machining, cooling, machine condition and suitability.

   It is very difficult under production rates to achieve a perfect cylinder from imperfectly formed stock. One way to reduce the possibility of error is by preliminary machining of component. Sometimes the production of error can be attributed to the component holding-device.

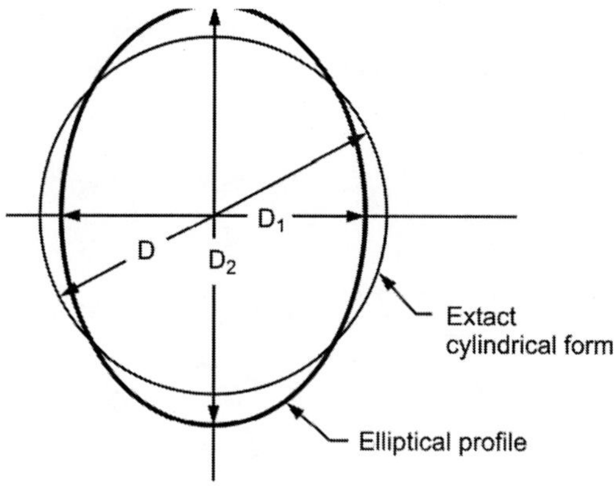

Fig. 11.12 Ovality (i.e., $D_1 \neq D_2 \neq D$).

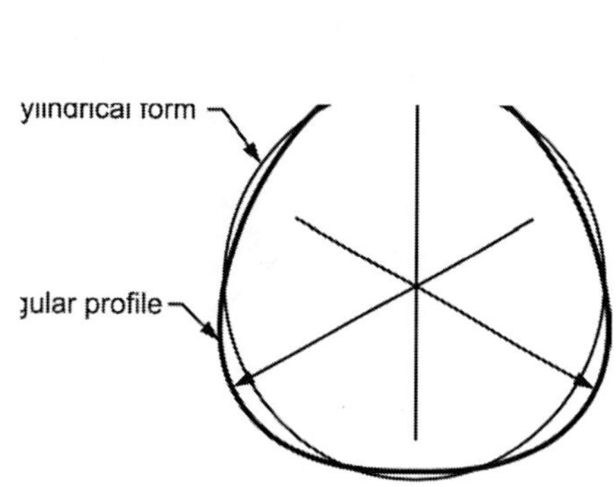

Fig. 11.13 Lobbing.

3. **Irregularities of no specific form.** Refer to Fig. 11.14. In this case irregularities are not of particular form.

Error of circularity. The ***error of circularity*** may be defined as follows:
"The radial distance between the minimum circumscribing circle and the maximum inscribing circle, which contain the profile of the surface at a section perpendicular to axis of rotation".

Evidently it is possible to draw several circles for a given profile and infer different results. Accordingly four types of reference circles (described below) have been standardised and the results should be specified indicating the reference circle adopted.

1. Maximum inscribed circle
2. Minimum circumscribed circle
3. Minimum zone circle
4. Least squares circle.

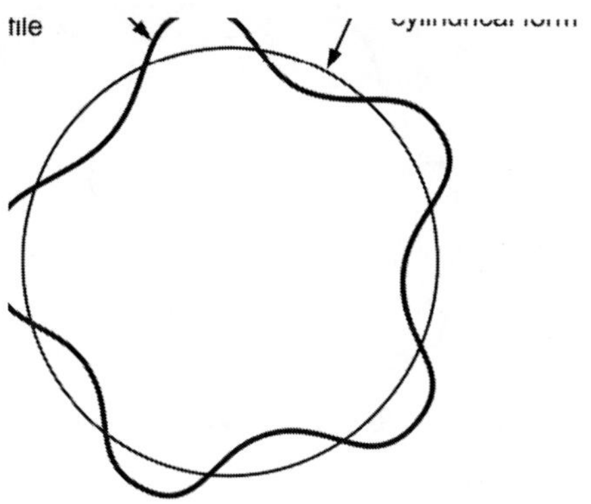

Fig. 11.14 Irregularities of no specific form.

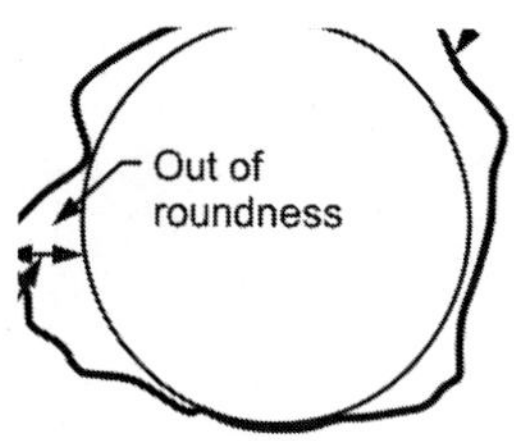

Fig. 11.15 Maximum inscribed circle.

1. **Maximum inscribed circle.** Refer to Fig. 11.15. The maximum inscribed circle is the *largest circle that can be drawn inside the profile without cutting it.* Its centre and radius can be found by trial-and-error using compasses, by template or computer. The *out-of-roundness is specified as the height of*

the largest peak above the circle. This value is nearly equivalent to : $S_p + S_y$ since S_v is zero, there being no valley inside the circle.

where, S_p = Radial distance of maximum peak from the circle, and
S_v = Radial distance of maximum valley from the circle.

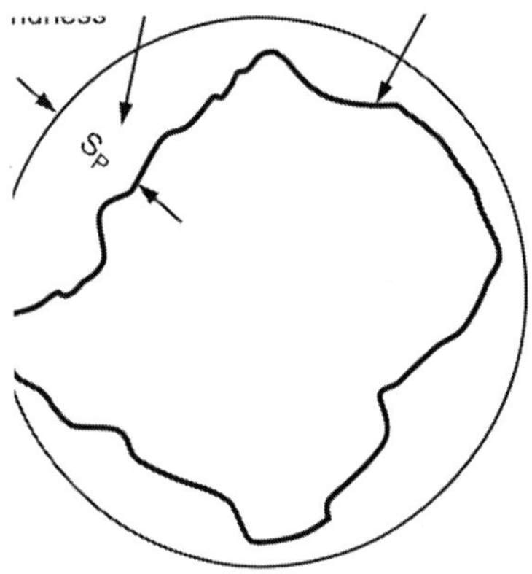

Fig. 11.16 Minimum circumscribed circle.

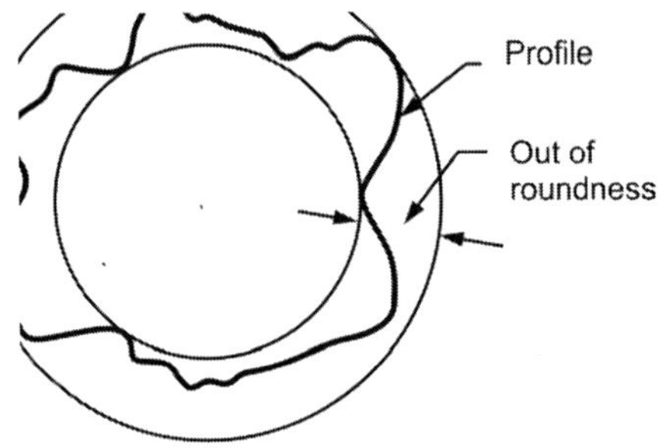

Fig. 11.17 Minimum zone circle.

2. **Minimum circumscribed circle.** Refer to Fig. 11.16. The minimum circumscribed circle is the *smallest circle that will completely enclose the profile without cutting it.* Its centre and radius can be found in a similar manner to that of the inscribed circle. This *out-of-roundness is the distance of the lowest valley from the circle, which is also equivalent to $S_p + S_v$, since S_p is zero, their being no peak outside the circle.*
3. **Minimum zone circle.** Refer to Fig. 11.17. There are two concentric circles that just enclose the profile and which have minimum radial separation. The value of *out-of-roundness is the radial distance between the two circles.* The centre of such a circle is termed as the *minimum zone centre.* These circles can be found by using a template, or computed by necessary accessories.
4. **Least squares circle.** Refer to Fig. 11.18. The least squares circle is the *circle which represents the average of all the peaks and valleys,* and can be defined mathematically as : *"The sum of squares of a sufficient number of equally spaced radial ordinates, measured from the circles to the profile has minimum value."* The centre of such a circle is referred to as the *'Least square centre'*. The value of out-of-roundness is the radial dista'nce of the maximum peak from this circle (S_p) plus the distance of the maximum valley from this circle (S_v), *i.e.*, $S_p + S_v$.

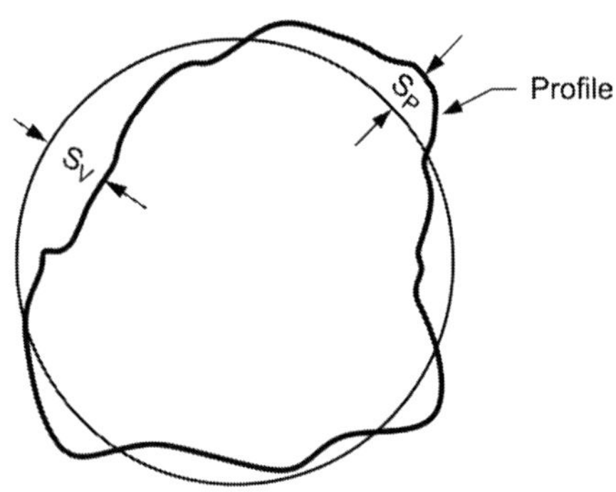

Fig. 11.18 Least squares circle.

It is very difficult to calculate the position of the least squares circle by manual methods but very easy by using a computer.

Determination of error of circularity. The *modern circularity measuring instruments* are used to determine error of circularity. These instruments make use of precise spindle on which the indicating means are carried and in which turntables are provided for carrying the work. The radial variations discovered by the instruments are generally displayed in the form of a *polar profile graph* or a meter reading or both. From the polar graph the departures from the ideal circle can be determined as "the *difference in radii between two best fitting concentric circles, within which profile is just con'tained*". Refer to Fig. 11.19. The *radial distance between the maximum circumscribing circle and the maximum inscribing circle is the measure of the error of circularity.*

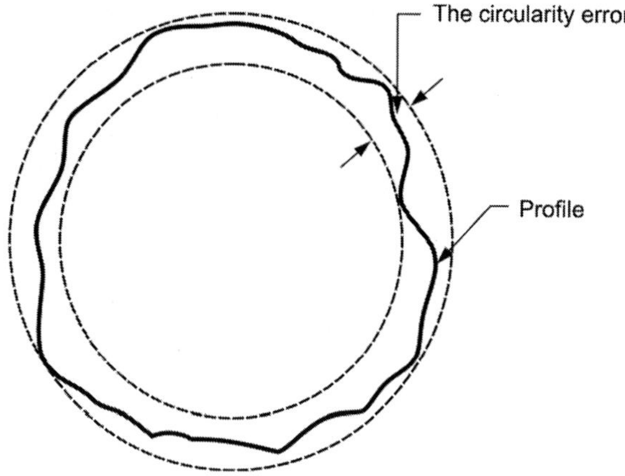

Fig. 11.19

11.4.2 Measurement of Circularity-Devices Used

The following *devices* are most commonly used for measurement of circularity error:

1. V-block and dial indicator
2. Precision measuring instrument (Talyrond).

11.4.2.1 V-block and dial indicator

The V-block method is quite accurate for workshop purpose. This method is capable of indicating *normal requirements of accuracy.*

In this method of measuring of circularity (lobbing) the following *procedure* is adopted : Refer to Fig. 11.20.

- The V-block is placed on a surface plate and the work to be checked is placed upon it.
- A sensitive dial indicator is firmly fixed in a stand and its feeler made to rest against the surface of work.
- The work is rotated to measure the rise and fall of the workpiece.
- In order to determine the number of lobes on the workpiece, the workpiece *is first tested in 60° V-block* and *then in 90° V-block.*

The number of lobes is then equal to the number of times the indicator pointer deflects during rotation of workpiece through 360°.

The idea of testing the work-piece in two V-blocks is that when an *elliptically shaped part* is rotated on a V-block of angle 60°, no change in reading is indicated, whereas if the same part is rotated on a 90° angle V-block, two maximum and minimum readings are indicated on the indicator.

The V-block method of measuring circularity has certain limitations, therefore, *following points need to be kept in mind*:

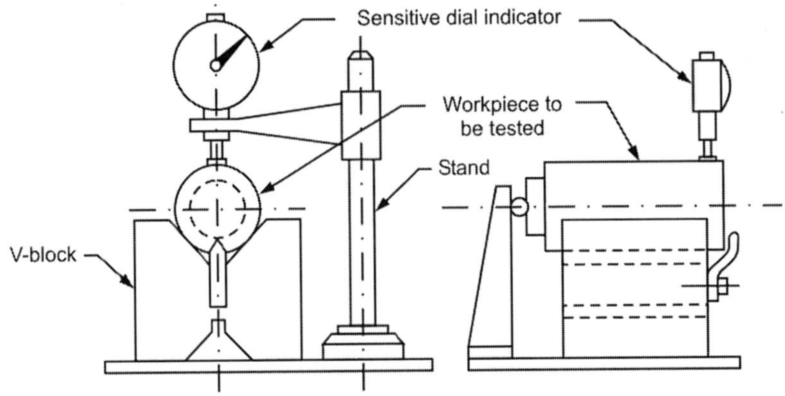

Fig. 11.20 V-block and dial indicator.

1. The circularity error measured on a V-block is greatly effected by the following *factors:*
   - The angle of V-block greatly influences the determination of error of circularity.
   - The position of the instrument, i.e., whether measured from top or bottom.
   - Number of lobes on the rotating part (e.g., triangular, quadrilateral, pentagonal, elliptical etc.)
2. The position of the instrument should be in the same vertical plane as the point of contact of the part with the V-block. If the error is measured at a point for from V-block, the error of circularity will be influenced by the radial run out of the part.
3. A leaf spring should always be kept below the indicator plunger and the surface of the part, otherwise, readings are likely to be affected by minute undulations of the surface, such as surface roughness.

Procedure to be adopted for the measurement and drawing of the '*Polar graph*':

Below is given the procedure to be adopted for the measurement and drawing *of the polar graph* and therefrom the *error of circularity*.

- To start with 12 markings equally spaced at angles of 30° are made on the face of the workpiece to be measured.
- The workpiece after proper cleaning is then placed on V-block against a rigid block with a steel ball in between as shown in Fig. 11.20.
- The dial indicator is placed just above the work-piece so that it is touching the workpiece, nearly at the centre of the V-block.
- The workpiece is then rotated such that the marking on the workpiece is below the indicator plunger.
- A small strip (leaf spring) is fixed between the indicator plunger and the surface of the part.
- The readings of the dial indicator are noted down for all markings. This is repeated three times to take the average value.

Plotting of polar graph:

In order to plot the polar graph, *a suitable scale is chosen depending on the maximum value of the reading.*

- Then a circle of diameter nearly 4 times the maximum reading of the dial indicator is drawn and is divided into 12 number of angular divisions as shown in Fig. 11.21.

- Inside the outer circle, another concentric circle of suitable diameter is drawn. Actually there are standard diameters used on polar graphs.
- Then the values of indicator are plotted in radial direction taking the smaller circle as the reference circle in order that both positive and negative readings are plotted within the prepared graph.
- Finally the individual points are joined by straight lines, thus obtaining the actual profile of the measured workpiece.
- *The error is measured as the radial distance between the minimum circumscribing circle and the maximum inscribing circle for the profile obtained. This is done by 'hit and trial method' such that the distance is minimum.*

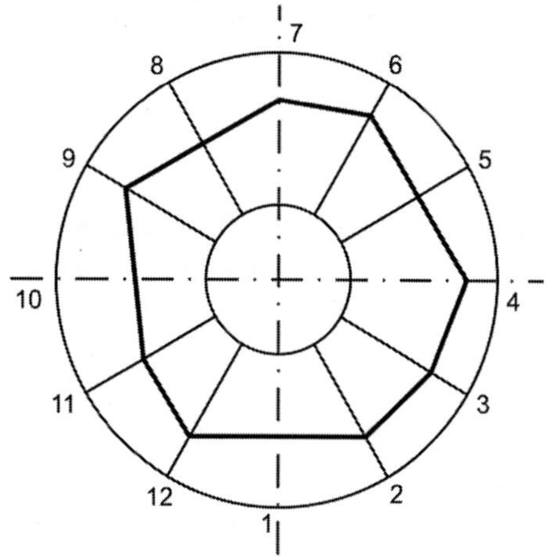

Fig. 11.21 Plotting of polar graph.

The actual value of circularity error is given by,

$$\text{Error} = \frac{\text{Error measured from polar graph}}{C}$$

where, C = a constant, the value of which depends upon the *shape of the workpiece and angle of V-block.*

11.4.2.2 Precision measuring instruments

The method involving the use of precision instrument is recommended for *very precise job where more reliable and more accurate results* (as compared to V-block method) *are desired.*

This method is *quicker and also eliminates the effects of angle of the block and the number of lobes on parts.* However, *this method is very costly.*

'The Talyrond'—*Roundness measuring machine:*

With the ever increasing importance of geometrical form in cylindrically generated components, as distinct from mere uniformity of diameter, it is becoming more and more necessary to be measure roundness accurately. The instrument to do this successfully and be available commercially was the Taylor-Hobson *"Talyrond".* This instrument was developed almost as a natural successor to the Talysurf surface texture measuring instrument and it utilises very similar principles of mechanical pick-up amplification and recording.

Construction and principle of working of a roundness measuring machine:

Construction: Refer to Fig. 11.22.

- The machine consists *of a precision spindle* of extreme accuracy that rotates about its axis. The permissible eccentricity should not exceed 0.01 μm.
- A *stylus arm* with a *diamond pointed stylus* is mounted on the spindle in such a way that the position of the stylus from

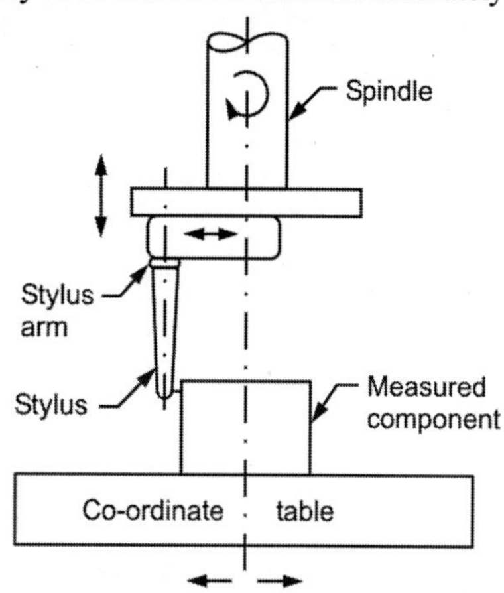

Fig. 11.22 Roundness measuring machine.

the axis of the rotation of the spindle can be varied according to the radius of *component* being measured.

- The spindle can also be moved upwards and downwards so that the measurement can be carried out at different heights along the component. This *enables the straightness of the component to be tested.*
- The component is mounted stationary on the *co-ordinate table which is provided with a centre device for initial centring. Final centring is done automatically by error signals* for driving the motors of the co-ordinate table.

Working:

- By rotating the spindle, the diamond stylus describes a true circle with reference to which the roundness error of component is measured. If the component is perfectly round, the stylus will not be radially displaced. Any deviation of actual contour from the true circle results in a radial movement of the stylus. This move of the stylus is converted into an electrical signal by means of two inductive elements mounted in the stylus arm. The electrical signal is amplified and then recorded on a *'polar co-ordinate recorder'* whose rotation is synchronised with that of the machine spindle. In this way a 'polar diagram' can be directly obtained showing the amount by which the periphery of the work being tested departs from the truly circular form, with an accuracy limited by the error in the spindle rotation only. The electrical magnification of the stylus movement can be increased upto 100000 times.
- Roundness tests can be carried out on the internal and external surfaces of rotation of the workpiece.
- As the stylus movement can be highly magnified, macro and micro-irregularities in relation to the axis of the workpiece can be seen on the recorded diagram. The roundness error of the component can be measured alone when the micro-irregularities are electrically filtered.

Note. Figure 11.23 shows typical Talyrond charts showing ovality and three lobbing. It is important to note that it is only the variations which are magnified; the actual trace on the chart may be either larger or smaller than work being measured.

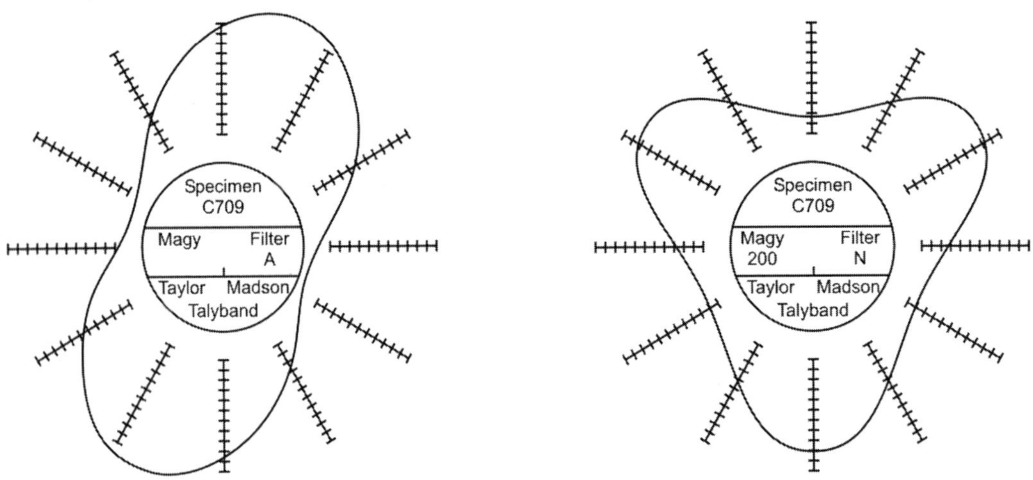

Fig. 11.23 Talyrond charts showing ovality and three lobbing.

"Modern roundness measuring instruments (machines)" are designed to make the numerical assessment of departures from ideal roundness from one of the following centres:

(*i*) Determination from minimum zone centre, *i.e.*, a centre for which the radial difference has a minimum value;

(*ii*) Determination from the least squares centre, *i.e.*, a centre defined on a "least squares" basis (the sum of squares of the errors is the minimum).

11.5 THE TOOLMAKER'S MICROSCOPE

The **toolmaker's microscope** is **an optical measuring** *machine equipped for external and internal length measurements as well as measurements on screw threads, profiles, curvatures and angles.* For these purposes, the microscope is provide with several measuring attachments such as:

(*i*) Centre stage for mounting of cylindrical components,

(*ii*) Revolving and angle measuring oculars,

(*iii*) Double image ocular,

(*iv*) Optical feeder, and

(*v*) Projection screen.

The *applications* of the instrument may be summarised broadly as follows :

1. The *determination of the relative position of various points on work* by measuring the travel necessary to bring a second point to the position previously occupied by the first, and so on.
2. *Measurement of angles* by using a protractor eye-piece.
3. *Comparison of thread forms with master profiles engraved in the eye-piece and measurement of pitch and effective diameter.*
4. *Comparison of an enlarged, projected image with a scale tracing fixed to the projection screen.*

Figure 11.24 shows a toolmaker's microscope. The main parts of this instrument are:

1. Rotatable table
2. Measuring stage
3. Swingable head
4. Ocular
5. Projection screen
6. Micrometers
7. Objective lens
8. Prism.

Construction. The microscope consists of a *rigid stand* on which a *swingable head* is mounted. The *measuring stage* moves on ball guideways by actuating two *measuring micrometers* arranged perpendicular to each other in the length and the cross-sections. The measuring range of each micrometer is 25 mm and the measuring capacity can be increased using slip gauges. A *rotatable table* is provided over the stage, on which the workpiece can be fixed either directly or between centres. This table can be rotated though 360° and the angular rotation can be read by a fixed vernier to a scale value of 3'.

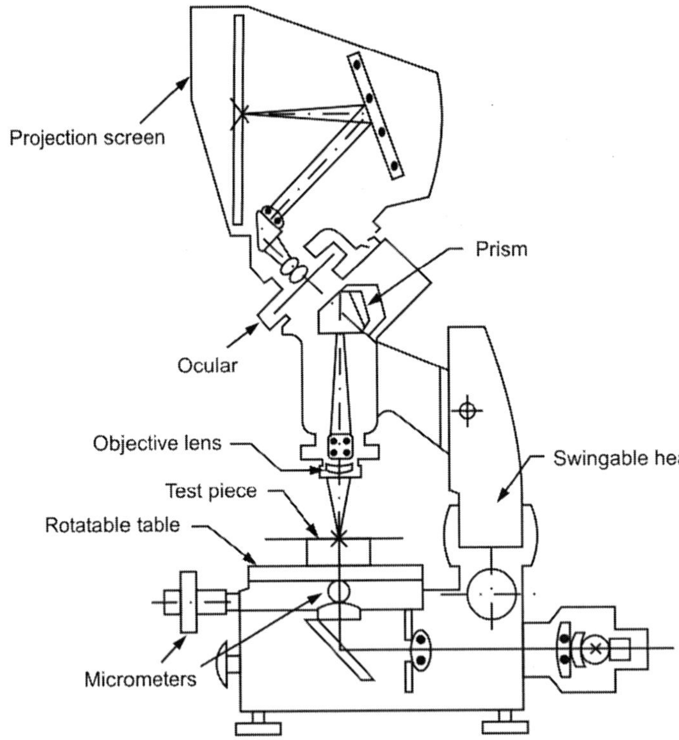

Fig. 11.24 Toolmaker's microscope.

Working. The component being measured is illuminated by the through light method. A parallel beam of light illuminates the lower side of workpiece which is then received by the objective lens in its way to a prism that deflects the light rays in the direction of the measuring ocular and the projection screen. Incident illumination can also be provided by an extra attachment. Exchangeable objective lens having magnification IX, 1.5X, 3X and 5X are available so that a total magnification of 10X, 15X, 30X and 50X can be achieved with an ocular of 10X. The direction of illumination can be tilted with respect to the workpiece by tilting the measuring head and the whole optical system. This inclined illumination is necessary in some cases as in screw thread measurements.

The scale value of this microscope:

- 0.01 mm for length measurement.
- 3' for angle measurement with rotatable table.
- 1' for angle measurement with the angle measuring ocular.

THEORETICAL QUESTIONS

1. Define 'Straightness'.
2. What is a straight edge ?
3. Name the materials of which the straight edges are made of ?
4. Enumerate the methods which are used for measuring Straightness.

5. Explain any two of the following methods employed for measuring Straightness:
   (*i*) The wedge method
   (*ii*) The level method
   (*iii*) The auto-collimator method.
6. Define 'Flatness'.
7. Name the methods which are used for the measurement flatness.
8. Explain any two of the following methods of measuring flatness:
   (*i*) Direct comparison
   (*ii*) Dial gauge
   (*iii*) Liquid method.
9. Explain how 'Beam comparator' is used for measuring flatness.
10. Explain 'Interference method' of measuring flatness of surface.
11. What are optical flats ? What is their field of application ?
12. Explain with the help of a neat diagram construction and working of a flatness inteferometer.
13. What is the difference between 'Roundness' and 'Circularity' ?
14. Explain briefly the types of irregularities of a circular part.
15. Explain briefly what you understand by:
    (*i*) Ovality
    (*ii*) Lobbing.
16. Define 'Error of circularity'.
17. What do you mean by the following ?
    (*i*) Maximum inscribed circle
    (*ii*) Minimum circumscribed circle
    (*iii*) Minimum zone circle
    (*iv*) Least squares circle.
18. Name the devices used for measurement of circularity and explain in detail any one of them.
19. Explain with neat sketch V-block method of measuring error of circularity. State its limitations.

❑❑❑

Measurement of Screw Threads and Gears

Chapter 12

12.1 SCREW THREAD MEASUREMENTS

12.1.1 Introduction

A **screw thread** *is helical ridge formed on uniform section round the curved surface.* The shape of the normal section of the thread depends upon the shape of the tool which produces its groove. A screw is a male threaded piece generally cylindrical in form, but sometimes, conical (or tapered), used in most cases as a temporary fastening; less frequently used as a means of transmitting motion or power.

The screw threads are applied to many devices for various *purposes* as follows:

1. To hold parts together as in the case of fastening.
2. To transmit power.
3. To control movement as in micrometer.
4. To increase the effect of applied effort as in auto jack.
5. To convey materials as in the case of fastening.

12.1.2 Classification of Threads

The threads may be *classified* as follows:

1. *According to the surface on which the threads are cut:*
   (*i*) External threads
   (*ii*) Internal threads.

   The external threads are cut into the surface of a *cylindrical bar.*

   The internal threads are cut into the surface of the *cylindrical hole* of a bar or cone.
2. *According to the direction of rotation of the threaded cylinder with respect to engagement or disengagement with the other part:*
   (*i*) Right handed thread
   (*ii*) Left handed thread.

   A *right handed thread* is one in which the nut must be turned in a right handed direction to screw it on (Fig. 12.1(*a*)). A left handed thread is one in which the nut would be screwed on by turning it to the left (Fig. 12.1(*b*)).

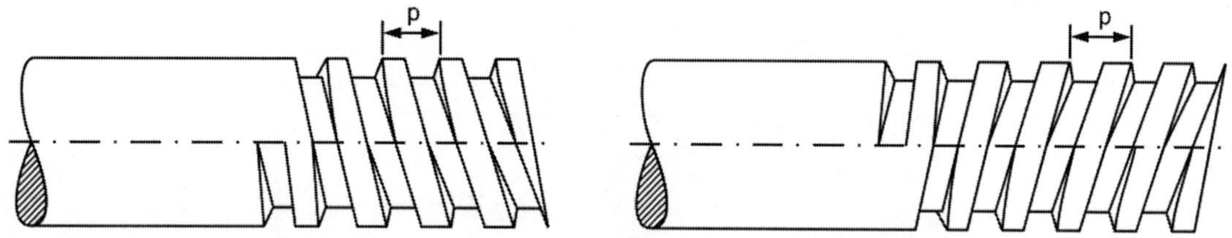

(a) Right handed thread. (b) Left handed thread.

Fig. 12.1

3. *According to number of starts:*
   (*i*) Single start threads
   (*ii*) Multi-start threads.

In a piece of work it is possible to have separate and independent threads running along it. Accordingly, there are single threaded screw and multiple or multi-start threaded screw. The independent threads are called starts and we may have single start, two start, three start etc. (Fig. 12.2).

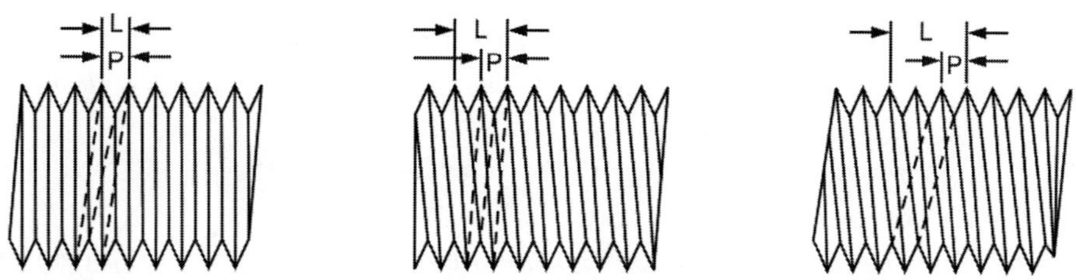

Fig. 12.2 (i) Single-start, (ii) Two-start, and (iii) Three-start threads.

A *single start threaded screw* is one in which there is a movement of one thread for one complete turn round the screw or bolt. In the *multi-start threaded screw* there is a movement of more than one thread. In the case of double start thread, for one complete turn, the thread advances two times as if it were a single thread.
Use of multi-start threads. Multi-start threads are used on those cases where *rapid movement or motion is required.*

Examples: *Fountain-pen cap, screw press, bottles, toothpaste etc.*

12.1.3 Elements of Screw Threads

To designate different parts of the screw threads the following *terms* are commonly employed :
Refers to Fig. 12.3.

1. **Major diameter.** It is the *largest diameter of a screw thread.* It is also termed as *outside* or *crest diameter.*
2. **Minor diameter.** It is the *smallest diameter of a screw thread.* It is also known as *root,* or *core diameter.*
3. **Effective or pitch diameter.** It is *an imaginary diameter in between the major and minor diameters, and is equal to the major diameter less than an amount equal to the single depth of a thread.*
4. **Axis of screw.** It is the *longitudinal central line through the screw.*
5. **Angle of thread.** It is the *angle between the two sides of a thread measured in an axial plane.*
6. **Depth of the thread.** It is the *angle between the crest and the root of a thread measured normal to axis.*

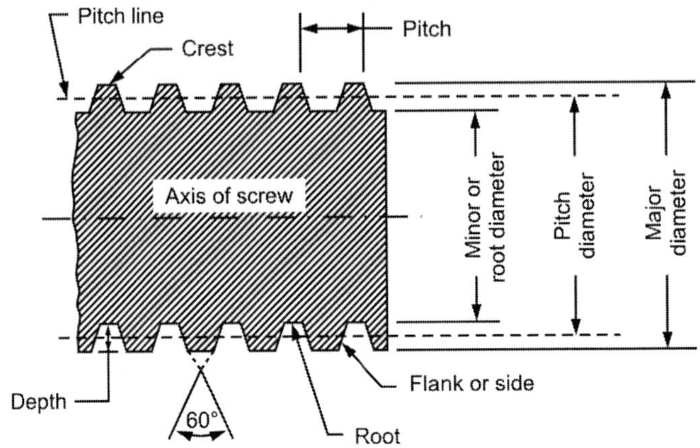

Fig. 12.3 Nomenclature of a screw thread.

7. **Thickness of the thread.** It is the *distance between the adjacent sides of the thread measured along or parallel to the pitch line.*
8. **Side.** It is the *slant surface of the thread which connects the crest with the root.*
9. **Helix angle.** It is the *angle made by helix of the thread at the pitch or effective diameter with the plane perpendicular to the axis.*
10. **Crest.** It is the top *surface joining the two sides of a thread.*
11. **Root** : It is the *bottom surface joining the two sides of a thread.*
12. **Pitch.** *It is the distance from a point on one thread to the corresponding point on the next thread measured parallel to the axis of the thread.* It is denoted by *'p'.*
13. **Lead.** *It is the distance a screw thread advances axially in one turn on a single thread screw, the lead is equal to pitch and for a double threaded screw, the lead becomes two times the pitch and so on.*

Note: A screw is specified by a *nominal diameter,* it is the diameter of the cylindrical piece on which the threads are cut.

12.1.4 Specifications of a Screw Thread

To specify a screw thread the following points are given due considerations:

1. Shape or form of thread
2. Pitch
3. Size (diameter)
4. Length
5. Number of starts
6. Material
7. Direction of threads
8. Internal or external threads.

12.1.5 Forms of Threads

Many different shapes of threads are employed but they can be satisfactorily grouped into three main classes :

1. V-threads
2. Square threads
3. Modification of both.

Comparison of 'V' and Square threads

S.No.	Particulars	V-threads	Square threads
1.	*Strength*	More strong	Less strong
2.	*Frictional resistance to motion*	More	Less
3.	*Cutting of threads*	Easy and hence cheaper	Difficult and hence costly
4.	*Suitability*	Suitable for fastening purposes.	Suitable for power transmission.

12.1.6 External Screw Thread Measurements

12.1.6.1 Measurement of Major Diameter (O_{major})

The major diameter of a screw thread can be determined by using an ordinary micrometer or, preferably, by using a bench micrometer.

Hand micrometer. For most applications a good hand micrometer is quite suitable for measurement of the major diameter of external thread. However *extreme care is necessary to ensure that only light pressure is applied because the anvils of the micrometer make contact only at the points of the screw.* Excessive pressure may lead to *elastic deformation* due to elastic compression need to be applied. It is, however, also desirable to check the micrometer reading on a cylindrical standard (setting gauge or standard) of approximately the same size, so that the zero error etc. might be eliminated.

Bench micrometer. *Greater accuracy is probably obtainable if the major diameter is measured by a 'bench micrometer'* (Fig. 12.4). A bench micrometer uses constant measuring pressure and with this machine the *error due to pitch error in the micrometer thread is avoided.* In order that all measurements be made at the same pressure, a fiducial indicator is used in place of the fixed anvil. In this machine *there is no provision for mounting the work piece between the centres and* it is to be held in hand. This is so, because, generally the *centres of the work piece are not true with its diameter. This machine is used as comparator in order to avoid any pitch error of micrometers, zero error setting etc.* A calibrated setting cylinder is used as the setting standard or gauge.

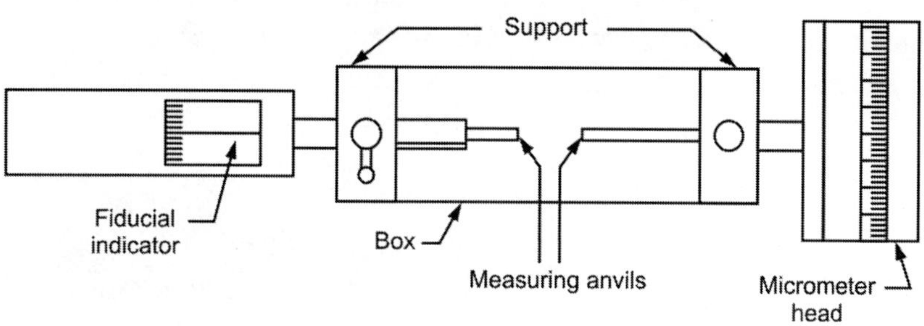

Fig. 12.4 Bench micrometer.

The *advantage of using cylinder as setting standard and not slip gauges etc., is that it gives greater similarity of contact at the anvils. The diameter of the setting cylinder must be nearly same as the major diameter.*

The *procedure of measuring major diameter* is as follows:

- The cylindrical standard (of known size) is held and reading of the micrometer is noted down [Fig.12.5(*a*)]

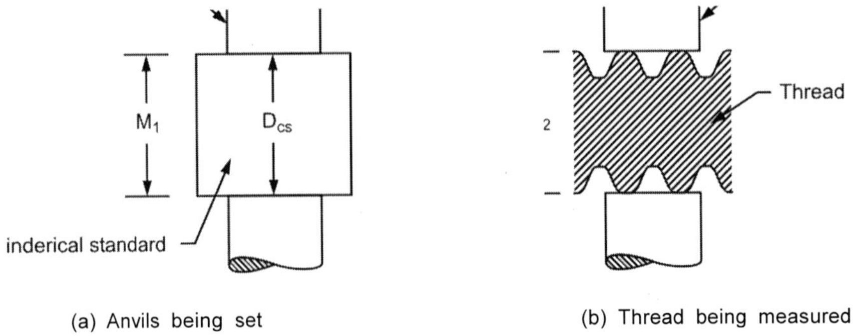

Fig. 12.5 Measurement of major diameter.

The cylindrical standard is then replaced by threaded workpiece and again micrometer reading is noted [Fig. 12.5(*b*)].

Let, M_1 = Micrometer reading over cylindrical standard,

M_2 = Micrometer reading over threaded workpiece major diameter,

D_{cs} = Diameter of cylindrical standard, and

D_{major} = Major diameter of thread.

Then, $D_{major} = D_{cs} \pm$ the difference between M_1, and M_2

The (+) or (–) is determined by whether the standard is a smaller or larger diameter than the thread major diameter.

Note. The setting gauge used for the initial reading should always be of similar geometrical form to that of the part to be measured. In this instance, a cylindrical form, such as a plug gauge, or roller, or other cylindrical standard, should be used.

12.1.6.2 Measurement of minor diameter (D_{minor})

The use of external hand micrometer is rarely made to measure the minor diameter screw threads because of *difficulty of ensuring that the measurement takes place at 90° to the thread axis.*

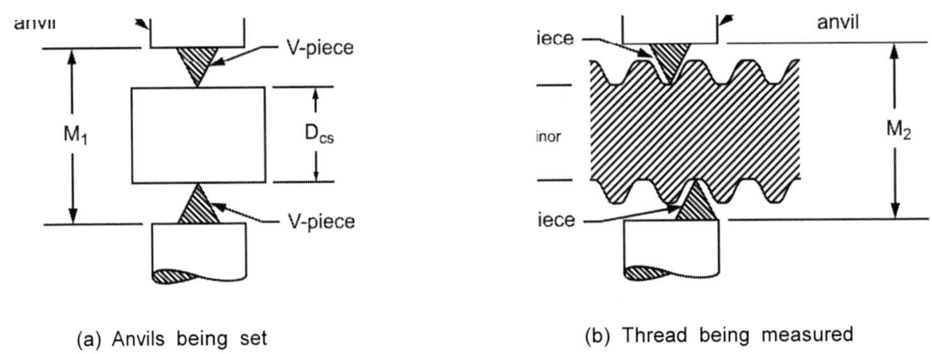

Fig. 12.6 Measurement of minor diameter.

For more accuracy and convenience the *bench micrometer* is used for this measurement. Here small Vee pieces which make contact with the root of the thread are used as shown in the Fig. 12.6. These Vee pieces (made of hardened steel) are made in several sizes, having suitable radii at the edges, and with included angles *less than the angle of the thread to be checked.* It is not necessary to know the dimension of the Vee pieces, since they are interpored between the micrometer faces and the cylinderical standard when the standard reading is taken [Fig. 12.6 (*a*)]. The thread being measured is shown in Fig. 12.6(*b*).

Refer to Fig. 12.6.

Let, M_1 = Micrometer reading over cylinderical and V-pieces,

M_2 = Micrometer reading over minor diameter and V-pieces,

D_{cs} = Diameter of the cylindrical standard, and

$D_{(minor)}$ = Minor diameter or the thread.

Then, $D_{(minor)} = D_{fy}$ difference between M_1, and M_2.

The (+) or (–) is determined by whether the standard is a small or larger diameter than the thread minor diameter.

This method of minor diameter measurement is *only applicable to threads such as Whitworth and B.A., where there is a definite radius at the root of the thread.*

Gauges for threads which are intended to make close contact at the minor diameter, which is not closely limited, are not normally measured by this method, but are checked visually on the projector or microscope.

12.1.6.3 Measurement of effective diameter (D_e)

For measuring the effective diameter, one of the following methods can be used:

1. Three and two wires method,
2. The thread micrometer method,
3. Optical measurement on tool-maker microscope,
4. Axial section method.

Three and two wires method. Effective diameter is measured by placing small cylinders or *wires of known diameter* in the Vees of the thread, and measuring over the tops of the wires with a micrometer or special machine. These *wires are made of hardened steel, and are lapped to sizes suitable for various pitches.* For each pitch of thread there is a "best size" wire; this is of such diameter that it makes contact with the flanks of the thread on the effective diameter or pitch line. Effective diameter may be measured with any diameter of wire which makes contact on the true flak of the thread, but values so obtained will differ from those obtained with *"best size"* wires if there is any error in angle or form of thread.

Three-wire method. Fig. 12.7 shows diagrammatically how three wires may be used for measuring effective diameter. If a hand micrometer is used, three wires must be used, so that the micrometer faces can be aligned parallel to the thread axis. With small diameter threads it is possible to hold gauge, with wires in place, with one hand, and the micrometer in the other, although one sometimes wishes for a third hand. Other effective methods of securing the wires include the use of grease in the threads, or sticking the ends of the wires into Plasticine or wax. In any case the wires must not be constrained, but must be free to adjust themselves under the micrometer pressure.

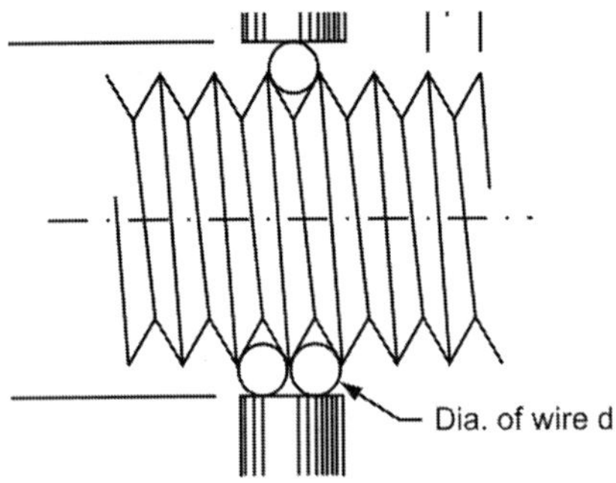

Fig. 12.7 Three-wire method of measuring effective diameter.

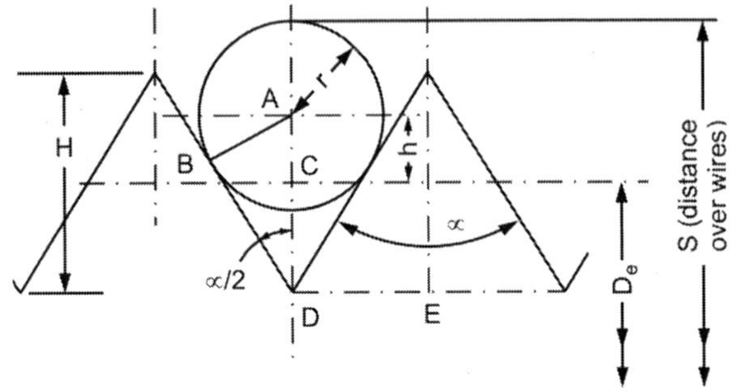

Fig. 12.8.

For screw gauge measurement, it is better to use a specially designed Diameter Measuring Machine incorporating a bench micrometer which maintains the axis of the micrometer spindle square to thread axis.

Checking the effective diameter when a screw is measured over three wires is given below for a general case. From the formula obtained particular values may then be inserted for any desired form of thread. One side of the thread is shown at Fig. 12.8.

Let S = Distance over the wires,

D_e = Effective diameter,

r = Radius of wire, and

d = Diameter of wire (= $2r$).

Referring to Fig. 12.8,

$$AD = AB \text{ cosec } \frac{\alpha}{2} = r \text{ cosec } \frac{\alpha}{2}$$

$$H = DE \cot \frac{a}{2} = \frac{p}{2} \cot \frac{\alpha}{2}$$

$$CD = \frac{1}{2} H = \frac{pd}{4} \cot \frac{\alpha}{2}$$

$$h = AD - CD = r \text{ cosec } \frac{\alpha}{2} - \frac{p}{4} \cot \frac{\alpha}{2}$$

and, distance over wires (S)

$$= D_e + 2h + 2r$$

$$= D_e + 2\left(r \text{ cosec} \frac{\alpha}{2} - \frac{p}{4} \cot \frac{\alpha}{2}\right) + 2r$$

$$= D_e + 2r \text{ cosec} \frac{\alpha}{2} - \frac{p}{4} \cot \frac{\alpha}{2} + 2r$$

$$= D_e + 2r\left(1 + \text{cosec} \frac{\alpha}{2}\right) - \frac{p}{4} \cot \frac{\alpha}{2}$$

$$= D_e + d\left(1 + \text{cosec} \frac{\alpha}{2}\right) - \frac{p}{4} \cot \frac{\alpha}{2} \quad (\because \; d = 2r)$$

Hence,

$$S = D_e + d\left(1 + \text{cosec} \frac{\alpha}{2}\right) - \frac{p}{2} \cot \frac{\alpha}{2} \qquad ...(12.1)$$

From this formula we may apply the special adaptation for common threads.

***(i)* Whitworth thread:** Refer to Fig. 12.9.

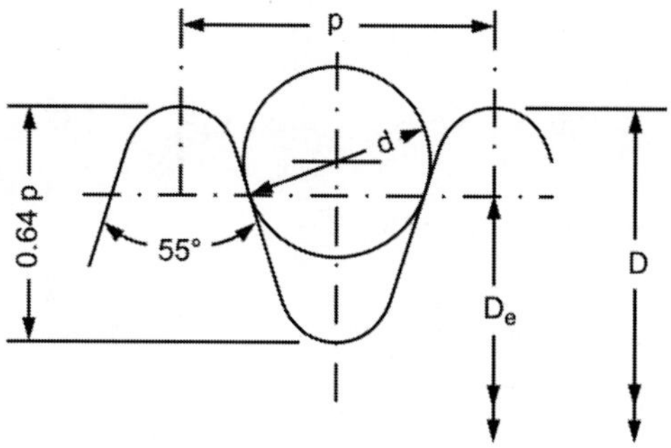

Fig. 12.9 Whitworth thread.

Depth of thread $= 0.64\,p$, so that $D_e = D - 0.64\,p$

$$\alpha = 55^\circ \text{ and } \operatorname{cosec}\frac{\alpha}{2} = 2.1657 \cos\frac{\alpha}{2} = 1.921$$

$$\text{Hence } S \text{ (over wires)} = D_e + d\,(1 + \operatorname{cosec}\frac{\alpha}{2}) - \frac{p}{2}\cot\frac{\alpha}{2}$$

or, $$S = D - 0.64\,p + d(3.1651) - \frac{p}{2}(1.921)$$

or, $$S = 0 + 3.1657\,d - 1.6\,p \qquad ...(12.2)$$

***(ii)* B.A. thread:** Refer to Fig. 12.10.

Depth of thread $= 0.6\,p$ and $D_e = D - 0.6\,p$

$$\theta = 47\frac{1}{2}^\circ;\ \operatorname{cosec}\frac{\alpha}{2} = 2.483;\ \cot\frac{\alpha}{2} = 2.2725$$

$$S = \text{(over wires)} = D_e + d\left(1 + \operatorname{cosec}\frac{\alpha}{2}\right) - \frac{p}{2}\cot\frac{\alpha}{2}$$

or, $$S = D - 0.6\,p + d(3.483) - \frac{p}{2}(2.2725)$$

or, $$S = D + 3.483\,d - 1.736\,p \qquad ...(12.3)$$

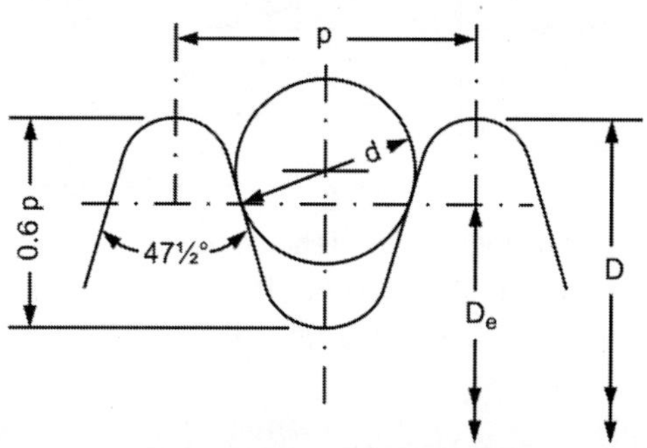

Fig. 12.10 B.A. thread.

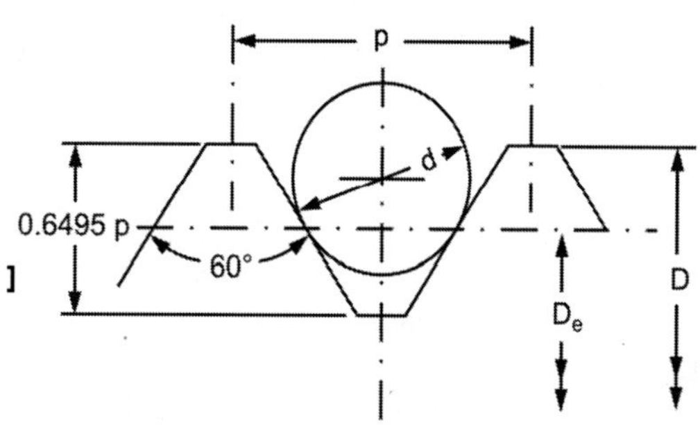

Fig. 12.11 Metric thread.

***(iii)* American National and Metric:**

Refer to Fig. 12.11.

$$\text{Depth of thread} = 0.6495\,p,$$

$$\text{so that } D_e = D - 0.6495\,p$$

$$\alpha = 60° \text{ and cosec } \frac{\alpha}{2} = 2;\ \cot \frac{\alpha}{2} = 1.732$$

$$S \text{ (over wires)} = D_e + d\left(1 - \text{cosec}\frac{\alpha}{2}\right) - \frac{p}{2}\cot\frac{\alpha}{2}$$

or,

$$S = D - 0.6495\,p + d(3) - \frac{p}{2}\,(1.732)$$

or,

$$S = D + 3d - 1.5155\,p \qquad ...(12.4)$$

(iv) Unified thread

The formula (12.4) for America National applies also to the unified thread which although differing from American thread in its root has the same effective diameter *($D - 0.6495\,p$)* and the same upper proportions.

Wire size. The above analysis for the effective diameter in its relation to the distance over wires holds good provided the wire touches the thread somewhere on its straight side, and provided the *thread angle is correct.* The extremes of wire sizes which touch on the straight sides and which can be measured are shown at (*a*) and (*c*), Fig. 12.12. For Whitworth, B.A. and International metric these limiting sizes are as follows:

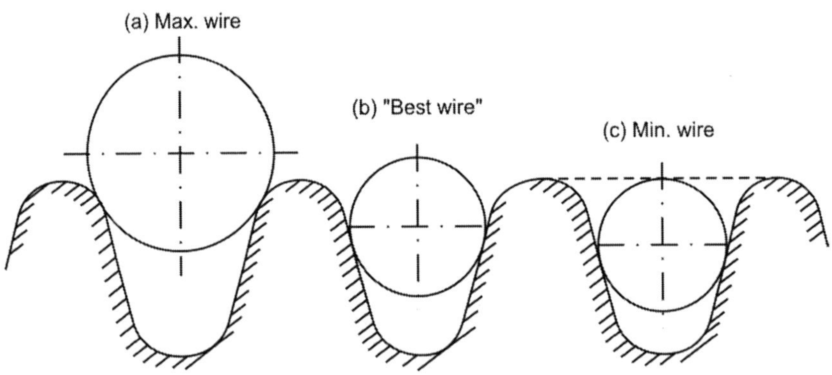

Fig. 12.12.

Thread form	*Maximum wire*	*Minimum wire*
Whitworth	0.853 *p*	0.506 *p*
B.A.	0.73 *p*	0.498 *p*
Metric, Unifiedand American	1.01 *p*	0.505 *p*

'Best' wire. If there is a possibility of the thread angle being incorrect the wire used should be such as to touch the thread exactly on the pitch line [Fig. 12.12(*b*)] and the inspection of the effective diameter from a reading over such wires will be independent of any error in the thread angle. Such wires are called the **"best"** wire sizes and are as follows:

Thread form	*"Best" wire dia*
Whitworth	0.564 p
B.A.	0.546 p
Metric, Unified and American	0.577 *p*

Two-wire method:

The two-wire method of measuring the effective diameter of a screw thread is illustrated in Fig. 12.13.

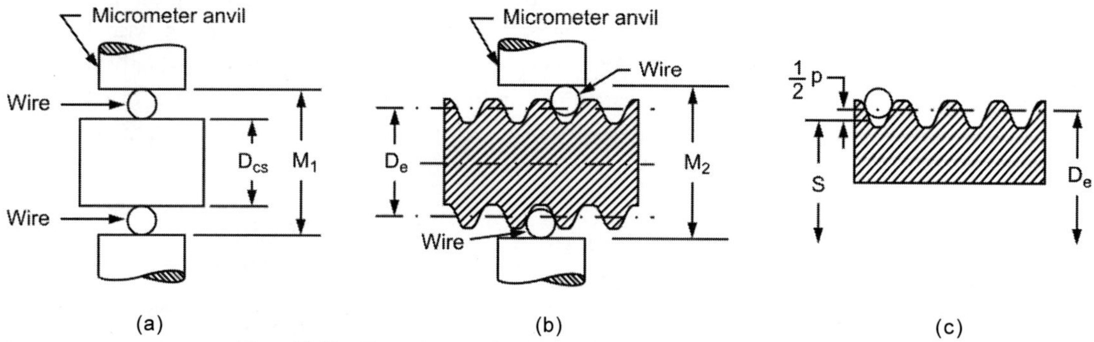

Fig. 12.13 Two-wire method of measuring effective diameter.

- Two wires/cylinders having been selected, a cylindrical standard is used and a micrometer reading is taken over the cylindrical standard, and the two wires.
- The cylindrical standard is then removed and is replaced by the screw, with the wires inserted in opposite grooves of the thread. The micrometer reading is then noted.

Let, M_1 = Micrometer reading over the cylindrical standard and wires,
M_2 = Micrometer reading over screw thread and wires,
D_{cs} = Diameter of cylindrical standard,
S_1 = Diameter under wires when the latter are in position on threads,
D_e = Effective diameter of thread, and
P = The difference between the effective diameter and the diameter under the wires.

The value of P is dependent on the diameter of the wires and form of the thread to be measured. Its value may be found as follows:

Let, p = Pitch of thread, and
d = Mean diameter of wires used.

For Whitworth threads (55°),

$$P = (0.96049\,p) - (1.16568\,d)$$

For B.A. threads (47°),

$$P = (1.13634\,p) - (1.48239\,d)$$

For Metric and Unified threads (60°),

$$P = (0.86603\,p) - d$$

Now, $$D_e = D_{cs} - M_1 + M_2 + P$$

When the effective diameters of screw plug gauges are being measured it is necessary to subtract the "rake correction" C and add the "compression correction" e to the micrometer reading to obtain accurate measurements. Both corrections are small, and they result from the rake angle of the thread and the compression of the wires respectively.

Thread micrometer method:

When the *threads are accurate the effective diameter of the thread will be equal to its pitch diameter.* In such cases, a specially designed thread micrometer is found to be most convenient way of measuring the effective diameter. *The thread micrometer* is shown in Fig. 12.14. The end of the spindle is pointed to a 60° cone for ISO thread and an accurate 60° Vee is grooved in the anvil. The anvil is free to rotate so as to adjust itself to the helix angle of the thread being measured. The sharp tip of the spindle point is ground off. Likewise, flats are ground on the peaks of the vee and the foot of the vee is ground out or cleared to ensure that only the pitch diameter is measured by the thread micrometer and not the root diameter or major diameter of the screw thread. Before using a thread micrometer, screw cone point of its spindle down to contact with the vee-anvil and check the micrometer thimble's zero reading.

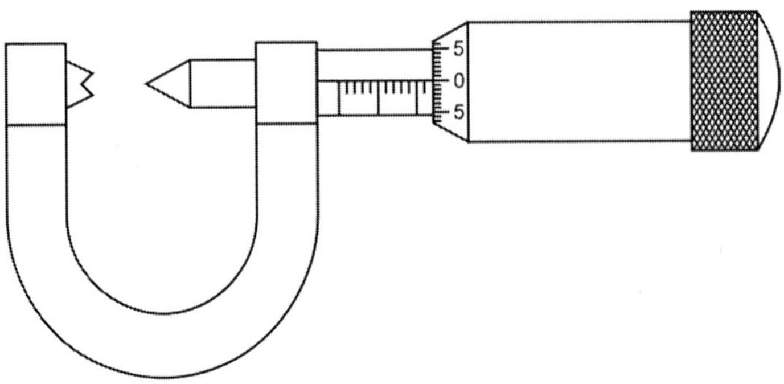

Fig. 12.14 Thread micrometer.

A thread micrometer is designed to measure threads within a certain range of thread pitches. Since any given thread micrometer is required to measure a range of threads of different pitches, each of which may cause a slight variation of anvil position on the thread, small errors in measurement are sometimes introduced. For this the *best pitch diameter to follow in using the thread micrometer is to first measure the pitch diameter of a standard thread plug gauge of the same size as the thread to be measured and then compesate for error.*

The description of remaining two methods, *viz.*, optical measurement of tool-maker microscope and axial section method is beyond the scope of this book.

12.1.6.4 Measurement of Pitch

The *pitch measurement should be carried out at the pitch line in the axial plane parallel to the axis of the screw thread.* Otherwise, the pitch errors cannot be correctly determined.

Rough measurement of pitch— *By thread pitch gauge:*
The *rough measurement* of the thread pitch can be done by means of a *thread pitch gauge* which consists of a number of threaded blades of different standard pitches.

Particularly for screw thread gauges, it is very important to measure the thread pitch to a high accuracy, preferable double the accuracy of measuring the effective diameter. This is due to the fact that *any pitch error is approximately doubled on the effective diameter.*

Accurate measurement of pitch—*By pitch measuring machine :*

The accurate measurement of pitch can be done on *a pitch measuring machine* (Fig. 12.15) as per the following procedure:

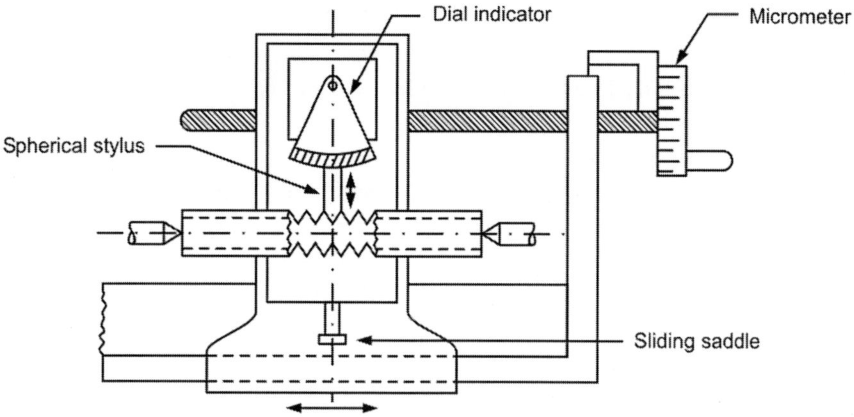

Fig. 12.15 Pitch measuring machine.

- The screw thread is mounted between fixed centres of the machine.
- The spherical stylus of the dial indicator is adjusted to make contact with the thread flanks at the pitch line.
- The dial indicator is mounted on a sliding saddle similar to that of a lathe, which can be moved by rotating the measuring micrometer. The saddle moves exactly parallel to the axis of the centres and the stylus falls in and out on the flanks of each thread.
- The micrometer and the dial indicator are set to read zero when the stylus engages the first thread space.
- The micrometer is then rotated until the stylus engages the next following thread space in such a position where the dial indicator returns to its initial zero reading. This ensures that the stylus is exactly in the same position relative to the axis of the screw thread as it was initially in the first threads space. Accordingly, the measuring direction is guaranteed parallel to the axis of the screw thread. The value of the thread pitch can then be read off on the micrometer.
- **Most accurate measurement of pitch**—*Axial section method :* The description of this method is beyond the scope of this book.

The thread pitch can also be measured on

- *Tool maker's microscope fitted* with a *screw thread template.*
- *Abbe horizontal measuring microscope using pitch measuring attachment.*

12.1.6.5 Measurement of thread angle and form

Thread angle and form are usually checked by optical means, using an *optical projector* which may be of horizontal or vertical form. Cabinet projectors are available which are compact and take up relatively little space. Also measuring microscopes may be used, one form of which is tool maker's microscope which can also be used to carry out a complete check of the thread elements.

The four essential elements of a *projection system are:*

1. Source of light — a lamp.
2. Collimating lens — gives a parallel beam of light.
3. Projection lens — combination lens forms a real image on the screen.
4. Screen — opaque or transluscent screen.

For a clear definition of a thread on the screen it *is necessary to project the light beam along the line of the thread helix angle.* This is done either by *inclining the work through the appropriate helix angle, or by doing the same with the lamp and collimating unit.*

The modern projectors are equipped with work holding fixtures, the projection lamp and the lenses situated on the top of the cabinet, and the screen at the front. The light rays from the lens are directed downwards into the cabinet, and hence to the screen by a system of prisms and mirrors, bringing everything within the reach of the operator. The enlarged image of the thread form appears on the ground glass screen on which is mounted the template or drawing of their form made to scale equal to the magnification of lens. This way the two forms, *i.e.,* ideal and projected are compared.

One of the difficulties in projecting screw thread is the fact that form is specified on an axial plane. Therefore correction for it must be considered.

12.1.7 Internal Screw Thread Measurements

The accurate measurement of internal threads, particularly the *screw thread ring gauges,* present more difficulties than external threads.

12.1.7.1 Measurement of major diameter (D_{major})

The major diameter of an internal thread is normally measured in some form of *Horizontal comparator* fitted with ball ended stylii of *radius less than the root radius of the thread to be measured.* When the gauge is mounted in the comparator with the stylii in opposite thread grooves, the gauge will align itself with the axis of the machine and measurement will be made along the helix of the thread as indicated in Fig. 12.16.

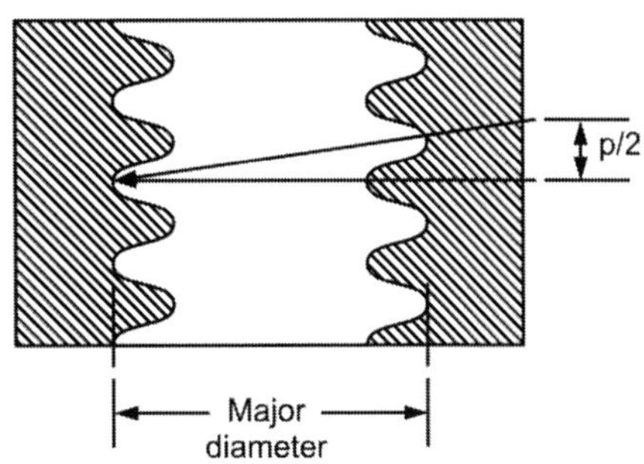

Fig. 12.16

The standard gap with which the comparator is set will have to be made to the dimension A and not to the nominal diameter. Thus,

$$A^2 = (\text{Nominal major diameter})^2 + \left(\frac{pitch}{2}\right)^2$$

or,

$$A = \left[(\text{Nominal major diameter})^2 + \left(\frac{pitch}{2}\right)^2\right]^{1/2}$$

12.1.7.2 Measurement of minor diameter (D_{minor})

(*i*) **When the value of minor diameter is large.** If the minor diameter is large, slip gauges and calibrated rollers provide an accurate method for measurement as shown in Fig. 12.17.

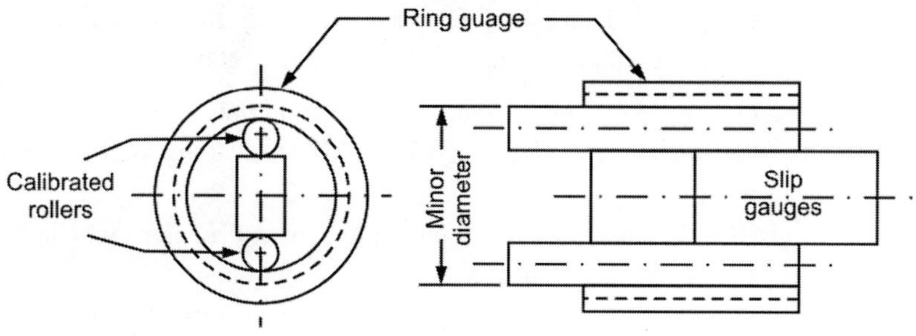

Fig. 12.17

(*ii*) **When the value of minor diameter is small.** For smaller diameter (less than $\frac{3}{4}$ inch) the use of taper parallels (consisting of a pair of wedges whose outer surfaces are curved and the radius of curvature is smaller than half the minor diameters, and micrometer is very common, Figure 12.18 illustrates this method.

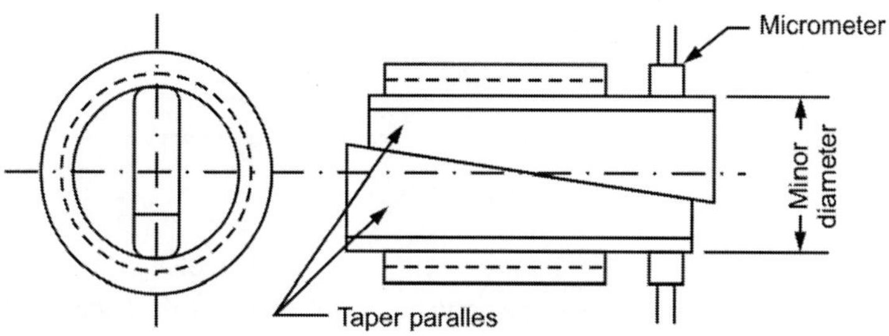

Fig. 12.18

12.1.7.3 Measurement of effective diameter (D_e)

The effective diameter of internal screw threads having a normal diameter above 50 mm, can be measured by *internal thread micrometers.*

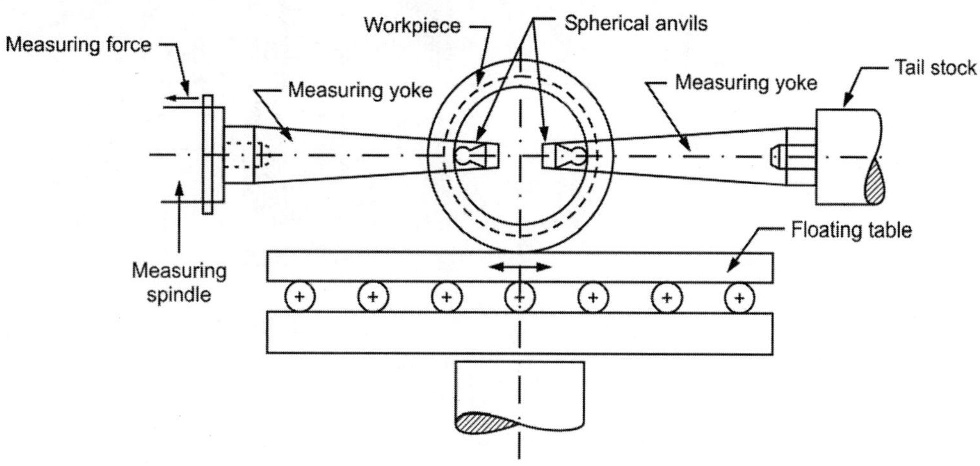

Fig. 12.19

The measurement can be easily carried out by a differential measuring attachment on the Abbe horizontal measuring microscope or any other length measuring machine as given below :

- The attachment (Fig. 12.19) consists of two internal measuring yokes, one is fixed to the *tail stock* while the *other yoke is mounted on the measuring spindle.* The measuring force is directed towards the back side of the headstock.
- Both yokes are provided with spherical anvils having the correct diameter that allows them to make contact with the thread flanks at the pitch line.
- The workpiece is mounted on the floating table so that it can adjust itself between measuring anvils.
- The anvils have to be initially set to the standard effective diameter by means of slip gauges or a screw thread ring gauge of known dimensions.

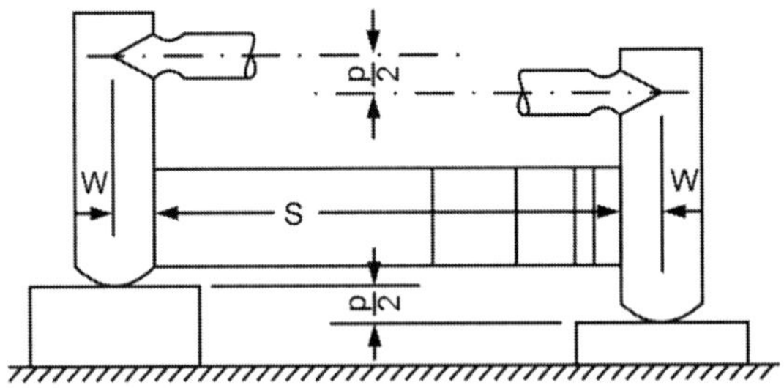

Fig. 12.20

- The slip gauge blocks are arranged as shown in Fig. 12.20. The 60° V-notches on the slip gauges represent the thread space which has to be vertically displaced by a distance equal to half the thread pitch.

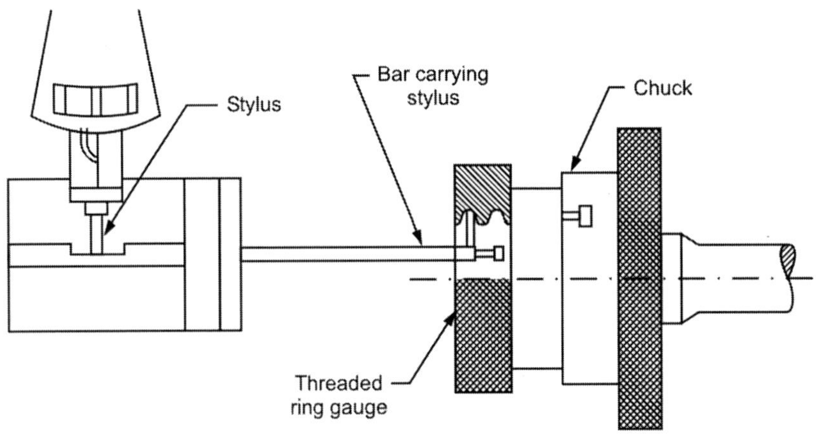

Fig. 12.21

The distance S, built up by the slip gauge block can be calculated from the following equation:

$$D_e = S + 2 - W \frac{p}{2} \cot \alpha/2 + k_1 + k_2$$

where, D_e = Effective diameter,
S = Distance built up by slip gauge blocks,
W = Depth of V-notches (known),
α = Thread angle, and
k_1, k_2 = Obliquity and the compression correction factors.

In case the thread angle α and notch angle α_0 are different, the equation for effective diameter is as follows:

$$D_e = S + 2 - W \frac{p}{2} . \cot \alpha/2 + dW (\text{cosec}\ \alpha/2 - \text{cosec}\ \alpha_0/2) + k_1 + k_2$$

12.1.7.4 Measurement of pitch

The pitch of an internal thread can be measured on any pitch measuring machine by *using an adaptor.* This adaptor carries a fine bar which can be inserted into the ring, the stylus being fitted to the bar end as shown in Fig. 12.21, engaging with the thread in the usual manner. The ring gauge is mounted on a face plate or the headstock of the machine, which will accommodate range of various sizes. A special set up on the face plate becomes essential for very large rings.

12.1.7.5 Measurement of thread angle and form

The thread angle of an internal screw thread can be measured as follows:

- A moulded replica is prepared whose thread angle can be measured as if it were a external screw thread. The materials used for moulds may be plastics, dental plaster or sulphur. Plastics have the disadvantage that the workpiece will be heated upto 100°C.
- To *ensure easy removal* of the moulded replica without screwing it out, the *height of the mould should be kept smaller than half the minor diameter of the nut.*

Since the replica is subjected to *shrinkage* and *distortion* therefore the *moulding method does not provide the accuracy* by which the thread pitch has to be measured.

12.2 GEAR MEASUREMENTS

12.2.1 Introduction

- A *gear is wheel provided with teeth which mesh with the teeth on another wheel, or on to a rack, so as to give a positive transmission of motion from one component to another.* Gears constitute the most commonly used device for power transmission or for changing power-speed ratios in a power system. They also afford a convenient way of changing the direction of motion. A number of devices such as differentials, transmission gear boxes, planetary drives etc. used in many construction machines employ gears as basic components.
- Gears belong to a category of machine elements that should be carefully measured and tested; as they almost effect the high accuracy demanded in the industry. This has necessitated the development of special measuring techniques to measure the different *parameters* which govern the gears, the position of two gears within a specific gear box arrangement. Within this scope a large variety of errors can take place which influence the specified accuracy and the interchangeability.

In the gear measurement it is *necessary to differentiate between the measurement of the individual parameters of a gear, i.e., their individual errors and the measurement of accumulative errors.* For measuring the individual gear parameters and their errors, it is necessary to know the errors of the gear cutting machine so that the gear production can be controlled and the machines can be reset. Moreover, individual measurements are also important for the inspection of high quality gears, such as, standard gears and gear cutting tools.

Because of the fact that in a gear box the errors of the individual gear parameters can counteract each other, it is usually sufficient, for testing the function of gear, to measure the accumulative errors and their effect on the function of the gear. The measurement of accumulative errors, is therefore, important for testing the gears by the consumers as well as by the producers so that the time required for inspection can be reduced.

12.2.2 Types of Gears

The different types of gears used are enumerated and described briefly below:

1. Spur gear
2. Helical gear
3. Bevel gear
4. Worm gear
5. Rack and pinion.

1. Spur gear. A spur gear is a gear wheel or pinion for transmitting motion between two *parallel shafts*. This is the simplest form of geared drive. The teeth are cast or machined parallel with the axis of rotation of the gear. Normally the teeth are of involute form. Figure 12.22 illustrates a spur gear drive, consisting of a pinion and a spur wheel.

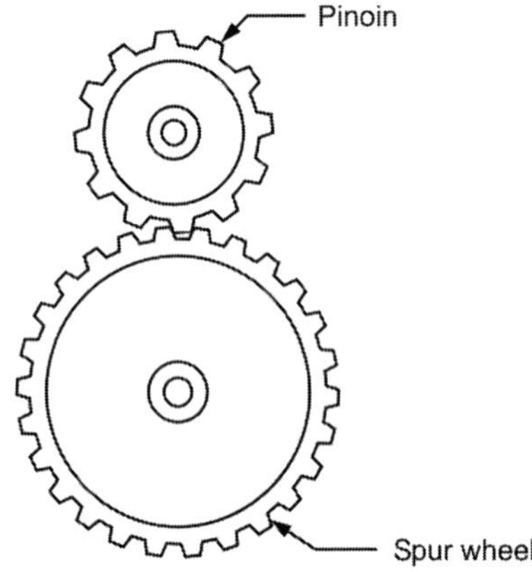

Fig. 12.22 Spur gear.

The efficiency of power transmission by these gears is very high and may be as much as 99% in case of high-speed gears with good material and workmanship of construction and good lubrication in operation. Under average conditions, efficiency of 96-98% are commonly attainable. The *disadvantages* are that they are liable to be *more noisy* in operation and may wear out and develop backlash more readily than the other types.

Fig. 12.23 Helical gears.

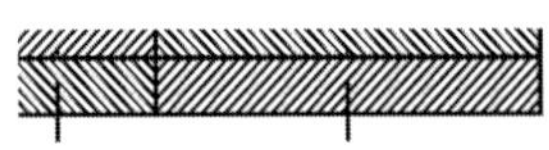

Fig. 12.24 Double helical gears.

2. Helical gear. Refer to Fig. 12.25. Helical gear is one in which teeth instead, of bearing parallel with shaft as in ordinary spur gears, are *inclined.* This ensures *smooth action and more accurate maintenance of velocity ratio.*

A *disadvantage* is that the inclination of the teeth *sets up a lateral thrust.* A method of neutralising this lateral or axial thrust is to use double-helical gears (also known as *Herring bone gears)* shown in Fig. 12.24.

3. Bevel gear. Refer to Fig. 12.25. A bevel gear transmits motion between two shafts which *intersect.* If the shafts are at right angles and wheels equal in size, they are called ***mitre gears***. If the shafts are not at right angles, they are sometimes called angle bevel gears. Spiral toothed bevel gears are preferred to straight toothed bevels in certain applications, because they will run more smoothly and make less noise at high speeds.

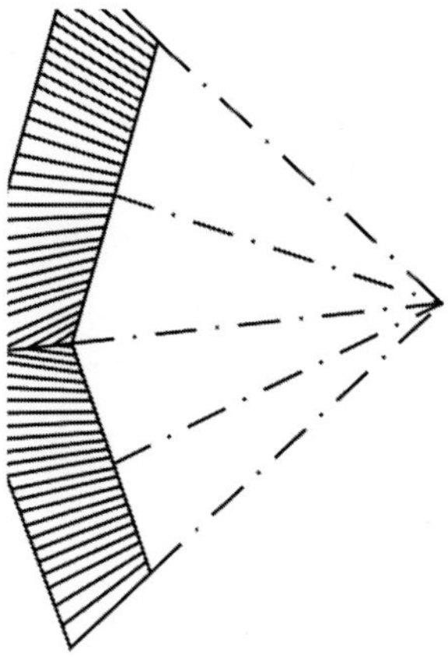

Fig. 12.25 Bevel gear.

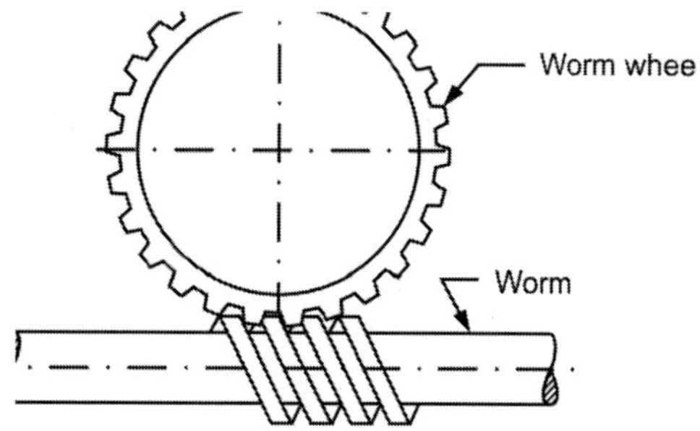

Fig. 12.26 Worm gear.

4. Worm gear. Refer to Fig. 12.26. Worm gears connect two *non-parallel, non-intersecting shafts which* are usually *at right angles*. One of the gears is called the *'worm'*. It is essentially part of a screw, meshing with the teeth on a gear wheel, called the *"worm wheel"*. The gear ratio is the ratio of number of teeth on the wheel to the number of threads on the worm.

One of the great advantages of worm gearing in that *high gear ratios (i.e.*, ratio of rotational speed of worm to that of worm wheel) are easily obtained. Worm gearing is *smooth and quiet*.

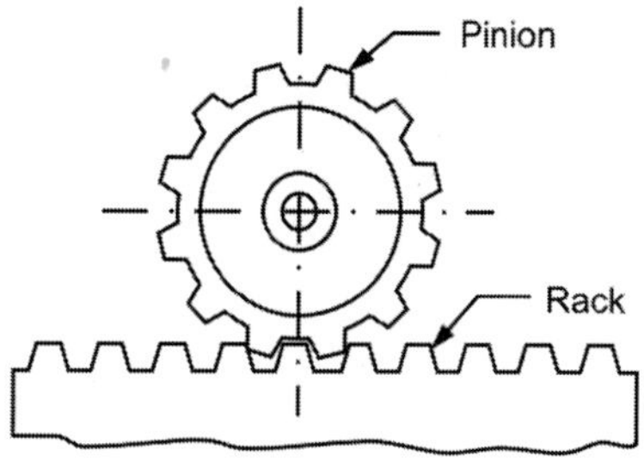

Fig. 12.27 Rack and pinion.

5. Rack and pinion. Refer to Fig. 12.27. A *rack* is a spur gear of infinite diameter, thus it assumes the shape of a straight gear. The rack is generally used with a pinion to *convert rotary motion into rectilinear motion*.

12.2.3 Measurement and Testing of Spur Gear

12.2.3.1 Measurement of tooth thickness

The following *methods* are used for measuring *tooth thickness:*

1. Gear tooth vernier.
2. The constant chord method.
3. The addendum comparator method.
4. The base tangent method.

1. Gear tooth vernier:

A gear tooth vernier is widely used to measure the tooth thickness. As the tooth thickness varies from top to the bottom any instrument for measuring on a single tooth must (*i*) measure the tooth thickness at a specified position on the tooth; (*ii*) fix that position at which the measurement is taken.

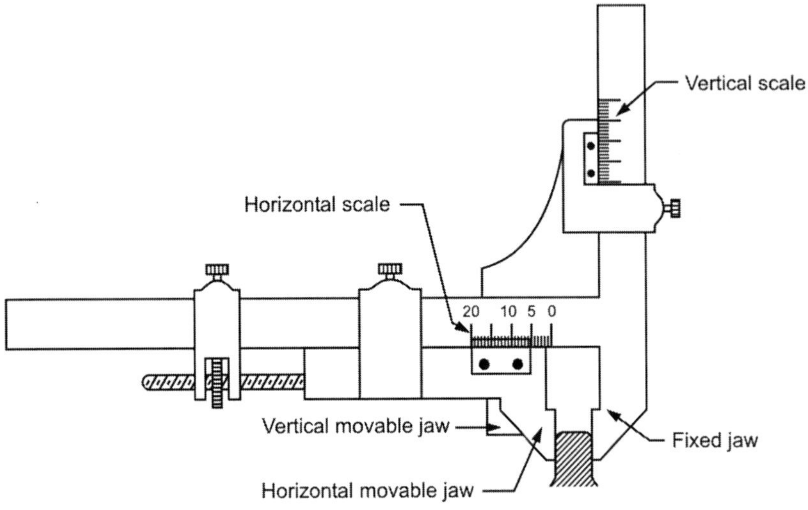

Fig. 12.28 Gear tooth vernier caliper.

Refer to Figs 12.28 and 12.29. The *"Gear tooth vernier caliper'* is an instrument similar to the ordinary vernier caliper but having a second beam at right angles to the main beam. This additional beam carries a tongue sliding between the jaws, which can be set on the vernier scale so that, when it rests on the top of a tooth, the tips of the jaws are at the correct distance from the tooth flanks for the required measurement.

For measuring the chordal thickness, the tongue of the gear tooth vernier is set so that the jaw will touch the flanks at the pitch circle. The thickness of the tooth is then measured on the main scale of the instrument. From the above it is obvious that the accuracy of the measurement by this instrument relies upon the 'feel' of operator and the resolution of the vernier scales of the caliper. Because of this the overall *efficiency of the determination is seldom better than 0.050 mm*. Usually the gear drawing will specify the chordal thickness at some selected tooth depth position, such as pitch circle, thus saving the operator from having to calculate the nominal values.

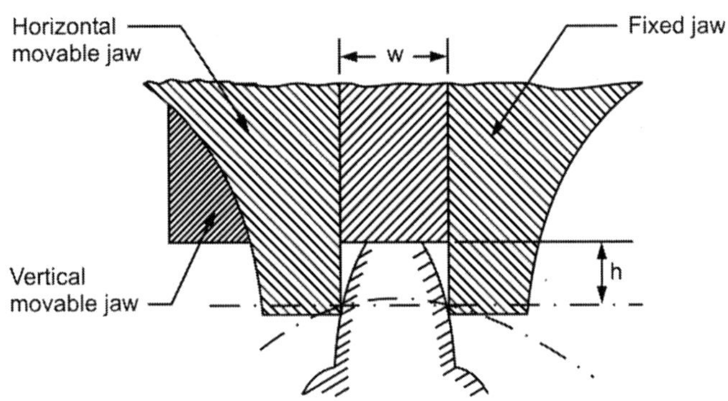

Fig. 12.29

Let
T = Number of teeth on the gear,
D = Pitch circle diameter,
R = Pitch circle radius,
w = Thickness of tooth at the pitch line,
p_d = Diametral pitch,
h = Depth from the top of tooth, and
m = Module.

To find *w*:

The angle subtended by a half tooth at the centre of a gear ($\angle B$ in Fig. 12.30)

$$\frac{1}{4} \text{ of } \frac{360°}{T} = \frac{90°}{T}$$

But diametral pitch, $p_d = \dfrac{\text{No. of teeth}(T)}{\text{Pitch diameter}(D)}$

i.e., $p_d = \dfrac{T}{D}$

or, $D = \dfrac{T}{p_d} = 2R$

or, $R = \dfrac{T}{2\,p_d}$

But, $AB = \dfrac{w}{2} = R \sin \dfrac{90°}{T}$

$\therefore$ $w = 2\,R \sin \dfrac{90°}{T}$

or, $w = 2 \times \dfrac{T}{2p_d} \sin \dfrac{90°}{T}$

or, $w = \dfrac{T}{p_d} \sin \dfrac{90°}{T}$ ...(12.5)

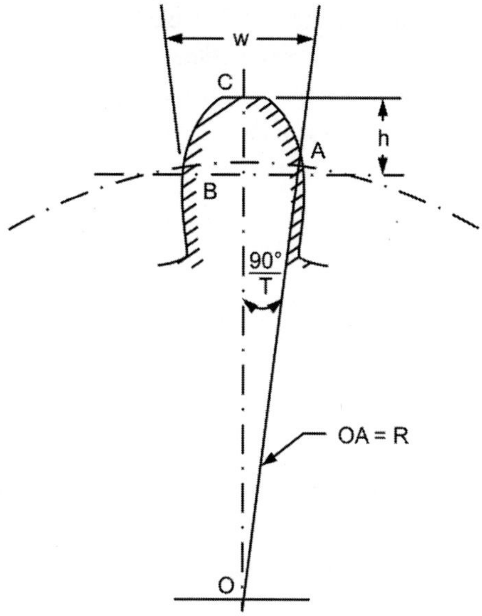

Fig. 12.30

or, $\omega = mT \sin \dfrac{90°}{T}$ $\left(\because m = \dfrac{1}{p_d}\right)$ ...(12.6)

To find *h*:

$$h = CB = OC - OB$$

But, $OC = R + \text{addendum} = \dfrac{T}{2p_d} + \dfrac{1}{p_d}$

and, $OB = R \cos \dfrac{90°}{T} = \dfrac{T}{2p_d} \cos \dfrac{90°}{T}$

Hence, $h = \dfrac{T}{2p_d} + \dfrac{1}{p_d} - \dfrac{T}{2p_d} \cos \dfrac{90°}{T}$

or, $h = \dfrac{1}{p_d} - \dfrac{T}{2p_d}\left(1 - \cos \dfrac{90°}{T}\right)$ ...(12.7)

or, $h = m + \dfrac{Tm}{2}\left(1 - \cos \dfrac{90°}{T}\right)$ $\left(\because m = \dfrac{1}{p_d}\right)$

or, $h = m\left[1 + \dfrac{T}{2}\left(1 - \cos \dfrac{90°}{T}\right)\right]$ (12.8)

2. The constant chord method:

In the tooth thickness expression at the pitch line it is seen that dimensions h and w are both dependent on the number of teeth. If a large number of gears for a set, each having different values of T, are to be tested, the separate calculations would become laborious. The constant chord is a useful dimension since it has the same nominal value for all gears of a common system, irrespective of the number of teeth, provided that the backlash allowance is the same on all the gears. *It is the chord between the points at which the tooth profile touches the basic rack of the system,* where the tangents to the flank lie at the pressure angle to the tooth centre line. The geometry is shown in Fig. 12.31.

The expressions for w and h in case of straight tooth spur gears are given below:

$$w = k\cos^2\alpha \qquad \text{where, } k \text{ is a constant} \qquad ...(12.9)$$

and,

$$h = m\left(1 - \frac{\pi}{8}\sin 2\alpha\right) \qquad ...(12.10)$$

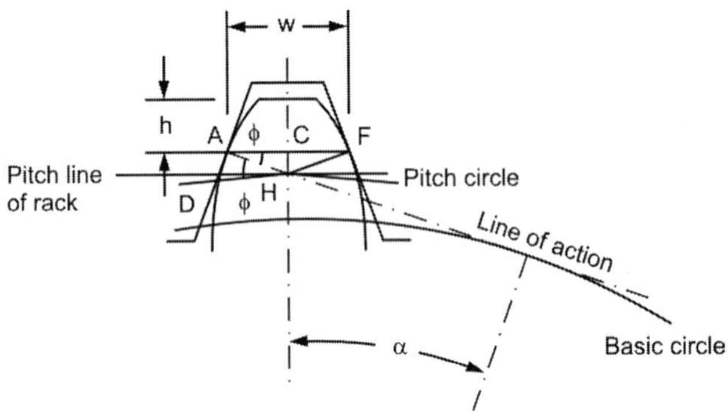

Fig. 12.31

The above expressions indicate that the values of w and h are quite independent of the number of teeth but in gear tooth vernier the corresponding expressions for both w and h depend upon the number of teeth. So if the gear of different number of teeth are to be measured, the calculations for each gear are to be made separate while such *laborious work is eliminated in constant chord method.*

The main feature of constant chord method is that it readily lends itself to a form of comparator which is more sensitive than tooth vernier.

3. The addendum comparative method:

The tooth thickness, using the addendum comparative method, is measured by *comparing the gear addendum with that of a basic rack.* The comparator jaws have the same angle as the normal pressure angle of the tooth form of the gear to be checked. A steel block corresponding to a rack tooth of the proper diametral pitch is used to set the comparator jaws to proper width. At this time the dial indicator is set to zero for standard addendum. When a gear is checked, a thin tooth will project further into the instrument and dial will read plus, while the reverse is also true. The change in tooth thickness may be computed by

$$\Delta w = 2 \times \textit{comparator reading} \times \textit{range of normal pressure angle.}$$

4. The base tangent method:

When the tooth thickness is measured by the vernier the results obtained are *not accurate* because of the following *reasons:*

(*i*) The measurement depends upon the vernier readings, each of which is a function of the other.

(*ii*) The measurement is made with an edge of the measuring jaws, not its face, which again does not lend itself to accurate measurement.

To overcome these problems, the span of a convenient number of teeth may be measured as shown in the Fig. 12.32. It can be seen that this uses a angle vernier caliper which overcomes the disadvantages mentioned above, except that inherent in the use of such an instrument, and this can be overcome by the use of more accurate equipment than a vernier caliper.

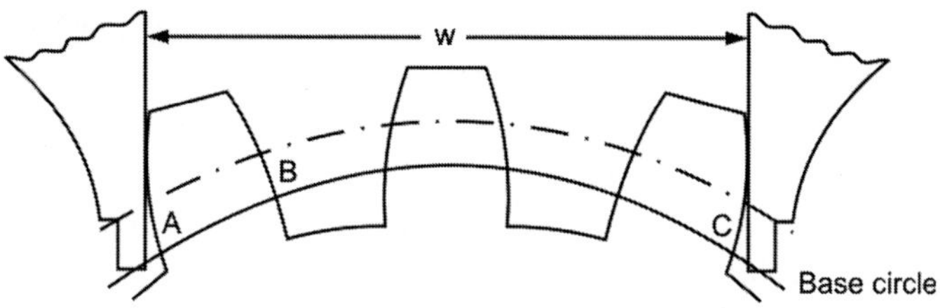

Fig. 12.32

Measurement by this method can be improved by the use of either a micrometer with flanged anvils or the *David Brown tangent comparator* as shown in Fig. 12.33. The instrument consists of a micrometer with limited movement on either side of a zero setting, the zero setting being made with gauge blocks or distant pieces. Further the method lends itself to the measurement of small tooth gears by projection, -either by direct measurement or by traversing the table of a toolmaker's microscope.

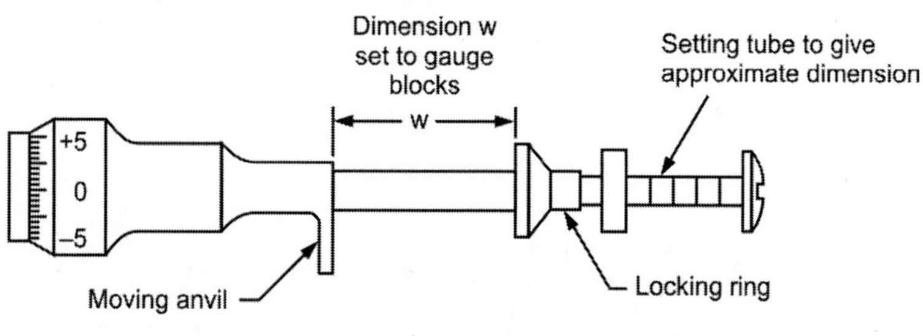

Fig. 12.33 David brown tangent comparator.

12.2.3.2 Measurement of tooth profile

To transmit power between two involute gears it is essential that the profile of the tooth flanks *maintain the exact involute form*. The involute profile of all teeth must be generated from the *same base circle* so that the *same pressure angle can be achieved.* The errors of the involute profile and those of the pressure angle should be exactly evaluated. These measurements for the determination of the actual tooth profile and its error, are among the most important individual measurements that should be carried out on gears. The measuring technique depends upon the measuring instrument used. The tooth profile measurement can be carried out by the following instruments :

1. Toolmaker's microscope or projector
2. Involute tester.

1. Toolmaker's microscope or projector:

The tooth flank profile can be tested by projection on the screen of the toolmaker's microscope or a projector. *The projected profile can be compared with the standard profile of the tooth which can be drawn on the projected screen.* This method can be used for testing the profile and the form of tooth on *gears having modulus smaller than one,* as the depth of the tooth flank is too small for any other method with reliable accuracy to be carried out. When the width of the gear blank is big, it is necessary to choose the correct method of illumination such that the projected tooth is free of shadows.

In order to conduct test on successive teeth, it is necessary to mount the gear on the circular table of the projector so that their axes coincide. To achieve the exact centring, a special fixture has to be used which has to be provided with a locating pin and a clamping element to prevent the rotation of the gear.

The use of this *projection method* can also be made to measure other *parameters of small gears,* such as *profile errors, tooth thickness and the base pitch* and their deviations from the theoretical values. For this, either an accurate scale or the micrometers of the co-ordinate table can be used. By using greatest possible magnification the measuring accuracy can be increased.

2. Involute tester:

Figure 12.34 shows diagrammatically. *'The Klingeluberg involute tester'* used to test gear tooth profile (an involute curve). Its description is given below :

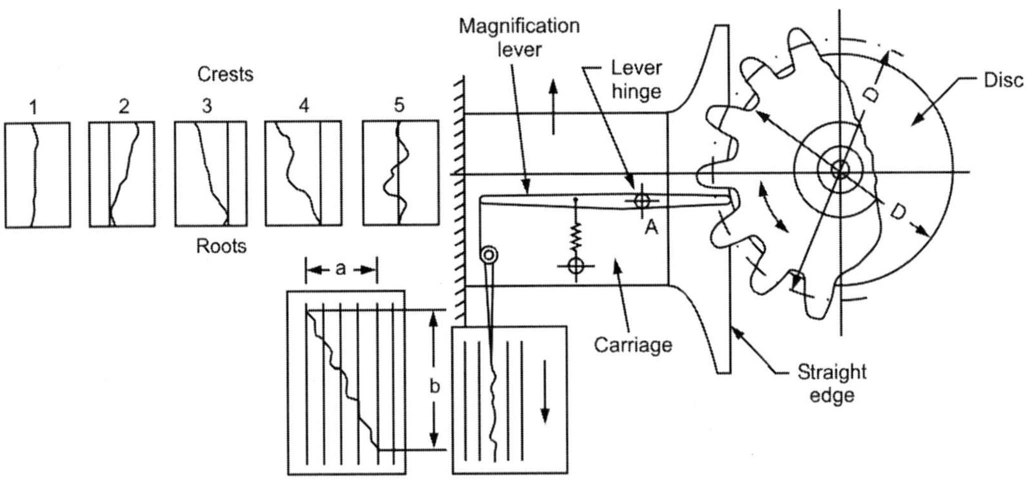

Fig. 12.34 Involute tester.

- It consists of a *disc whose diameter must be equal to* the base *circle diameter* of the gear under test.
- The gear being tested is mounted on the same spindle on which the disc is fixed.
- A *straight-edge* is made to contact the disc and to move tangentially to it so that the disc will be rotated without slip.
- A *magnification lever,* fitted with a diamond stylus that contacts the tooth flank at the base circle, is hinged at A.
- The movement of the stylus is magnified (250 ×) through a lever and gear mechanism and then recorded.

- Due to the linear motion of the straight-edge and the rotational motion of the gear tested, the stylus, which will trace an *involute curve,* will slide over the tooth flank.
- If *the profile of the flank is exactly involute, the lever will remain stationary.* In this case, the recorded curve is a straight line parallel to the direction of motion of the flat surface. Irregularities in the straightness of the recorded line indicate errors in the gear tooth profile while the inclination of the recorded line to the direction of motion of the recording paper will indicate the errors of the base circle, *i.e., errors of the pressure angle.* The actual base circle diameter D_b can be calculated by adding to or substracting its error e_b from the theoretical base circle diameter D_b

$\therefore$
$$D_b = D'_b \pm e_b$$

and
$$e_b = D'_b \frac{a \times M_a}{b \times M_b}$$

where, M_a = Magnification of the stylus lever and gear mechanism,
M_b = Tangential magnification of the recorder, and
a = Deviation of the recorded line over the recorded length b.

From the recorded diagrams shown in Fig. 12.34 the following results can be obtained.

Diagram	*Tooth profile*	*Base circle diameter*
1.	Correct	Correct
2.	Correct	Small
3.	Correct	Great
4.	Not correct	Great
5.	Not correct	Correct

12.2.3.3 Measurement of pitch

- To measure the circular pitch (p_c) is practically impossible. However, it can be easily substituted by measuring the chordal pitch (p_{ch}) which is directly proportional to the circular pitch. The theoretical value of the circular pitch, $p_r = \pi m$.
- The chordal pitch (p_{ch}) *is the length of the chord of the pitch circle connecting two similarly located points on two continuous teeth.* It can be calculated from:

  $p_{ch} = D \sin \dfrac{180°}{T}$, where D and T are the pitch diameter and number of teeth respectively.

  Therefore, the chordal pitch should be measured at the pitch circle.
- It is also possible to measure the angular pitch p_a instead of measuring the chordal pitch where 360° $p_a = \dfrac{360°}{T}$. The relative chordal pitch errors can be measured with reference to the axis of the gear if the gear blank is mounted and located on mandrel that can be fixed between centres. The permissible concentricity error between the gear blank and mandrel should not exceed 1 μm. This arrangement can be made by using one of the following *gear measuring machines.*

  (*i*) Klingeluberg pitch measuring machine
  (*ii*) Zeiss gear measuring machine
  (*iii*) Theodolite or alignment telescope

(*iv*) Keilpart base pitch tester
(*v*) Zeiss base pitch tester
(*vi*) Zeiss constant tangent base pitch tester
(*vii*) Gear measuring machine.

Circular pitch measuring machine. This instrument is used for checking the circular pitch of gear tooth. The two measuring contact tips are applied on the same side of adjacent teeth of the gear. The left-hand tip is first set up to the required module by means of some suitable arrangement. The right hand tip is a two armed lever whose one end contacts the gear tooth and the other one actuates the contact point of the dial indicator. Two guide points are also provided for the stability of the instrument.

Figure 12.35 shows an instrument by which pitch variations can be also be measured. It employs a fixed finger and stop for consistent positioning on successive pairs of teeth, and a movable finger whose movement can be sensed by a dial indicator. It may be noted that readings obtained by this instrument will be affected by profile variations and run out of the gear.

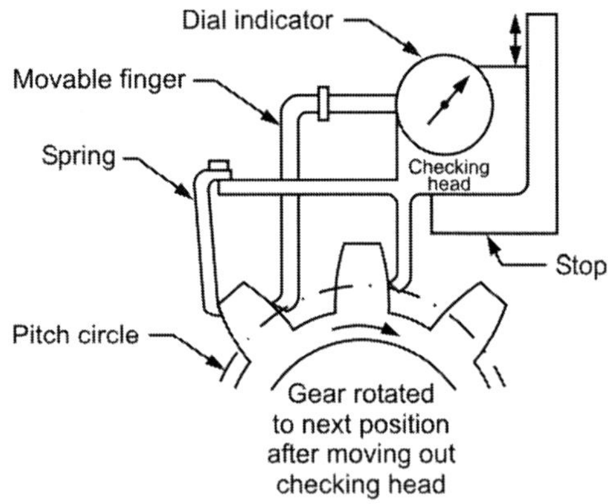

Fig. 12.35.

THEORETICAL QUESTIONS

1. Define a screw thread.
2. Enumerate the purposes for which screw threads are used.
3. How are threads classified ?
4. Explain any four of the following elements of a screw threads:

   (*i*) Major diameter (*ii*) Minor diameter
   (*iii*) Effective material (*iv*) Helix angle
   (*v*) Thickness of thread (*vi*) Axis of the screw
   (*vii*) Crest.

5. Enumerate the points which should be given due consideration while giving specifications of a screw thread.
6. Give the comparison between Vee and square threads.
7. Give the procedure of measuring 'Major diameter' of a screw thread by using a bench micrometer.
8. Explain with neat sketches the method of measuring minor diameter of a screw thread.
9. Explain two and three wires methods of measuring effective diameter of a screw thread.
10. Explain briefly 'Thread micrometer method' of measuring effective diameter of a screw thread.
11. Explain how can the pitch of a screw thread be measured on a pitch measuring machine.
12. Describe briefly the method of measuring thread angle of a screw thread.
13. Explain briefly how the measurements of the following elements of internal screw threads can be done:

   (*i*) Major diameter (*ii*) Minor diameter
   (*iii*) Effective diameter (*iv*) Pitch
   (*v*) Thread angle and form.

Gear Measurement

14. What is a gear?
15. Enumerate the uses of gears.
16. Explain briefly any two of the following gears:

    (*i*) Spur gear
    (*ii*) Helical gear
    (*iii*) Bevel gears
    (*iv*) Worm gears
    (*v*) Rack and pinion.

❑❑❑

Surface Texture/Finish

Chapter 13

13.1 INTRODUCTION

- Each of the methods employed for producing a surface imparts some characteristic quality to the structure of the surface layer of metal and it will be readily appreciated that a chip-producing formation method will have a different effect from such treatments as planishing, rolling or drawing where the surface layer is caused to slide over the material beneath. Certain qualities in the surface layer are important for various classes of service; for example, the conditions imposed on the races of a ball or roller bearing where heavy loads must be carried by very small areas, whilst under the action of rolling friction demand that the surfaces must possess a high degree of homogeneity, elasticity and hardness.
- The finish and texture of lifting surfaces have an important effect on bearing friction, rate of wear, initial toteralnces etc. Every surface, more or less, is composed of minute hills and valleys and when the conditions of service require that such surfaces shall operate one against the other the early stages of action will result in the levelling down of the "hills" on each member. If the combination has a bearing this initial wear will result in an increased clearance, so that the eventual conditions, after the bearing has "settled down", will depend on both the initial fit and quality of the finish. The reliability of a press or force fit and its approach to theoretical conditions is largely dependent on surface quality and finish. Even for surfaces which do not fit or serve as bearings smoothness is often important since in highly stressed parts fatigue cracks originate from surface blemishes and under corrosive influence a high class surface may be more durable than one less finished.

13.2 SURFACE TEXTURE

The *surface texture* may be defined as:

"The characteristic quality of an actual surface due to small departures from its general geometrical form which, occurring at regular or irregular intervals, tend to form a pattern or texture on the surface".

Surface textures vary according to the machining processes used in producing it. This is certainly true in case of metal machining. These differences are apparent by visual examination and can be felt readily by passing a finger nail over the surface. The texture of surfaces may be regular or irregular in character and may lie in a particular direction or be non-directional in character. Additional factors producing surface irregularities may include faulty tools, inherent imperfections in the machine tools used and, of course, errors due to the personal element. The problem of the measurement of surface texture is complex, owing to the number of possible variables. Therefore, in practice, for the normal control of manufactured components, a complete analysis is not possible, or even desirable. The difficulty is overcome mainly by specifying the finishing process (such as grinding, lapping, etc.) to be used and by stating, in standard units, the quality of finish required.

- The *problem of the measurement of surface* texture is *basically geometrical.* Although, fundamentally, the problem is three dimensional in character, in practice it is conveniently reduced to one of two-dimensional geometry. This is accomplished by limiting individual measurements to the examination of profiles of plane sections taken through the surface being measured. It is important to make the measurement in the *"correct"* plane, which is usually in a plane approximately at right angles to the "lay" (or direction of the predominant markings) of the surface and, generally, it furnishes the most efficient results. However, other planes may have to be developed in special cases.

13.3 PRIMARY AND SECONDARY TEXTURE

Any material being machined by chip removal process can't be finished perfectly due to some departures from ideal conditions as envisaged by the designer. Due to conditions not being ideal, the surface produced will have some irregularities; and these geometrical irregularities could be classified into following four categories:

First order. *Irregularities arising out of inaccuracies in the machine tool itself (e.g.,* lack of straightness of guideways on which toolpost is moving).

Surface irregularities arising due to deformation of work under the action of cutting forces and the height of the material itself are also induced under this head.

Second order. *Irregularities caused due to vibrations of any kind such as chatter marks.*

Third order. *Irregularities caused by a machining itself due to characteristics of the process.* This also includes the feed marks of the cutting tool.

Fourth order. Irregularities arising from the rupture of the material during the separation of the chip.

These irregularities of four orders can be grouped into the following *two* groups :

1. Primary texture (or Roughness)
2. Secondary texture.

Primary texture (or roughness). In this group are included *irregularities of small wavelength caused by direct action of the cutting element on the material or by some other disturbance such as friction, wear, or corrosion.* These errors are chiefly caused due to *tool chatter, i.e.,* it includes irregularities of third and fourth order and constitutes that *microgeometrical errors.*

Secondary texture. In this group are included *irregularities of considerable wavelength of a periodic character resulting from mechanical disturbances in the generating set up.* These errors are termed as *macrogeometrical errors* and include irregularities of first and second order and are mainly due to *misalignment of centres, lack of straightness of guideways and non-linear feed motion.*

Any surface could be considered to be combination of two forms of wavelengths *(large wavelength for waviness and smaller wavelength for roughness)* superimposed upon each other.

One of the problems in measuring surfaces finish is to *separate the waviness from the roughness.*

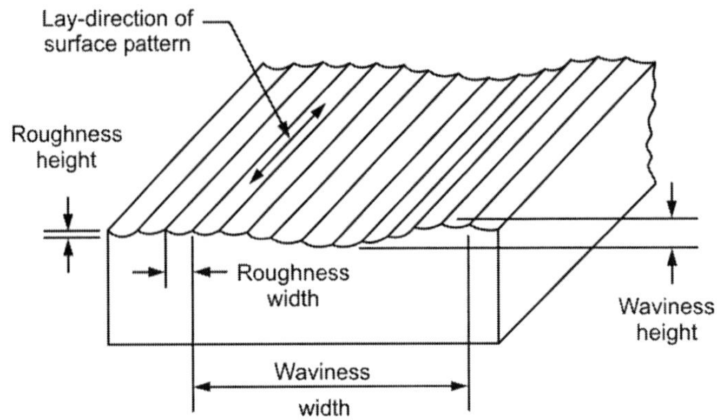

Fig. 13.1

Figure 13.1 shows the various terms used in connection with surface finish.

13.4 METHODS OF MEASURING SURFACE FINISH

The surface finish of machined part can be measured by the following *two* methods:

1. Surface inspection by comparison methods
2. Direct instrument measurements.

13.4.1 Surface Inspection by Comparison Methods

In comparative methods, the surface texture is assessed by observation of the surface. But these methods are not reliable as they can be misleading if comparison is not made with surface produced by same techniques. The various methods available are:

1. Touch inspection
2. Visual inspection
3. Scratch inspection
4. Microscopic inspection
5. Surface photographs
6. Micro-interferometer
7. Wallace surface dynamometer
8. Reflected light intensity
9. Comparison with standard specimens.

13.4.1.1 Touch inspection

One can feel the distance between the peaks of the surface irregularities by touching it with finger, on conditions that this distance is greater than 3 mm. The feeling of the human being is very sensitive to vibrations, and the maximum sensitivity can be achieved by a frequency of 360 c/s. Therefore the surface finish can be tested when the finger nail or any metallic edge slides over the surface. Surfaces having a maximum depth of roughness of about 0.1 μm can be compared when the sliding speed of the edge is between 3 and 5 cm/s. The judgement, whether the surface is rough or smooth, must be done in comparison with specimens having standard surface roughness.

The main *limitation* of this method is that the *degree of surface roughness can't be assessed.* Also *minute flows cannot be detected.* This *method can simply tell which surface is more rough.*

13.4.1.2 Visual inspection

Under normal conditions, one can visualize and differentiate between surfaces having a roughness variation of not less than 5% due to the fact that the brightness of these surfaces varies proportionally. When a sharp straight edge is placed over the surface under test, the edge can, by good illumination of 100 to 150 lux identify from a distance of 250 mm an air gap of not less than 0.5 μm and a distance between the peaks of about 70 μm.

Visual inspection by naked eye is always likely to be misleading particularly when surfaces having high degree of surfaces finish are inspected. The method is, therefore, *limited to rougher surfaces and results vary from person-to-person.* More accurate inspection can be done by illuminated magnifiers.

13.4.1.3 Scratch inspection

In scratch inspection method, a soft material like lead, babbit or plastic is rubbed over the surface to be inspected. By doing so it carries the impression of the scratches on the surfaces which can be easily visualised.

13.4.1.4 Microscopic inspection

This method of measuring surface finish is perhaps the best method but suffers due to *limitations that only a small portion of the surface can be inspected at a time*. Thus several readings are required to get an average value.

- In this method a master finished surface is placed under the microscope and compared with the surface under inspection.

In another method a straight edge is placed on the surface to be inspected and a beam of light projected at about 60° to the work. Thus the shadows cast into the surface scratches are magnified and surface irregularities can be studied.

13.4.1.5 Surface photographs

In this method *magnified photographs* of the surface are taken with *difference types of illumination.*

- In case we use vertical *illumination* then *defects* like *irregularities and scratches appear as dark spots and flat portion of the surface appears as bright area.*
- In case of *oblique illumination,* reverse is the case. Photographs with different illumination are compared and the results assessed.

13.4.1.6 Micro-interference

In micro-interference method of measuring finish an *optical flat is placed on the surface to be inspected and illuminated by a monochromatic source of light. Interference bands are studied through a microscope.* Defects, *i.e.,* scratches in the surfaces *appear as interference lines extending from the dark bands into the bright bands. The depth of the defects is measured in terms of the fraction of the interference band.*

13.4.1.7 Wallace surface dynamometer

Wallace surface dynamometer is a sort of friction meter. It consists of a pendulum in which the testing shoes are clamped to the bearing surface and a predetermined spring pressure can be applied. The pendulum is lifted to its initial starting position and allowed to swing over the surface to be tested. If *the surface is smooth, then there will be less friction and pendulum swings for a longer period.* Thus *time of swing is a direct measure of surface finish.*

13.4.1.8 Reflected light intensity

In this method a beam of light of known quantity is projected upon the surface. This light is reflected in several directions as beams of lesser intensity and the *change in light intensity in different directions is measured by a photo cell.* The measured intensity changes are already calibrated by means of readings taken from the surface of known roughness by some other suitable method.

13.4.1.9 Comparison with standard specimens

For qualitative assessment of the surface roughness, the roughness comparison specimens are used. In this qualitative assessment of the roughness of the finished product is evaluated by visual or tactile comparison (*i.e.,* lightly moving a finger nail across the lay of the roughness with roughness comparison specimens of known roughness value finished by similar machining process. The roughness comparison specimens are produced in a variety of materials (steel having a hardness of about 405 HB) by direct application of the production process or by copying techniques and are available in various grades of roughness. The copying technique consists of (*i*) electro-forming positive replicas of master surfaces and (*ii*) making positive replicas of master surfaces in plastic materials by coating or otherwise and having the feel and appearances of the naturally machined metal surface. It is general practice to reproduce the specimen surfaces as replicas of master surfaces in a hard wearing material like nickel alloy or metallised plastic.

The specimens used for roughness comparison are available either in flat form or segmented cylindrical form. The flat specimens have the size: *length* = 25 mm; *width* = 16 mm and *thickness* = 1.6 to 10 mm. The specimen is rigidly supported on a base to prevent it from being warped or distorted during use. The segmented cylindrical specimens will have either convex surface or concave surface having a surface radius about 125 mm. *These specimens have also proved to be very good device for training the personnel to judge the roughness value by feel and appearance of the surfaces.*

While these specimens are being used for direct comparison purposes, it is worth noting that: (*i*) two surfaces being compared should resemble each other in material, shape and method of finish; (*ii*) two surfaces must be clearly alike in general character in sense that the cutting technique and machine feeds used for making the specimen should resemble those used for the product; (*iii*) surfaces being compared should be clean and dry.

13.4.1.10 Some other methods of measuring surface roughness

(*i*) Glass measurement
(*ii*) Diffraction techniques
(*iii*) Taper sectioning
(*iv*) Tactile tests
(*v*) Pneumatic gauging
(*vi*) Optical secorns
(*vii*) Thermal comparator
(*viii*) Capacitance gauges etc.

Methods (*i*), (*ii*) and (*iii*) are discussed below:

Glass measurement. There is a variation in the optical properties of different rough surfaces as judged selectively by the human eye and various instruments have been developed to try and quality this variation more objectively. In one instrument the reflection of a series of vertical bars in the surface is reflected as the specular angle. This angle is gradually decreased until the bars can no longer the distinguished. The *roughness is inversely proportional to the cosine of the angle of extinction.*

Diffraction techniques. Speckle is produced when a rough surface is illuminated with spatially coherent light and observed with an optical system of the finite aperture. Light is received at each point in the image from several different points on the object because of the resolution limitations of the system. The path length of the light from each point to the surface depends on the height of the surface contour at that point. If this height varies significantly across the width of the point spread function, interference effects will appear as speckle. From measurements of speckle pattern surface roughness can be deduced.

Taper sectioning. It is a very accurate method but is a destructive one. A section is cut through the surface to be examined at a shallow angle α thus magnifying height variations by a factor of cot α, and the section is examined by *optical microscope.*

13.4.2 Direct instrument measurement

The quantitative measurement of surface roughness can be done by the following methods:

1. **Intersection method:**
   (*i*) Cross-section method
   (*ii*) Light-section method.
2. **Interface method.**
3. **Stylus method:**
   (*i*) Stylus instruments having *mechanical magnification:*
   - Mikrokarato surface roughness tester.
   - Tomlinson tester

   (*ii*) Stylus instrument having *optical magnification:*
   - Forster surface roughness tester
   - Kiesemetter tester

   (*iii*) Stylus instruments having *electrical magnification:*
   - Talysurf surface roughness tester
   - Phillips roughness tester
   - Abbott profilograph
4. **Pneumatic measuring method.**
5. **Perthen condenser method.**

13.4.2.1 Intersection method

The actual surface texture can be observed in a plane that intersects the component perpendicular to its surface.

(i) Cross-section method:

- When a specimen/component is cut perpendicular to its surface, the surface texture gets destroyed due to deformation that takes place during the cutting operation. This can be avoided when a protective layer of approximately the same hardness is applied to the surface before cutting, either by chromate plating or by casting of a thin layer of white metal on the surface.

- The specimen, after cutting, has to be polished and the actual surface texture can be observed under a measuring microscope.

Although *this method is accurate yet its troublesome procedure makes it inconvenient for practical measurement of the surface roughness.*

- The destruction of a component for measuring its surface hardness can be avoided *by making a surface replica from celluloid or gelatine. After drying the replica,* it can be *sliced* and observed under the microscope. This method is now finding wide application in the investigation of microstructure by the *electronic microscope.*

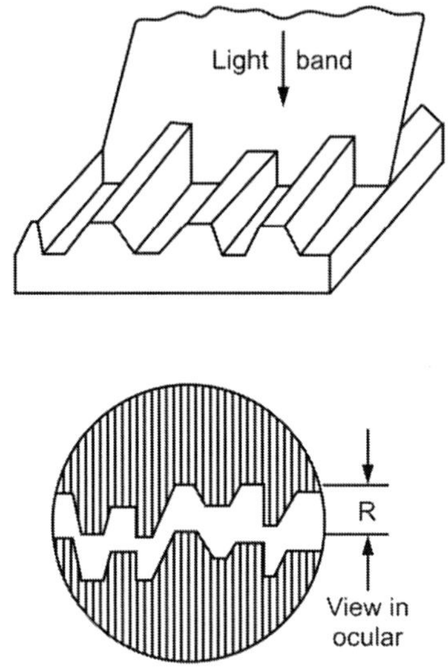

Fig. 13.2 Measuring principle.

(ii) Light-section method

- This method is a non-destructive way of rendering the surface texture visible.
- A narrow light band is made to intersect the test surface and the intersected contour of the surface can be observed by a microscope as shown in Fig. 13.2. Figures 13.3 and 13.4 show two possible methods of projecting and observing the surface texture.

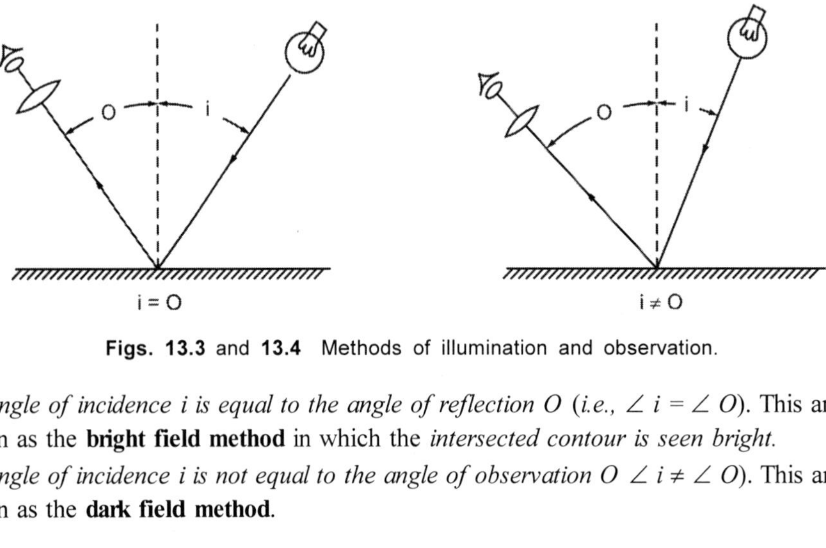

Figs. 13.3 and **13.4** Methods of illumination and observation.

1. The *angle of incidence i is equal to the angle of reflection O* (*i.e.,* $\angle i = \angle O$). This arrangement is known as the **bright field method** in which the *intersected contour is seen bright.*
2. The *angle of incidence i is not equal to the angle of observation O* $\angle i \neq \angle O$). This arrangement is known as the **dark field method.**

13.4.2.2 Interference method (Interference microscopes)

- *Interference microscope* can be used to examine surface texture. It enables the depths of surface irregularities to be measured in *terms of wavelengths of monocromatic light.*

  A beam of monochromatic light of known wavelength is used together with an optical flat placed over the test surface. When interference occurs between the incident and reflected rays of light, the contour pattern or interference fringes can be observed through microscope. *From the shape of the interference fringes the maximum depth of roughness of the surface can be measured.* A highly accurate optical flat should be used and it must also be of approximately the reflectivity as that of the surface being examined. The *optical flat is usually given a silver coating to suit highly reflective test surfaces.* Ordinary machined surfaces should be covered with a *very thin layer of vaseline* equally distributed over the surface to *increase the reflectivity and to reduce the absorption of the surface.*

- The *measuring range* of surface roughness by the interference method, using a monochromatic light of a wavelength equal to 0.6 mm, *does not exceed* 2 μm. The low limit is about one-fifth the distance between two successive fringes which is equal to 0.05 μm.
- *Bad reflecting surfaces having surface roughness greater than 1* μm, *cannot be examined by the interference method.*

13.4.2.3 Stylus method

- This method is largely used for measuring the surface feature by using a very fine needle which can trace the actual surface texture and *penetrate to the full depth of its valleys.*
- *The stylus is made to slide over the surface being tested or it can receive low frequency impulses while the surface moves relative to it.* In both cases, the vertical movement of the needle is highly magnified and can be read off on the instrument indicator or on a recorder graph whose horizontal movement can also be magnified.
- Vertical magnifications are usually of the order of 1000 to 100000 times while the horizontal magnification ranges between 20 and 200.

The following can affect the measuring accuracy of the stylus method.

(*i*) Stylus hinge
(*ii*) Stylus point
(*iii*) Measuring force
(*iv*) Speed of traverse
(*v*) Stylus skid.

Some instruments are described below:

Talysurf Surface Roughness Tester:

Refer Fig. 13.5. The various parts of the Talysurf surface roughness tester are:

(*i*) E-shape soft iron head
(*ii*) Armature
(*iii*) Stylus
(*iv*) Cylindrical skid.

Construction: The *stylus* is mounted on the *armature,* which is provided at the central limb of an *E-shaped soft iron head.* The outer limbs of the head are provided with two induction coils and a small air gap is left between the armature and outer limbs of the head.

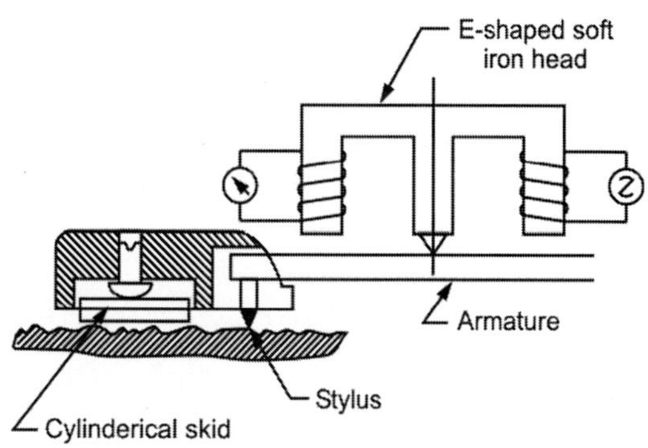

Fig. 13.5 Talysurf surface roughness tester.

Working: A downward movement of the stylus results in decreasing the air gap of the primary coil and in an equal increase of the air gap at the secondary coil. The *impedance of the coils will be changed according to the variation of air gap and an additional alternating current flows in the secondary coil whose magnitude is governed by the variation of the impedance of the coil and is proportional to the displacement of the stylus. The position of the stylus,* therefore, *controls or modulates the carrier.*

- This method of carrier modulation *enables true graphs of the surface texture to be obtained.*
- A cylindrical *skid* is provided as a datum and a pick-up unit is moved across the test surface by means of a motor and a gear box. This provides different measuring speeds. The maximum length that can be traced is 10 mm.
- The skid length is made greater than twice the pitch of the waviness and a diamond stylus is used having a *four-sided pyramid shape.* The pyramid angle is 90° and the nose radius of the stylus is equal to 2 μm.
- The measuring force of the stylus is 0.1 g.
- The vertical magnification can be varied in steps from 1000 to 50000 times.

THEORETICAL QUESTIONS

1. Define 'Surface texture'.
2. Explain briefly the following:
   (*i*) Primary texture (*ii*) Secondary texture.
3. How can geometrical irregularities be classified ?
4. Differentiate between first, second, third and fourth order irregularities.
5. Enumerate various methods of measuring surface finish.
6. Explain any two of the following methods of measuring surface finish.
   (*i*) Touch inspection (*ii*) Visual inspection
   (*iii*) Scratch inspection (*iv*) Microscopic inspection
   (*v*) Reflected light intensity.
7. Describe any one of the following intersection method:
   (*i*) Cross-section method (*ii*) Light-section method.

❑❑❑

B. Tech

(SEMESTER-IV) THEORY EXAMINATION, 2011-12

MEASUREMENT AND METROLOGY

EME-403

Time : 3 Hours *Total Marks : 100*

Note: Attempt **all** Sections

SECTION-A

1. Answer **all** the questions in **50 – 75** words. Each question carries **2** marks.

(a) Define measurement and explain applications of measurement.

(b) Explain the functions of a sensor in a measurement system.

(c) Differentiate between gauge pressure and static pressure.

(d) Differentiate between thermocouple and thermistor.

(e) Explain different types of seismic instruments.

(f) Explain the purpose of a surface plate in metrology.

(g) Explain different types of fits.

(h) Define and list out various types of comparators.

(i) Differentiate between primary and secondary texture.

(j) Explain the importance of measurement and inspection in the field of metrology.

SECTION-B

2. Answer any **three** questions. Each question carries **10** marks.

(a) Classify different types of errors and explain how these errors can be removed or rectified in an instrument.

(b) Explain the working of a bimetallic thermometer with a neat sketch.

(c) Differentiate between uni-lateral and bi-lateral tolerances. Why is unilateral tolerance preferred over bilateral tolerance?

(d) Define optical flat and explain its uses in detail.

(e) Define straightness, flatness and roundness and also explain how autocollimator is useful in finding these parameters.

SECTION-C

Answer **all** the questions. Each question carries **10** marks.

3. Classify transducers and explain capacitive and piezoelectric transducer with neat sketches.

OR

Explain in detail the generalized measurement system and its functional elements with a block diagram.

4. Explain the McLeod pressure gauge with neat sketches.

OR

Classify different types of resistance gauges and explain the working of Wheatstone bridge under balanced and unbalanced conditions.

5. Differentiate between Hole basis and Shaft basis system with neat diagrams.

OR

Explain how sine bar is used for measuring the angle of a taper plug gauge with neat diagram.

6. Define interferometry. Explain the working principle and applications of Tool Makers's microscope with a neat diagram.

OR

Differentiate between Roughness and Waviness. Explain different numerical assessment methods to find surface roughness.

7. Explain the importance of thread micrometer and explain how screw thread is measured by using two wire and three wire method.

OR

Explain working of a Stroboscope with a neat diagram.

❑❑❑

B. Tech

(SEMESTER-IV) Theory Examination, 2012-13

MEASUREMENT AND METROLOGY

EME-403

Time : 3 Hours *Total Marks : 100*

SECTION-A

1. Attempt **all** question parts: **10 × 2 = 20**

(a) State the difference between accuracy and precision of a measurement.

(b) Give any four methods of measurement.

(c) What is meant by analog and digital transducer? Give example.

(d) What is thermocouple and thermopile?

(e) With a aid of sketch explain any one type of elastic pressure transducer.

(f) What is strain gauge and strain rosettes?

(g) What are the various types of linear measuring instruments?

(h) How the following are designated?

(i) Standard tolerance grade

(ii) Tolerance size

(i) Name the different types of interferometer.

(j) What are the various methods used for measuring the gear tooth thickness?

SECTION-B

2. Attempt any **three** question parts: **3 × 10 = 30**

(a) Discuss in detail:

(i) the various dynamic characteristics of a measuring system

(ii) primary and secondary transducer with neat sketch

(b) Briefly explain various methods of measuring torque.

(c) Discuss the construction and working of Optical Pyrometer with neat sketch.

(d) With the aid of neat sketch explain the working of Johansson 'Mikrokator'.

(e) Describe how the straightness error of a Lathe bed is checked using an Auto-collimator.

SECTION-C

Attempt **all** questions: **5 × 10 = 50**

3. Attempt any **two** parts: **2 × 5 = 10**

(a) Draw a sketch showing the movement of sigma comparator.

(b) What is surface texture and give its representation with a sketch? Also explain about primary and secondary texture.

(c) With neat sketches, explain the significance of some important parameters used for measuring surface roughness. Why so many parameters are needed?

4. Attempt any **one** part: **1 × 10 = 10**

(a) What are the basic blocks of a generalized instrumentation system? Draw the various blocks and explain their functions.

(b) Discuss in detail various types of errors associated in measurement and how these errors can be minimized.

5. Attempt any **one** part: **1 × 10 = 10**

(a) Write short notes on any two of the following:

(i) Stroboscope

(ii) Vibration pickups and decibel meters

(iii) Different types of I6ad cells

(b) Explain the theory of strain gauges and also discuss the temperature compensation in strain gauge.

6. Attempt any **one** part: **1 × 10 = 10**

(a) Explain how the component shown in Fig. 1 is checked for its dimensional and form accuracy. Justify the selection of instruments after considering various alternatives possible.

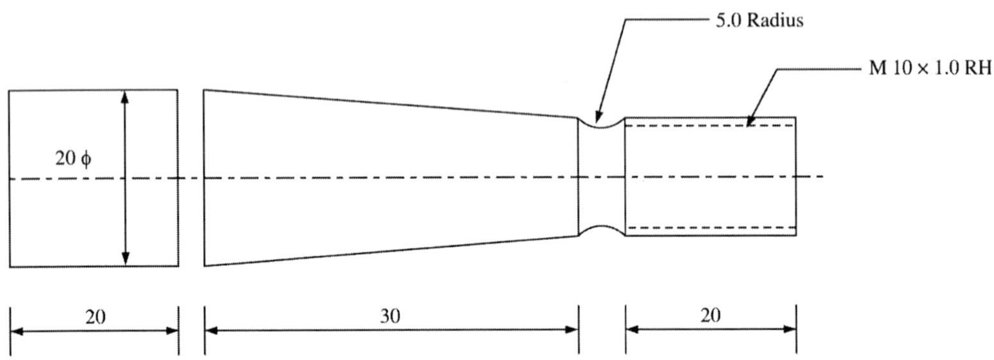

Fig. 1

(b) Explain Taylor's principle of gauge design. Determine the dimensions of hole and shaft for a fit 30 H_7/hg. Also determine the allowance and maximum clearance.

7. Attempt any **two** parts: **2 × 5 = 10**

(a) Derive the expression for best size wire.

(b) Calculate the length of base tangent over two teeth for a helical gear having the following specification : width of gear = 30 mm, No. of teeth = 10, Helix angle = 30°, pressure angle = 20°, module = 4.

(c) In the measurement of surface roughness, heights of 20 successive peaks and troughs were measured from a datum and were 35, 25, 40, 22, 35, 18, 42, 25, 35, 22, 36, 18, 42, 22, 32, 21, 37, 18, 35, 20 microns. If the measurements were obtained over length of 20 mm, determine the C.L.A and R.M.S value of the rough surface.

❑❑❑

Index

Index

A

C

D

E

F

G

H

I

L

M

O

P

R

S

T

U

V

W

X

Win Prizes!

ATTENTION READERS

We request you for your frank assessment regarding some of the aspects of the book given as under:

A TEXT BOOK OF MEASUREMENT AND METROLOGY

By

R.K. RAJPUT

Please fill up the given space in neat capital letters.

(i) What topics of your syllabus that are important from you examination point of view are not covered in the book ?

....

....

....

....

(ii) What are the chapters wherein the treatment of the subject matter is not systematic or organised or updated ?

....

....

....

....

(iii) Have you come across misprints/mistakes/factual inaccuracies in the book? Please specify the chapters and the page numbers.

....

....

....

....

(iv) What other books on the same subject have you found/heard better than the present book? Please specify in terms of price and quality.

....

....

....

....

(v) Further suggestions and comments for the improvement of the book:

....

....

....

....

(PS: If need be please attach a separate sheet to write your views.)

Other Details :

(i) Who recommended you the book ? (Please tick in the box near the option relevant to you.)

☐ Teachers ☐ Friends ☐ Bookseller

(ii) Name of the recommending teacher, his designation and address.

....

....

....

....

(iii) Name and address of the bookseller you purchased the book from:

....

....

....

....

(iv) Name and address of your institution (Please mention the University or Board, as the case may be)

....

....

....

....

(v) Your name and complete postal address:

....

....

....

....

(vi) Course you are enrolled in:

....

....

....

....

The best assessment will be awarded each month. The award will be in the form of our publications as decided by the Editorial Board.

Please mail the filled up coupon at your earliest to:

S.K. KATARIA & SONS®

Publisher of Engineering Books

4885/109, Prakash Mahal, Dr. Subhash Bhargav Lane,

Opposite Delhi Medical Association, Daryaganj, New Delhi–110002 (INDIA)

Phone: +91-11-43551243; *Mobile:* +91-9871775858, 7678425209

e-mail: katariabook@gmail.com, katariabooks@yahoo.com

Website: www.skkatariaandsons.com